Applied Geophysics: Modeling and Simulation

Applied Geophysics: Modeling and Simulation

Edited by Karl Seibert

SYRAWOOD
PUBLISHING HOUSE
New York

Published by Syrawood Publishing House,
750 Third Avenue, 9th Floor,
New York, NY 10017, USA
www.syrawoodpublishinghouse.com

Applied Geophysics: Modeling and Simulation
Edited by Karl Seibert

International Standard Book Number: 978-1-68286-642-9 (Hardback)

Cataloging-in-Publication Data

Applied geophysics : modeling and simulation / edited by Karl Seibert.
    p. cm.
Includes bibliographical references and index.
ISBN 978-1-68286-642-9
1. Geophysics. 2. Geophysics--Simulation methods. 3. Geological modeling. I. Seibert, Karl.
QE511 .A67 2019
550--dc23

TABLE OF CONTENTS

PREFACE

It is often said that books are a boon to mankind. They document every progress and pass on the knowledge from one generation to the other. They play a crucial role in our lives. Thus I was both excited and nervous while editing this book. I was pleased by the thought of being able to make a mark but I was also nervous to do it right because the future of students depends upon it. Hence, I took a few months to research further into the discipline, revise my knowledge and also explore some more aspects. Post this process, I begun with the editing of this book.

Applied geophysics is concerned with the implementation of geophysical theories and concepts to practical problems and tasks of civil engineering such as groundwater mapping, ore and mineral prospecting, etc. It studies physical phenomena like magnetism, electricity, radioactivity, etc. It also encompasses geological concepts to understand and analyze dynamics of plate tectonics, volcanism, rock formation, Earth's gravitational and magnetic fields, etc. The principles of applied geophysics are also significant to a number of prominent disciplines such as Earth systems science, climatology, earthquake research, etc. This book is compiled to provide in-depth knowledge about the theory and practice of geophysics. It strives to provide a fair idea about this discipline and to help develop a better understanding of the latest advances within this field. The content included herein is appropriate for students seeking detailed information in this area as well as for experts.

I thank my publisher with all my heart for considering me worthy of this unparalleled opportunity and for showing unwavering faith in my skills. I would also like to thank the editorial team who worked closely with me at every step and contributed immensely towards the successful completion of this book. Last but not the least, I wish to thank my friends and colleagues for their support.

Editor

Laboratory experimental investigation of heat transport in fractured media

Claudia Cherubini[1,2], **Nicola Pastore**[3], **Concetta I. Giasi**[3], **and Nicoletta Maria Allegretti**[3]

[1]Department of Mechanical, Aerospace & Civil Engineering, Brunel University London, Uxbridge, UB8 3PH, UK
[2]School of Civil Engineering, The University of Queensland, Queensland, Australia
[3]DICATECh, Department of Civil, Environmental, Building Engineering, and Chemistry, Politecnico di Bari, Bari, Italy

Correspondence to: Claudia Cherubini (claudia.cherubini@brunel.ac.uk) and Nicola Pastore (nicola.pastore@poliba.it)

Abstract. Low enthalpy geothermal energy is a renewable resource that is still underexploited nowadays in relation to its potential for development in society worldwide. Most of its applications have already been investigated, such as heating and cooling of private and public buildings, road defrosting, cooling of industrial processes, food drying systems or desalination.

Geothermal power development is a long, risky and expensive process. It basically consists of successive development stages aimed at locating the resources (exploration), confirming the power generating capacity of the reservoir (confirmation) and building the power plant and associated structures (site development). Different factors intervene in influencing the length, difficulty and materials required for these phases, thereby affecting their cost.

One of the major limitations related to the installation of low enthalpy geothermal power plants regards the initial development steps that are risky and the upfront capital costs that are huge.

Most of the total cost of geothermal power is related to the reimbursement of invested capital and associated returns.

In order to increase the optimal efficiency of installations which use groundwater as a geothermal resource, flow and heat transport dynamics in aquifers need to be well characterized. Especially in fractured rock aquifers these processes represent critical elements that are not well known. Therefore there is a tendency to oversize geothermal plants.

In the literature there are very few studies on heat transport, especially on fractured media.

This study is aimed at deepening the understanding of this topic through heat transport experiments in fractured networks and their interpretation.

Heat transfer tests have been carried out on the experimental apparatus previously employed to perform flow and tracer transport experiments, which has been modified in order to analyze heat transport dynamics in a network of fractures. In order to model the obtained thermal breakthrough curves, the Explicit Network Model (ENM) has been used, which is based on an adaptation of Tang's solution for the transport of the solutes in a semi-infinite single fracture embedded in a porous matrix.

Parameter estimation, time moment analysis, tailing character and other dimensionless parameters have permitted a better understanding of the dynamics of heat transport and the efficiency of heat exchange between the fractures and the matrix. The results have been compared with the previous experimental studies on solute transport.

1 Introduction

An important role in transport of natural resources or contaminant transport through subsurface systems is given by fractured rocks. Interest in the study of dynamics of heat transport in fractured media has grown in recent years because of the development of a wide range of applications, including geothermal energy harvesting (Gisladottir et al., 2016).

Quantitative geothermal reservoir characterization using tracers is based on different approaches to predicting thermal breakthrough curves in fractured reservoirs (Shook, 2001; Kocabas, 2005; Read et al., 2013).

The characterization and modeling of heat transfer in fractured media are particularly challenging as open and well-connected fractures can induce highly localized pathways which are orders of magnitude more permeable than the rock matrix (Klepikova et al., 2016; Cherubini and Pastore, 2011).

The study of solute transport in fractured media has recently become a widespread research topic in hydrogeology (Cherubini, 2008; Cherubini et al., 2008, 2009, 2013d; Masciopinto et al., 2010), whereas the literature about heat transfer in fractured media is somewhat limited.

Hao et al. (2013) developed a dual continuum model for the representation of discrete fractures and the interaction with the surrounding rock matrix in order to give a reliable prediction of the impacts of fracture–matrix interaction on heat transfer in fractured geothermal formations.

Moonen et al. (2011) introduced the concept of a cohesive zone which represents a transition zone between the fracture and undamaged material. They proposed a model to adequately represent the influences of fractures or partially damaged material interfaces on heat transfer phenomena.

Geiger and Emmanuel (2010) found that matrix permeability plays an important role in thermal retardations and attenuation of thermal signals. At high matrix permeability, poorly connected fractures can contribute to the heat transport, resulting in heterogeneous heat distributions in the whole matrix block. For lower matrix permeability heat transport occurs mainly through fractures that form a fully connected pathway between the inflow and outflow boundaries, which results in highly non-Fourier behavior characterized by early breakthrough and long tailing.

Numerous field observations (Tsang and Neretnieks, 1998) show that flow in fractures is being organized in channels due to the small-scale variations in the fracture aperture. Flow channeling causes dispersion in fractures. Such channels will have a strong influence on the transport characteristics of a fracture, such as, for instance, its thermal exchange area, crucial for geothermal applications (Auradou et al., 2006). Highly channelized flow in fractured geologic systems has been credited with early thermal breakthrough and poor performance of geothermal circulation systems (Hawkins et al., 2012).

Lu et al. (2012) conducted experiments of saturated water flow and heat transfer in a regularly fractured granite at meter scale. The experiments indicated that the heat advection due to water flow in vertical fractures nearest to the heat sources played a major role in influencing the spatial distributions and temporal variations of the temperature, impeding heat conduction in the transverse direction; such an effect increased with larger water fluxes in the fractures and decreased with a higher heat source and/or a larger distance of the fracture from the heat source.

Neuville et al. (2010) showed that fracture–matrix thermal exchange is highly affected by the fracture wall roughness. Natarajan et al. (2010) conducted numerical simulation of thermal transport in a sinusoidal fracture–matrix coupled system. They affirmed that this model presents a different behavior with respect to the classical parallel plate fracture–matrix coupled system. The sinusoidal curvature of the fracture provides high thermal diffusion into the rock matrix.

Ouyang (2014) developed a three-equation local thermal non-equilibrium model to predict the effective solid-to-fluid heat transfer coefficient in geothermal system reservoirs. They affirmed that due to the high rock-to-fracture size ratio, the solid thermal resistance effect in the internal rocks cannot be neglected in the effective solid-to-fluid heat transfer coefficient. Furthermore the results of this study show that it is not efficient to extract the thermal energy from the rocks if fracture density is not large enough.

Analytical and semi-analytical approaches have been developed to describe the dynamics of heat transfer in fractured rocks. Such approaches are amenable to the same mathematical treatment as their counterparts developed for mass transport (Martinez et al., 2014). One of these is the analytical solution derived by Tang et al. (1981).

While the equations of solute and thermal transport have the same basic form, the fundamental difference between mass and heat transport is that (1) solutes are transported through the fractures only, whereas heat is transported through both fractures and matrix, and (2) the fracture–matrix exchange is large compared with molecular diffusion. This means that the fracture–matrix exchange is more relevant for heat transport than for mass transport. Thus, matrix thermal diffusivity strongly influences the thermal breakthrough curves (BTCs) (Becker and Shapiro, 2003).

Contrarily, since the heat capacity of the solids will retard the advance of the thermal front, the advective transport for heat is slower than for solute transport (Rau et al., 2012).

The quantification of thermal dispersivity in terms of heat transport and its relationship with velocity has not been properly addressed experimentally and has conflicting descriptions in the literature (Ma et al., 2012).

Most studies neglect the hydrodynamic component of thermal dispersion because of thermal diffusion being more efficient than molecular diffusion by several orders of magnitude (Bear, 1972). Analysis of heat transport under natural gradients has commonly neglected hydrodynamic dispersion (e.g., Bredehoeft and Papadopulos, 1965; Domenico and Palciauskas, 1973; Taniguchi et al., 1999; Reiter, 2001; Ferguson et al., 2006). Dispersive heat transport is often assumed to be represented by thermal conductivity and/or to have little influence in models of relatively large systems and modest fluid flow rates (Bear, 1972; Woodbury and Smith, 1985).

Some authors suggest that thermal dispersivity enhances the spreading of thermal energy and should therefore be part of the mathematical description of heat transfer in analogy to solute dispersivity (de Marsily, 1986), and have incorporated this term into their models (e.g., Smith and Chapman, 1983; Hopmans et al., 2002; Niswonger and Prudic, 2003). In the same way, other researchers (e.g., Smith and Chapman, 1983; Ronan et al., 1998; Constanz et al., 2002; Su et al., 2004) have included the thermomechanical dispersion tensor representing mechanical mixing caused by unspecified heterogeneities within the porous medium.

By contrast, some other researchers argue that the enhanced thermal spreading is either negligible or can be described simply by increasing the effective diffusivity; thus, the hydrodynamic dispersivity mechanism is inappropriate (Bear, 1972; Bravo et al., 2002; Ingebritsen and Sanford, 1998; Keery et al., 2007). Constantz et al. (2003) and Vandenbohede et al. (2009) found that thermal dispersivity was significantly smaller than the solute dispersivity. Others (de Marsily, 1986; Molina-Giraldo et al., 2011) found that thermal and solute dispersivity was on the same order of magnitude.

Tracer tests of both solute and heat were carried out at Bonnaud, Jura, France (de Marsily, 1986), and the thermal dispersivity and solute dispersivity were found to be of the same order of magnitude.

Bear (1972), Ingebritsen and Sanford (1998), and Hopmans et al. (2002), among others, concluded that the effects of thermal dispersion are negligible compared to conduction and set the former to zero.

However, Hopmans et al. (2002) showed that dispersivity is increasingly important at higher flow water velocities, since it is only then that the thermal dispersion term is of the same order of magnitude or larger than the conductive term.

Sauty et al. (1982) suggested that there was a correlation between the apparent thermal conductivity and Darcy velocity; thus, they included the hydrodynamic dispersion term in the advective–conductive modeling.

Other similar formulations of this concept are present in the literature (e.g., Papadopulos and Larson, 1978; Smith and Chapman, 1983; Molson et al., 1992). Such treatments have not explicitly distinguished between macrodispersion, which occurs due to variations in permeability over larger scales, and the components of hydrodynamic dispersion that occur due to variations in velocity at the pore scale.

One group of authors have utilized a linear relationship to describe the thermal dispersivity and the relationship between thermal dispersivity and fluid velocity (e.g., de Marsily, 1986; Anderson, 2005; Hatch et al., 2006; Keery et al., 2007; Vandenbohede et al., 2009; Vandenbohede and Lebbe, 2010; Rau et al., 2012), while others have identified the possibility of a nonlinear relationship (Green et al., 1964).

The present study is aimed at providing a better understanding of heat transfer mechanisms in fractured rocks. Laboratory experiments on mass and heat transport in a fractured rock sample have been carried out in order to analyze the contribution of thermal dispersion in heat propagation processes, the influence of nonlinear flow dynamics on the enhancement of thermal matrix diffusion and finally the optimal conditions for thermal exchange in a fractured network.

Section 1 shows a short review of mass and heat transport in fractured media highlighting what is still unresolved or contrasting in the literature.

In Sect. 2 the theoretical background related to nonlinear flow and solute and heat transport behavior in fractured media has been reported.

A better development of the Explicit Network Model (ENM), based on Tang's solution developed for solute transport in a single semi-infinite fracture inside a porous matrix, has been used for the fitting of the thermal BTCs. The ENM model explicitly takes the fracture network geometry into account and therefore permits one to understand the physical meaning of mass and heat transfer phenomena and to obtain a more accurate estimation of the related parameters. In an analogous way, the ENM has been used in order to fit the observed BTCs obtained from previous experiments on mass transport.

Section 3 shows the thermal tracer tests carried out on an artificially created fractured rock sample that has been used in previous studies to analyze nonlinear flow and non-Fickian transport dynamics in fractured formations (Cherubini et al., 2012, 2013a, b, c, 2014).

In Sect. 4 have been reported the interpretation of flow and transport experiments together with the fitting of BTCs and interpretation of estimated model parameters. In particular, the obtained thermal BTCs show more enhanced early arrival and long tailing than solute BTCs.

The travel time for solute transport is an order of magnitude lower than for heat transport experiments. Thermal convective velocity is thus more delayed with respect to solute transport. The thermal dispersion mechanism dominates heat propagation in the fractured medium in the carried out experiments and thus cannot be neglected.

For mass transport the presence of the secondary path and the nonlinear flow regime are the main factors affecting non-Fickian behavior observed in experimental BTCs, whereas for heat transport the non-Fickian nature of the experimental BTCs is governed mainly by the heat exchange mechanism between the fracture network and the surrounding matrix. The presence of a nonlinear flow regime gives rise to a weak growth on heat transfer phenomena.

Section 5 reports some practical applications of the knowledge acquired from this study on the convective heat transport in fractured media for exploiting heat recovery and heat dissipation. Furthermore the estimation of the average effective thermal conductivity suggests that there is a solid thermal resistance in the fluid-to-solid heat transfer processes due to the rock–fracture size ratio. This result matches previous analyses (Pastore et al., 2015) in which a lower heat dissipation with respect to Tang's solution in correspondence to the single fracture surrounded by a matrix with more limited heat capacity has been found.

2 Theoretical background

2.1 Nonlinear flow

With few exceptions, any fracture can be envisioned as two rough surfaces in contact. In cross section the solid areas representing asperities might be considered the grains of porous media.

Therefore, in most studies examining hydrodynamic processes in fractured media, the general equations describing flow and transport in porous media are applied, such as Darcy's law, which depicts a linear relationship between the pressure gradient and fluid velocity (Whitaker, 1986; Cherubini and Pastore, 2010).

However, this linearity has been demonstrated to be valid in low flow regimes ($Re < 1$). For $Re > 1$ a nonlinear flow behavior is likely to occur (Cherubini et al., 2013d).

When $Re \gg 1$, a strong inertial regime develops that can be described by the Forchheimer equation (Forchheimer, 1901):

$$-\frac{dp}{dx} = \frac{\mu}{k} \cdot u_f + \rho\beta \cdot u_f^2, \qquad (1)$$

where x (m) is the coordinate parallel to the axis of the single fracture (SF), p ($ML^{-1}T^{-2}$) is the flow pressure, μ ($ML^{-1}T^{-1}$) is the dynamic viscosity, k (L^2) is the permeability, u_f (LT^{-1}) is the convective velocity, ρ (ML^{-3}) is the density and β (L^{-1}) is called the inertial resistance coefficient, or non-Darcy coefficient.

It is possible to express the Forchheimer law in terms of hydraulic head h (L):

$$-\frac{dh}{dx} = a' \cdot u_f + b' \cdot u_f^2. \qquad (2)$$

The coefficients a' (TL^{-1}) and b' (TL^{-2}) represent the linear and inertial coefficient, respectively, equal to

$$a' = \frac{\mu}{\rho g k}; b' = \frac{\beta}{g}. \qquad (3)$$

The relationship between hydraulic head gradient and flow rate Q ($L^3 T^{-1}$) can be written as

$$-\frac{dh}{dx} = a \cdot Q + b \cdot Q^2. \qquad (4)$$

The coefficients a (TL^{-3}) and b (T^2L^{-6}) can be related to a' and b':

$$a = \frac{a'}{\omega_{eq}}; b = \frac{b'}{\omega_{eq}^2}, \qquad (5)$$

where ω_{eq} (L^2) is the equivalent cross-sectional area of SF.

2.2 Heat transfer by water flow in single fractures

Fluid flow and heat transfer in a single fracture (SF) undergo advective, diffusive and dispersive phenomena. Dispersion is caused by small-scale fracture aperture variations.

Flow channeling is one example of macrodispersion caused by preferred flow paths, in that mass and heat tend to migrate through the portions of a fracture with the largest apertures.

In fractured media another process is represented by diffusion into the surrounding rock matrix. Matrix diffusion attenuates the mass and heat propagation in the fractures.

According to the boundary layer theory (Fahien, 1983), solute mass transfer q_M (ML^{-2}) per unit area at the fracture–matrix interface (Wu et al., 2010) is given by

$$q_M = \frac{D_m}{\delta} (c_f - c_m), \qquad (6)$$

where c_f (ML^{-3}) is the concentration across fractures, c_m (ML^{-3}) is the concentration of the matrix block surfaces, D_m (LT^{-2}) is the molecular diffusion coefficient, and δ (m) is the thickness of the boundary layer (Wu et al., 2010). For small fractures, δ may become the aperture w_f (m) of the SF.

In an analogous manner, the specific heat transfer flux q_H (MT^{-3}) at the fracture–matrix interface is given by

$$q_H = \frac{k_m}{\delta} (T_f - T_m), \qquad (7)$$

where T_f (K) is the temperature across fractures, T_m (K) is the temperature of the matrix block surfaces, and k_m ($MLT^{-3}K^{-1}$) is the thermal conductivity.

The continuity conditions at the fracture–matrix interface require a balance between mass transfer rate and mass diffused into the matrix described as

$$q_M = -D_e \frac{\partial c_m}{\partial z}\bigg|_{z = w_f/2}, \qquad (8)$$

where z (m) is the coordinate perpendicular to the fracture axis and w_f is the aperture of the fracture.

In the same way, the specific heat flux must be balanced by heat diffused into the matrix described as

$$q_H = -k_e \frac{\partial T_m}{\partial z}\bigg|_{z = w_f/2}. \qquad (9)$$

The effective diffusion coefficient takes into account the fact that diffusion can only take place through pore and fracture openings because mineral grains block many of the possible pathways. The effective thermal conductivity of a formation consisting of multiple components depends on the geometrical configuration of the components as well as on the thermal conductivity of each.

The effective terms (D_e instead of D_m and k_e instead of k_m) have been introduced in order to include the effect of various system parameters such as fluid velocity, porosity, surface area, and roughness that may enhance the mass and heat transfer effect. For instance, when large flow velocity occurs, convective transport is stronger along the center of the fracture, enhancing the concentration or temperature gradient at the fracture–matrix interface. As is known, roughness plays an important role in increasing mass or heat transfer because of increasing turbulent flow conditions.

According to Bodin (2007) the governing equation for the 1-D advective–dispersive transport along the axis of a semi-infinite fracture with 1-D diffusion in the rock matrix, in perpendicular direction to the axis of the fracture, is

$$\frac{\partial c_f}{\partial t} + u_f \frac{\partial c_f}{\partial x} = \frac{\partial}{\partial x}\left(D_f \frac{\partial c_f}{\partial x}\right) - \frac{D_e}{\delta}\left.\frac{\partial c_m}{\partial z}\right|_{z=w_f/2}, \quad (10)$$

where D_f (L^2T^{-1}) is the dispersion in the fracture. The latter mainly depends on two processes: Aris–Taylor dispersion and geometrical dispersion. Previous experiments (Cherubini et al., 2012a, b, c, 2014) show that, due to the complex geometrical and topological characteristics of the fracture network that create tortuous flow paths, Aris–Taylor dispersion may not develop. A linear relationship has been found between velocity and dispersion, so geometrical dispersion is mostly responsible for the mixing process along the fracture:

$$D_f = \alpha_{LM} u_f, \quad (11)$$

where α_{LM} (L) is the dispersivity coefficient for mass transport.

Assuming that fluid flow velocity in the surrounding rock matrix is equal to zero, the equation for the conservation of heat in the matrix is given by

$$\frac{\partial c_m}{\partial t} = D_a \frac{\partial^2 c_m}{\partial z^2}, \quad (12)$$

where D_a is the apparent diffusion coefficient of the solute in the matrix expressed as a function of the matrix porosity θ_m, $D_a = D_e/\theta_m$ (Bodin et al., 2007).

Tang et al. (1981) presented an analytical solution for solute transport in a semi-infinite single fracture embedded in a porous rock matrix with a constant concentration at the fracture inlet ($x = 0$) equal to c_0 (ML^{-3}) and with an initial concentration equal to zero. The solute concentration in the fracture $\bar{c}_f$ and in the matrix $\bar{c}_m$ are as follows:

$$\bar{c}_f = \frac{c_0}{s} \exp(vL)\exp\left[-vL\left\{1 + \beta^2\left(\frac{s^{1/2}}{A} + s\right)\right\}^{1/2}\right], \quad (13)$$

$$\bar{c}_m = \bar{c}_f \exp\left[-Bs^{1/2}(z - w_f/2)\right], \quad (14)$$

where s is the integral variable of the Laplace transform and L (L) is the length of SF; the v, A, β^2 and B coefficients are expressed as follows:

$$v = \frac{u_f}{2D_f}, \quad (15)$$

$$A = \frac{\delta}{\sqrt{\theta_m D_e}}, \quad (16)$$

$$\beta^2 = \frac{4D_f}{u_f^2}, \quad (17)$$

$$B = \frac{1}{\sqrt{D_e}}, \quad (18)$$

whereas the gradient of $\bar{c}_m$ at the interface $z = w_f/2$ is

$$\left.\frac{d\bar{c}_m}{dx}\right|_{x=w_f/2} = -\bar{c}_f B s^{1/2}. \quad (19)$$

Defining the residence time as the average amount of time that the solute spends in the system, on the basis of these analytical solutions the probability density function (PDF) of the solute residence time in the single fracture in the Laplace space can be expressed as

$$\bar{\Gamma}(s) = \exp(vL)\exp\left[-vL\left\{1 + \beta^2\left(\frac{s^{1/2}}{A} + s\right)\right\}^{1/2}\right]. \quad (20)$$

Assuming that density and heat capacity are constant in time, the heat transport conservation equation in SF can be expressed as follows:

$$\frac{\partial T_f}{\partial t} + u_f \frac{\partial T_f}{\partial x} = \frac{\partial}{\partial x}\left(D_{fH} \frac{\partial T_f}{\partial x}\right) - \frac{k_e}{\rho_w C_w \delta}\left.\frac{\partial T_m}{\partial z}\right|_{z=w_f/2}, \quad (21)$$

where ρ_w (ML^{-3}) and C_w ($L^2T^2K^{-1}$) represent the density and the specific heat capacity of the fluid within SF, respectively. D_f for heat transport assumes the following expression:

$$D_{fH} = \frac{\lambda_L}{\rho_w C_w}, \quad (22)$$

where λ_L is the thermodynamic dispersion coefficient ($MLT^{-3}K^{-1}$). Sauty et al. (1982) and de Marsily (1986) proposed an expression for the thermal dispersion coefficient where the thermal dispersion term varies linearly with velocity and depends on the heterogeneity of the medium, as for solute transport:

$$\lambda_L = k_0 + \alpha_{LH}\rho_w C_w u_f, \quad (23)$$

where k_0 is the bulk thermal conductivity ($MLT^{-3}K^{-1}$) and α_{LH} (L) is the longitudinal thermal dispersivity.

The heat transport conservation equation in the matrix is expressed as follows:

$$\rho_m C_m \frac{\partial T_m}{\partial t} = k_e \frac{\partial^2 T_m}{\partial z^2}. \quad (24)$$

Note that the governing equations of heat and mass transport highlight similarities between the two processes; thus, Tang's solution can also be used for heat transport.

In terms of heat transport, the coefficients v, A, β^2 and B are expressed as follows:

$$v = \frac{u_f}{2D_{fH}}, \quad (25)$$

$$A = \frac{\delta}{\sqrt{\theta D_e}}, \quad (26)$$

where $\theta = \rho_m C_m / \rho_w C_w$ and $D_e = k_e / \rho_w C_w$.

$$\beta^2 = \frac{4 D_f}{u_f^2} \tag{27}$$

$$B = \frac{1}{\sqrt{D_e}} \tag{28}$$

Three characteristic timescales can be defined:

$$t_u = \frac{L}{u_f}; t_d = \frac{L^2}{D_f}; t_e = \frac{\delta^2}{D_e}, \tag{29}$$

where L (L) is the characteristic length; t_u (T), t_d (T) and t_e (T) represent the characteristics timescales of convective transport, dispersive transport and loss of the mass or heat into the surrounding matrix.

The relative effect of dispersion, convection and matrix diffusion on mass or heat propagation in the fracture can be evaluated by comparing the corresponding timescale.

Peclet number P_e is defined as the ratio between dispersive (t_d) and convective (t_u) transport times:

$$Pe = \frac{t_d}{t_u} = \frac{u_f L}{D_f}. \tag{30}$$

At high Peclet numbers transport processes are mainly governed by convection, whereas at low Peclet numbers it is mainly dispersion that dominates.

Another useful dimensionless number, generally applied in chemical engineering, is the Damköhler number that can be used in order to evaluate the influence of matrix diffusion on convection phenomena. Da relates the convection timescale to the exchange timescale:

$$Da = \frac{t_u}{t_e} = \frac{\alpha L}{u_f}, \tag{31}$$

where α (T^{-1}) is the exchange rate coefficient corresponding to

$$\alpha = \frac{D_e}{\delta^2}. \tag{32}$$

Note that the inverse of t_e has the same meaning as the exchange rate coefficient α (T^{-1}).

When t_e values are of the same order of magnitude as the transport time t_u ($Da \cong 1$), diffusive processes in the matrix are more relevant. In this case concentration or temperature distribution profiles are characterized by a long tail.

When $t_e \gg t_u$ ($Da \ll 1$), the fracture–matrix exchange is very slow and it does not influence mass or heat propagation. By contrast, when $t_e \ll t_u$ ($Da \gg 1$), the fracture–matrix exchange is rapid, there is instantaneous equilibrium between the fracture and the matrix, and they have the same concentration or temperature. These two circumstances close the standard advective–dispersive transport equation.

The product between Pe and Da represents another dimensionless group which is a measure of transport processes:

$$Pe \times Da = \frac{t_d}{t_e} = \frac{\alpha L^2}{D_f}. \tag{33}$$

When $Pe \times Da$ increases, t_e decreases more rapidly than t_d, and subsequently the mass or heat diffusion into the matrix may be dominant on the longitudinal dispersion.

2.3 Explicit Network Model (ENM)

The 2-D Explicit Network Model (ENM) depicts the fractures as 1-D pipe elements forming a 2-D pipe network, and therefore expressly takes the fracture network geometry into account. The ENM permits one to understand the physical meaning of flow and transport phenomena and therefore to obtain a more accurate estimation of flow and transport parameters.

With the assumption that a jth SF can be schematized by a 1-D pipe element, the Forchheimer model can be used to write the relationship between head loss Δh_j (L) and flow rate Q_j (L^{3}T^{-1}) in finite terms:

$$\frac{\Delta h_j}{L_j} = a Q_j + b Q_j^2 \Rightarrow \Delta h_j = \left[L_j \left(a + b Q_j \right) \right] Q_j, \tag{34}$$

where L_j (L) is the length of jth SF, and a (TL^{-3}) and b (T^{2}L^{-6}) represent the Forchheimer parameters written in finite terms. The term in the square brackets constitutes the resistance to flow $R_j \left(Q_j \right)$ (TL^{-2}) of jth SF.

In case of steady-state conditions and for a simple 2-D fracture network geometry, a straightforward manner can be applied to obtain the solution of a flow field by applying the first and second Kirchhoff laws.

In a 2-D fracture network, fractures can be arranged in series and/or in parallel. Specifically, in a network in which fractures are set in a chain, the total resistance to flow is calculated by simply adding up the resistance values of each single fracture. The flow in a parallel fracture network breaks up, with some flowing along each parallel branch and recombining when the branches meet again. In order to estimate the total resistance to flow, the reciprocals of the resistance values have to be added up, and then the reciprocal of the total has to be calculated. The flow rate Q_j across the generic fracture j of the parallel network can be calculated as (Cherubini et al., 2014)

$$Q_j = \sum_{i=1}^{n} Q_i \left[\frac{1}{R_j} \left(\sum_{i=1}^{n} \frac{1}{R_i} \right)^{-1} \right], \tag{35}$$

where $\sum_{i=1}^{n} Q_i$ (LT^{-3}) is the sum of the mass flow rates at fracture intersections in correspondence to the inlet bond of j fracture, whereas the term in square brackets represents the probability of water distribution of j fracture $P_{Q,j}$.

Once the flow field in the fracture network is known, to obtain the PDF at a generic node, the PDFs of each elementary path that reaches the node have to be summed up. They can be calculated as the convolution product of the PDFs of each single fracture composing the elementary path.

Definitely, the BTC describing the concentration in the fracture as a function of time at the generic node, using the convolution theorem, can be obtained as follows:

$$c_f(t) = c_0 + c_{inj}(t) \cdot \mathcal{L}^{-1} \left[\sum_{i=1}^{N_p} \prod_{j=1}^{n_{f,i}} P_{M,j} \overline{\Gamma}_j(s) \right], \quad (36)$$

where c_0 (ML^{-3}) is the initial concentration and c_{inj} (ML^{-3}) is the concentration injection function, $*$ is the convolution operator, $\mathcal{L}^{-1}$ represents the inverse Laplace transform operator, N_p is the number of the paths reaching the node, $n_{f,i}$ is the number of the SF belonging to the elementary path ith, and $P_{M,j}$ and $\overline{\Gamma}(s)$ are the mass distribution probability and the PDF in the Laplace space of the generic jth SF, respectively. The inverse Laplace transform $\mathcal{L}^{-1}$ can be solved numerically using the Abate and Ward (2006) algorithm.

In the same way the BTC T_f which describes the temperature in the fracture as a function of time at the generic node can be written as

$$T_f(t) = T_0 + T_{inj}(t) \cdot \mathcal{L}^{-1} \left[\sum_{i=1}^{N_p} \prod_{j=1}^{n_{f,i}} P_{H,j} \overline{\Gamma}_j(s) \right], \quad (37)$$

where T_0 (K) is the initial temperature, T_{inj} (K) is the temperature injection function and $P_{H,j}$ is the heat distribution probability.

$P_{M,j}$ and $P_{H,j}$ can be estimated as the probabilities of the mass and heat distribution at the inlet bond of each individual SF, respectively. The mass and heat distribution is proportional to the correspondent flow rates:

$$P_{M,j} = P_{H,j} = \frac{Q_j}{\sum_{i=1}^{n} Q_i}. \quad (38)$$

Note that if Eq. (38) is valid, the probability of water distribution is equal to the probabilities of mass and heat distribution (term in square brackets in Eq. 34). Therefore, the ENM model regarding each SF can be described by four parameters $(u_{f,j}, D_{f,j}, \alpha_j, P_{Q,j})$.

3 Material and methods

3.1 Description of the experimental apparatus

The heat transfer tests have been carried out on the experimental apparatus previously employed to perform flow and tracer transport experiments at bench scale (Cherubini et al. 2012, 2013a, b, c, 2014). However, the apparatus has been modified in order to analyze heat transport dynamics. Two thermocouples have been placed at the inlet and the outlet of a selected fracture path of the limestone block with parallelepiped shape ($0.6 \times 0.4 \times 0.08$ m^3) described in previous studies. A TC-08 Thermocouple Data Logger (pico Technology) with a sampling rate of 1 s has been connected to the thermocouples. An extruded polystyrene panel with thermal conductivity equal to 0.034 Wm^{-1} K^{-1} and thickness 0.05 m has been used to thermally insulate the limestone block which has then been connected to a hydraulic circuit. The head loss between the upstream tank connected to the inlet port and the downstream tank connected to the outlet port drives flow of water through the fractured block. An ultrasonic velocimeter (DOP3000 by Signal Processing) has been adopted to measure the instantaneous flow rate that flows across the block. An electric boiler with a volume of 10^{-2} m^3 has been used to heat the water. In a flow cell located in correspondence to the outlet port, a multiparametric probe is positioned for the instantaneous measurement of pressure (dbar), temperature (°C) and electric conductivity (μS cm^{-1}). Figure 1a shows the fractured block sealed with epoxy resin and Figure 1b shows the thermal insulated fractured block connected to the hydraulic circuit, whereas the schematic diagram of the experimental apparatus is shown in Fig. 2.

3.2 Flow experiments

The average flow rate through the selected path can be evaluated as

$$\overline{Q} = \frac{S_1}{t_1 - t_0}(h_1 - h_0), \quad (39)$$

where S_1 (L^2) is the cross-sectional area of the flow cell, and $\Delta t = t_1 - t_0$ is the time for the flow cell to be filled from h_0 (L) and h_1 (L). To calculate the head loss between the upstream tank and the flow cell, the following expression is adopted:

$$\Delta h = h_c - \frac{h_0 + h_1}{2}, \quad (40)$$

where h_c is the hydraulic head measured in the upstream tank. Several tests have been carried out varying the control head, and in correspondence to each value of the average flow rate and head loss, the average resistance to flow has been determined as

$$\overline{R}(Q) = \left[\frac{S_1}{t_1 - t_0} \ln\left(\frac{h_0 - h_c}{h_1 - h_c} \right) \right]^{-1}. \quad (41)$$

3.3 Solute and temperature tracer tests

Solute and temperature tracer tests have been conducted through the following steps.

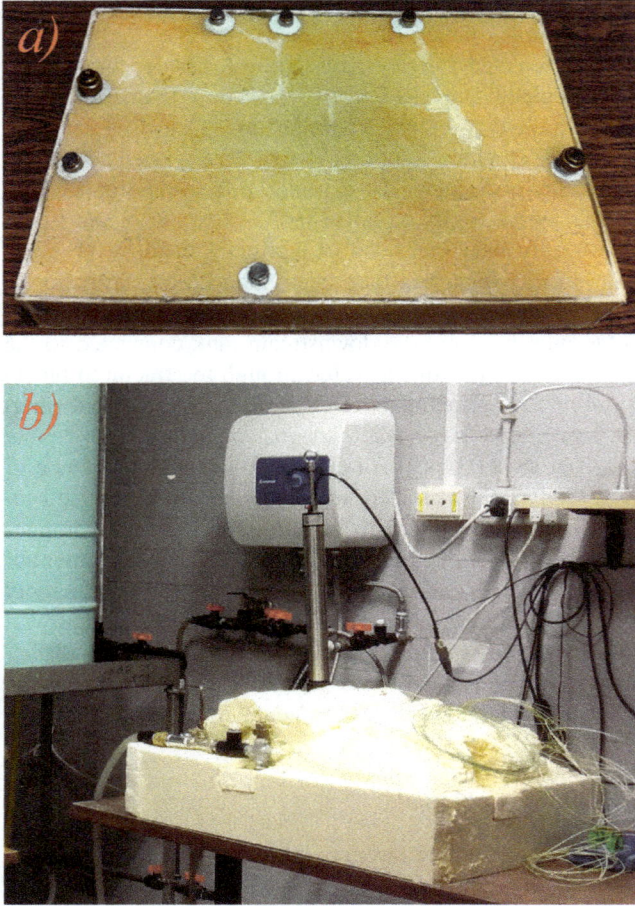

Figure 1. (a) Fractured block sealed with epoxy resin. (b) Thermal insulated fracture block connected to the hydraulic circuit.

As an initial condition, a specific value of the hydraulic head difference between the upstream tank and the downstream tank has been assigned. At $t = 0$, valve a is closed so that the hydrostatic head inside the block assumes the same value as the one in the downstream tank. At $t = 10$ s, valve a is opened.

For the solute tracer test at time $t = 60$ s by means of a syringe, a mass of 5×10^{-4} kg sodium chloride is injected into the inlet port. Due to the very short source release time, the instantaneous source assumption can be adopted which assumes the source of a solute as an instantaneous injection (pulse). The multiparametric probe located within the flow cell measures the solute BTC.

As concerns thermal tracer tests at the time $t = 60$ s, the valve d is opened, while the valve c is closed. In such a way a step temperature function in correspondence to the inlet port $T_{\text{inj}}(t)$ is imposed and measured by the first thermocouple. The other thermocouple located inside the outlet port is used to measure the thermal BTC.

The ultrasonic velocimeter is used in order to measure the instantaneous flow rate, whereas a multiparametric probe located at the outlet port measures the pressure and the electric conductivity.

4   Results and discussion

4.1   Flow characteristics

The Kirchhoff laws have been used in order to estimate the flow rates flowing in each single fracture. In Fig. 3 a sketch of the 2-D pipe conceptualization of the fracture network is reported.

The resistance to flow of each SF can be evaluated as the square bracket in Eq. (34). For simplicity the linear and nonlinear terms have been considered constant and equal for each SF.

The resistance to flow for the whole fracture network $\overline{R}(Q)$ can be evaluated as the sum of the resistance to flow of each SF arranged in a chain and the total resistance of the parallel branches equal to the reciprocal of the sum of the reciprocal of the resistance to flow of each parallel branch:

$$\overline{R}(Q) = R_1(Q_0) + R_2(Q_0)$$
$$+ \left(\frac{1}{R_6(Q_1)} + \frac{1}{R_3(Q_2) + R_4(Q_2) + R_5(Q_2)} \right)^{-1}$$
$$+ R_7(Q_0) + R_8(Q_0) + R_9(Q_0), \qquad (42)$$

where R_j with $j = 1$–9 represents the resistance to flow of each SF, Q_0 is the injection flow rate, and Q_1 and Q_2 are the flow rates flowing in parallel branches 6 and 3–5, respectively.

The flow rate Q_1 is determined in an iterative manner using the following iterative equation derived by Eq. (35) at node 3:

$$Q_1^{k+1} = Q_0 \cdot \left[\frac{1}{R_6(Q_1^k)} \left(\frac{1}{R_3(Q_0 - Q_1^k) + R_4(Q_0 - Q_1^k) + R_5(Q_0 - Q_1^k)} \right. \right.$$
$$\left. \left. + \frac{1}{R_6(Q_1^k)} \right)^{-1} \right], \qquad (43)$$

whereas the flow rate Q_2 is determined merely as

$$Q_2 = Q_0 - Q_1. \qquad (44)$$

The linear and nonlinear terms representative of the whole fracture network have been estimated by matching the average experimental resistance to flow resulting from Eq. (41) with resistance to flow estimated from Eq. (42).

The linear and nonlinear terms are equal, respectively, to $a = 7.345 \times 10^4$ sm^{-3} and $b = 11.65 \times 10^9$ s^2 m^{-6}. Inertial forces dominate viscous ones when the Forchheimer number (Fo) is higher than one. Fo can be evaluated as the ratio between the nonlinear loss (bQ^2) and the linear loss (aQ). The critical flow rate Q_{crit} which represents the value of the

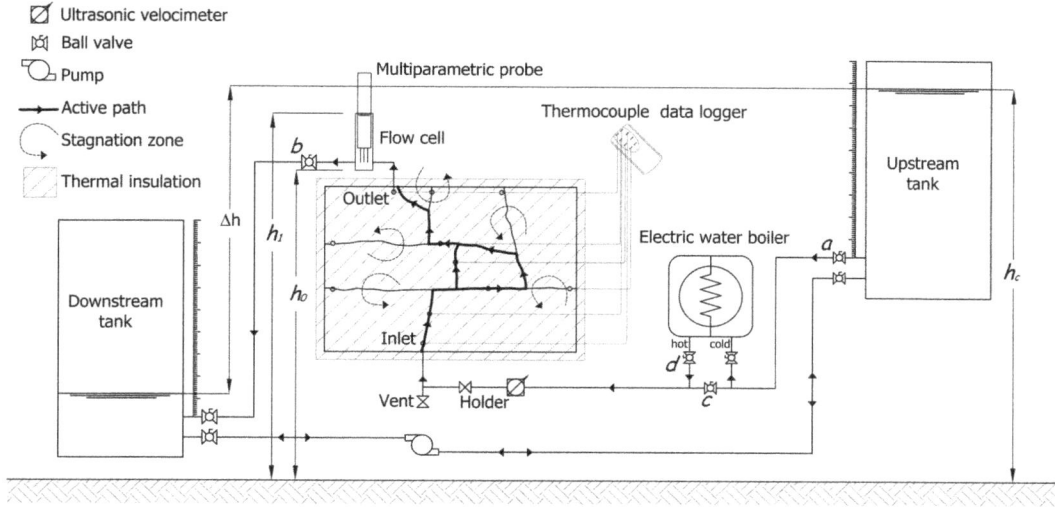

Figure 2. Schematic diagram of the experimental setup.

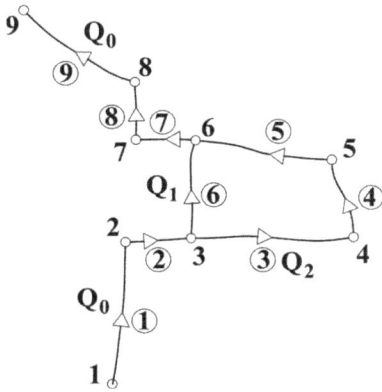

Figure 3. Two-dimensional pipe network conceptualization of the fracture network of the fractured rock block in Fig. 1. Q_0 is the injection flow rate; Q_1 and Q_2 are the flow rates that flow in parallel branches 6 and 3–5, respectively.

flow rate for which $Fo = 1$ is derived as the ratio between a and b resulting in $Q_{crit} = 6.30 \times 10^{-6} \, \mathrm{m^3 \, s^{-1}}$.

Because of the nonlinearity of flow, varying the inlet flow rate Q_0, the ratio between the flow rates Q_1 and Q_2 flowing, respectively, in branches 6 and 3–5 is not constant. When Q_0 increases, Q_2 increases faster than Q_1. The probability of the water distribution of the branch 6 $P_{Q,6}$ is evaluated as the ratio between Q_0 and Q_1, whereas the probability of the water distribution of branches 3–5 is equal to $P_{Q,3-5} = 1 - P_{Q,6}$.

4.2 Fitting of breakthrough curves and interpretation of estimated model parameters

The behavior of mass and heat transport has been compared by varying the injection flow rates. In particular, 21 tests in

the range 1.83×10^{-6}–$1.26 \times 10^{-5} \, \mathrm{m^3 \, s^{-1}}$ (Re in the range 17.5–78.71) for heat transport have been performed and compared with the 55 tests in the range 1.32–$8.34 \times 10^{-6} \, \mathrm{m^3 \, s^{-1}}$ (Re in the range 8.2–52.1) for solute transport presented in previous studies.

The observed heat and mass BTCs for different flow rates have been individually fitted using the ENM approach presented in Sect. 2.3. For simplicity, the transport parameters u_f, D_f and α are assumed equal for all branches of the fracture network. The probability of mass and heat distribution is assumed equal to the probability of water distribution.

The experimental BTCs are fitted using Eqs. (36) and (37) for mass and heat transport, respectively. Note that, for mass transport, $c_{inj}(t)$ supposing the instantaneous injection condition becomes a Dirac delta function.

The determination coefficient (r^2) and the root mean square error (RMSE) have been used in order to evaluate the goodness of fit.

Tables 1 and 2 show the values of transport parameters, the RMSE and r^2 for mass and heat transport, respectively. Furthermore Figs. 4 and 5 show the fitting results of BTCs for some values of Q_0.

The results presented in Tables 1 and 2 highlight that the estimated convective velocities u_f for heat transport are lower than for mass transport, whereas the estimated dispersion D_f for heat transport is higher than for mass transport. Regarding the transfer rate coefficient α, it assumes very low values for mass transport relative to the convective velocity. Instead, for heat transport the exchange rate coefficient is on the same order of magnitude of the convective velocity and, considering a characteristic length equal to $L = 0.601$ m corresponding to the length of the main path of the fracture network, the effect of dual porosity is very strong and cannot be neglected relative to the investigated injection flow range.

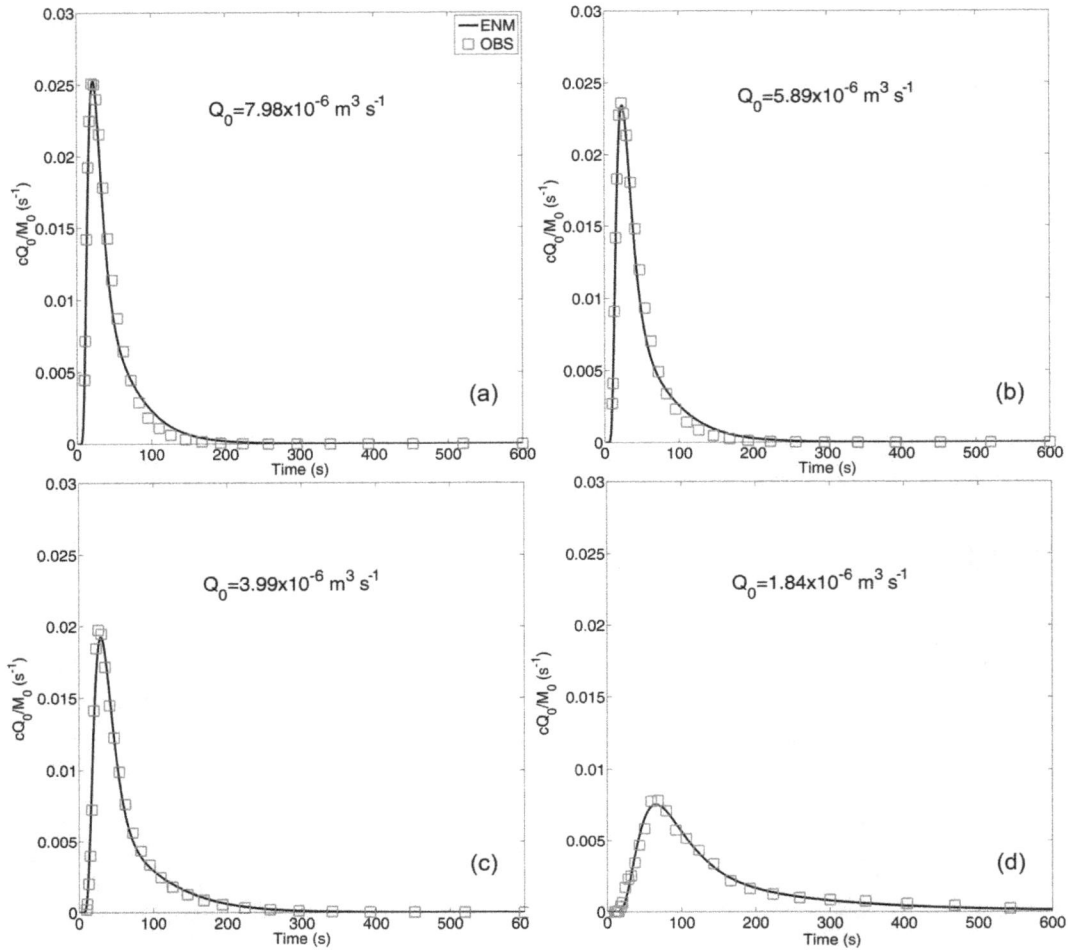

Figure 4. Fitting of BTCs at different injection flow rates using the ENM with Tang's solution for mass transport. The green square curve is the observed specific mass flux at the outlet port; the continuous black line is the simulated specific mass flux.

Both mass and heat transport show a satisfactory fitting. In a particular manner, *RMSE* varies in the range 0.0015–0.0180 for mass transport and in the range 0.0030–0.236 for heat transport, whereas r^2 varies in the range 0.9863–0.9987 for mass transport and in the range 0.0963–0.9998 for heat transport.

In order to investigate the different behavior between mass and heat transport, the relationships between injection flow rate and the transport parameters have been analyzed. In Fig. 6 the relationship between u_f and Q_0 is reported, whereas in Figs. 7 and 8 the dispersion coefficient D_f and the exchange term α as a function of u_f, respectively, are reported. The figures show a very different behavior between mass and heat transport.

Regarding mass transport experiments according to previous studies (Cherubini at al., 2013a, b, c, 2014), Fig. 5 shows that for values of Q_0 higher than $4 \times 10^{-6}\,\mathrm{m}^3\,\mathrm{s}^{-1}$ u_f increases less rapidly. This behavior was due to the presence

of inertial forces that gave rise to a retardation effect on solute transport.

Instead, Fig. 7 shows a linear relationship between u_f and D_f, suggesting that inertial forces did not exert any effect on dispersion and that geometrical dispersion dominates the Aris–Taylor dispersion.

In the same way as for mass transport, for heat transfer a linear relationship is evident between dispersion and convective velocity. Even if heat convective velocity is lower than solute advective velocity, the longitudinal thermal dispersivity assumes higher values than the longitudinal solute dispersivity. Also, for heat transport experiments, a linear relationship between u_f and D_f has been found.

Figure 8 shows the exchange rate coefficient α as a function of the convective velocity u_f for both mass and heat transport.

Regarding the mass transport, the estimated exchange rate coefficient α is much lower than the convective velocity. These results suggest that in the case study fracture–matrix

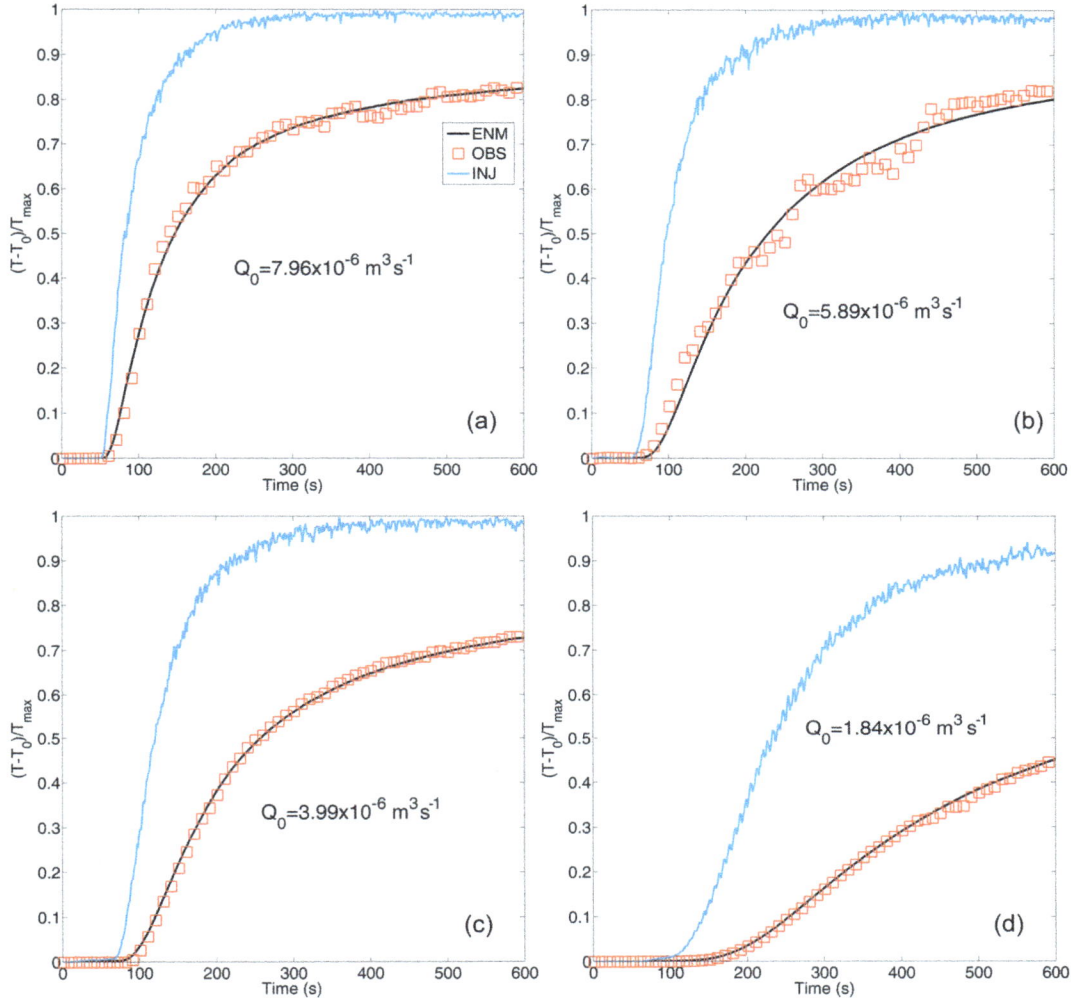

Figure 5. Fitting of BTCs at different injection flow rates using the ENM with Tang's solution for heat transport. The blue curve is the temperature observed at the inlet port used as the temperature injection function, the red square curve is the observed temperature at the outlet port, and the black continuous curve is the simulated temperature at the outlet port.

exchange is very slow and that it may not influence mass transport. Non-Fickian behavior observed in the experimental BTCs is therefore dominated mainly by the presence of inertial forces and the parallel branches.

A very different behavior is observed for heat transport. Heat convective velocity does not seem to be influenced by the presence of the inertial force, whereas u_f is influenced by fracture–matrix exchange phenomena resulting in a significant retardation effect. Once the model parameters for each flow rate have been determined, the unit response function (f_{URF}), corresponding to the PDF obtained from impulsive injection of both solute and temperature tracers, is obtained. The unit response function can be characterized using the time moments and tail character analysis.

The mean residence time t_m assumes the following expression:

$$t_m = \frac{\int_0^\infty t\, f_{URF}(t)\, dt}{\int_0^\infty f_{URF}(t)\, dt}, \tag{45}$$

whereas the nth normalized central moment of distribution of f_{URF} versus time can be written as

$$\mu_n = \frac{\int_0^\infty (t - t_m)^n f_{URF}(t)\, dt}{\int_0^\infty f_{URF}(t)\, dt}. \tag{46}$$

The second moment μ_2 can be used in order to evaluate the dispersion relative to t_m, whereas the skewness is a measure

Table 1. Estimated values of parameters, RMSE, and determination coefficient r^2 for the ENM with Tang's solution at different injection flow rates for mass transport.

Injection flow rate Q_0 (m^3 s^{-1}) $\times 10^{-6}$	Convective velocity u_f (ms^{-1}) $\times 10^{-3}$	Dispersion D_f (ms^{-2}) $\times 10^{-3}$	Exchange rate coefficient (s^{-1}) $\times 10^{-6}$	RMSE	r^2
1.319	4.38 ÷ 4.47	0.68 ÷ 0.70	4.80 ÷ 5.06	0.0053	0.9863
1.843	6.21 ÷ 6.28	0.57 ÷ 0.58	2.86 ÷ 3.01	0.0026	0.9954
2.234	6.54 ÷ 6.59	0.66 ÷ 0.67	3.09 ÷ 3.13	0.0017	0.9976
2.402	7.64 ÷ 7.68	0.67 ÷ 0.67	2.65 ÷ 2.68	0.0015	0.9983
2.598	9.88 ÷ 9.94	0.80 ÷ 0.82	2.76 ÷ 2.84	0.0015	0.9987
2.731	8.27 ÷ 8.35	0.75 ÷ 0.76	2.80 ÷ 2.91	0.0018	0.9977
2.766	8.35 ÷ 8.41	0.84 ÷ 0.85	2.65 ÷ 2.69	0.0021	0.9978
3.076	11.33 ÷ 11.43	0.89 ÷ 0.91	2.53 ÷ 2.59	0.0029	0.9982
3.084	10.86 ÷ 10.95	0.87 ÷ 0.89	3.11 ÷ 3.18	0.0022	0.9982
4.074	15.88 ÷ 16.02	1.19 ÷ 1.21	2.89 ÷ 2.94	0.0048	0.9979
4.087	15.07 ÷ 15.20	1.11 ÷ 1.13	3.75 ÷ 3.83	0.0045	0.9976
4.132	14.71 ÷ 14.82	1.08 ÷ 1.09	2.93 ÷ 2.98	0.0028	0.9985
4.354	15.63 ÷ 15.77	1.14 ÷ 1.16	3.24 ÷ 3.30	0.0052	0.9979
4.529	17.05 ÷ 17.21	1.30 ÷ 1.32	2.88 ÷ 2.94	0.0055	0.9978
5.852	19.26 ÷ 19.38	1.44 ÷ 1.46	4.21 ÷ 4.25	0.0042	0.9983
5.895	19.38 ÷ 19.54	1.37 ÷ 1.39	3.77 ÷ 3.82	0.0058	0.9981
6.168	18.98 ÷ 19.17	1.36 ÷ 1.39	2.87 ÷ 2.92	0.0091	0.9973
7.076	20.64 ÷ 20.86	1.36 ÷ 1.39	3.33 ÷ 3.39	0.0123	0.9963
7.620	20.47 ÷ 20.75	1.52 ÷ 1.55	2.33 ÷ 2.39	0.0180	0.9951
7.983	21.33 ÷ 21.58	1.61 ÷ 1.64	2.92 ÷ 2.98	0.0137	0.9965
8.345	21.71 ÷ 21.97	1.65 ÷ 1.68	2.81 ÷ 2.86	0.0136	0.9964

Table 2. Estimated values of parameters, RMSE, and determination coefficient r^2 for the ENM with Tang's solution at different injection flow rates for heat transport.

Injection flow rate Q_0 (m^3 s^{-1}) $\times 10^{-6}$	Convective velocity u_f (ms^{-1}) $\times 10^{-3}$	Dispersion D_f (ms^{-2}) $\times 10^{-3}$	Exchange rate coefficient (s^{-1}) $\times 10^{-3}$	RMSE	r^2
1.835	2.20 ÷ 2.91	1.91 ÷ 1.95	6.27 ÷ 6.59	0.0065	0.9997
2.325	1.74 ÷ 2.73	1.82 ÷ 1.91	5.39 ÷ 9.26	0.0098	0.9992
2.462	0.35 ÷ 0.52	2.42 ÷ 2.57	2.25 ÷ 2.33	0.0138	0.9984
2.605	0.44 ÷ 0.54	2.33 ÷ 2.40	0.74 ÷ 0.77	0.0073	0.9995
2.680	2.18 ÷ 2.95	1.77 ÷ 1.83	5.68 ÷ 8.31	0.0030	0.9998
2.800	0.36 ÷ 0.79	2.53 ÷ 2.68	3.54 ÷ 3.72	0.0213	0.9982
2.847	1.73 ÷ 3.16	1.98 ÷ 2.06	4.95 ÷ 13.45	0.0283	0.9978
3.003	2.34 ÷ 2.87	2.24 ÷ 2.32	5.33 ÷ 6.55	0.0033	0.9998
3.998	2.56 ÷ 2.75	6.63 ÷ 6.80	2.05 ÷ 2.11	0.0150	0.9993
4.030	2.60 ÷ 2.83	7.18 ÷ 7.36	1.42 ÷ 1.52	0.0147	0.9993
4.217	3.85 ÷ 4.56	8.92 ÷ 9.29	4.86 ÷ 5.77	0.0228	0.9945
4.225	2.43 ÷ 2.64	7.53 ÷ 7.84	1.64 ÷ 1.80	0.0251	0.9987
4.471	2.30 ÷ 3.13	9.18 ÷ 9.50	1.06 ÷ 1.33	0.1115	0.9957
5.837	3.51 ÷ 4.13	4.95 ÷ 5.36	0.61 ÷ 0.79	0.2360	0.9872
5.880	2.71 ÷ 3.10	4.23 ÷ 4.60	0.04 ÷ 0.05	0.1997	0.9926
6.445	4.71 ÷ 5.12	6.18 ÷ 6.81	1.49 ÷ 1.54	0.2156	0.9863
7.056	8.15 ÷ 8.46	10.05 ÷ 10.74	5.63 ÷ 6.00	0.0694	0.9951
7.959	9.64 ÷ 10.11	18.40 ÷ 19.47	10.92 ÷ 11.55	0.0662	0.9971
8.971	13.40 ÷ 13.79	24.57 ÷ 25.82	15.35 ÷ 15.85	0.0303	0.9985
12.364	11.01 ÷ 11.67	21.97 ÷ 22.63	5.23 ÷ 5.25	0.0631	0.9939
12.595	13.71 ÷ 14.26	26.65 ÷ 27.61	9.26 ÷ 9.41	0.0426	0.9955

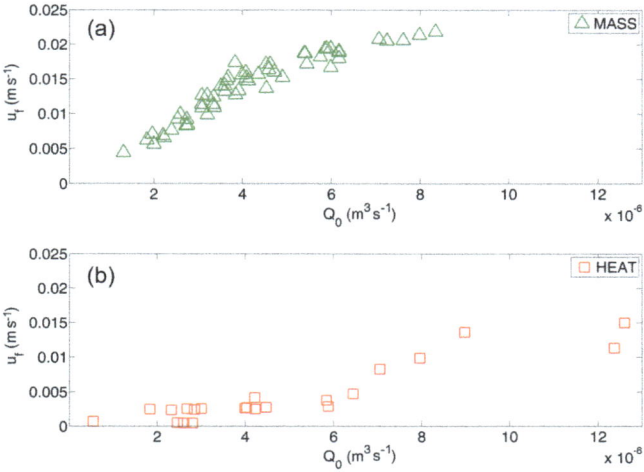

Figure 6. Velocity u_f (m s^{-1}) as a function of the injection flow rate Q_0 (m^3 s^{-1}) for the ENM with Tang's solution for both mass and heat transport.

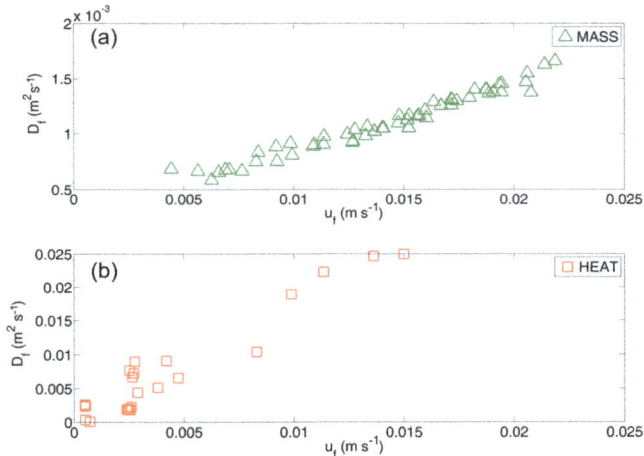

Figure 7. Dispersion D_f (m s^{-2}) as a function of velocity u_f (m s^{-1}) for the ENM with Tang's solution for both mass and heat transport.

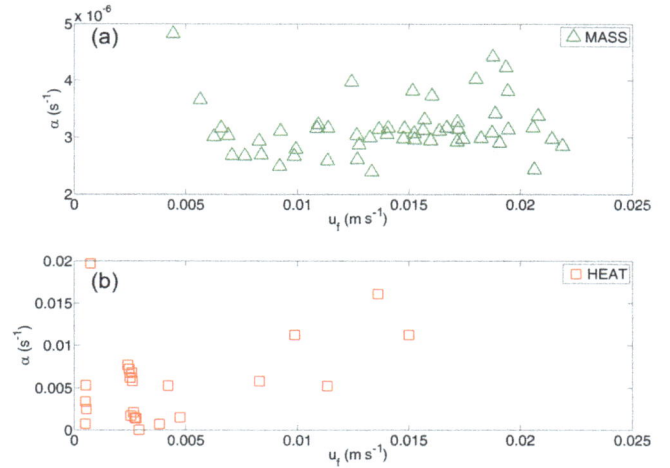

Figure 8. Transfer coefficient α (s^{-1}) as a function of velocity u_f (m s^{-1}) for both mass and heat transport.

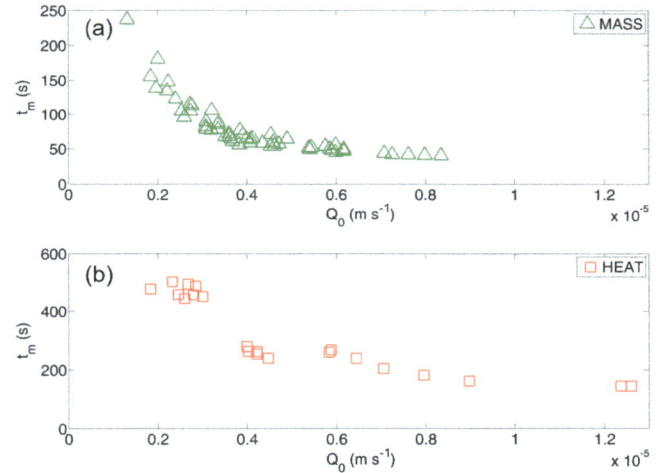

Figure 9. Mean travel time t_m (s) as a function of the injection flow rate for both mass and heat transport.

of the degree of asymmetry and is defined as follows:

$$S = \mu_3/\mu_2^{3/2}. \tag{47}$$

The tailing character t_c can be described as

$$t_c = \frac{\Delta t_{\text{fall}}}{\Delta t_{\text{rise}}}, \tag{48}$$

where Δt_{fall} denotes the duration of the falling limb defined as the time interval from the peak to the tail cutoff, which is the time when the falling limb first reaches a value that is 0.05 times the peak value. Δt_{rise} is defined as the time interval from the first arrival to the peak. This quantity provides a measure of the asymmetry between the rising and falling limbs. A value of t_c significantly higher than 1 indicates an elongated tail compared to the rising limb (Cherubini et al., 2010).

In Fig. 9 is reported the residence time versus the injection flow rates. The figure highlights that t_m for heat transport is about 3 times higher than for mass transport. In a particular way, t_m varies in the range 40.3–237.1 s for mass transport and in the range 147.8–506.9 s for heat transport. This result still highlights that heat transport is more delayed than mass transport.

In the same way the skewness S (Fig. 10) and the tailing character t_c (Fig. 11) are reported as a function of Q_0.

A different behavior for heat and mass transport is observed for the skewness coefficient. For heat transfer the skewness shows a growth trend which seems to decrease af-

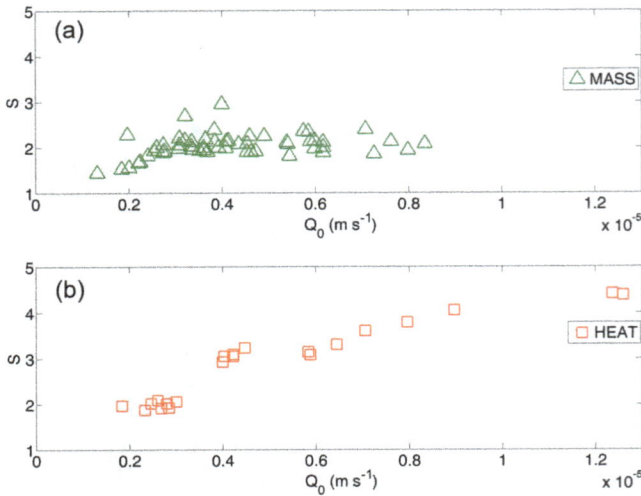

Figure 10. Skewness as a function of the injection flow rate for both mass and heat transport.

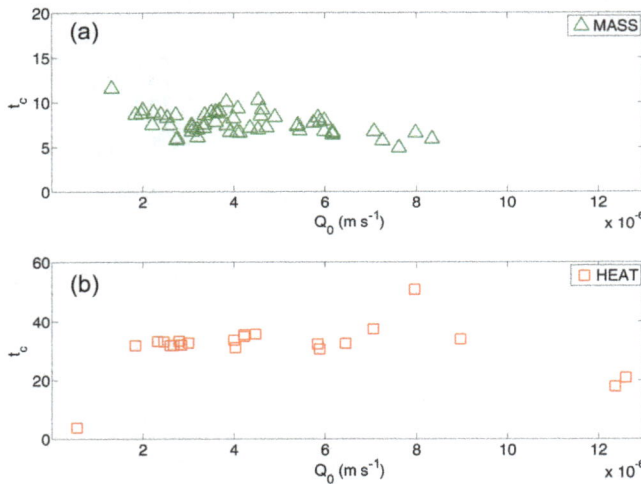

Figure 11. Tailing character t_c as a function of the injection flow rate for both mass and heat transport.

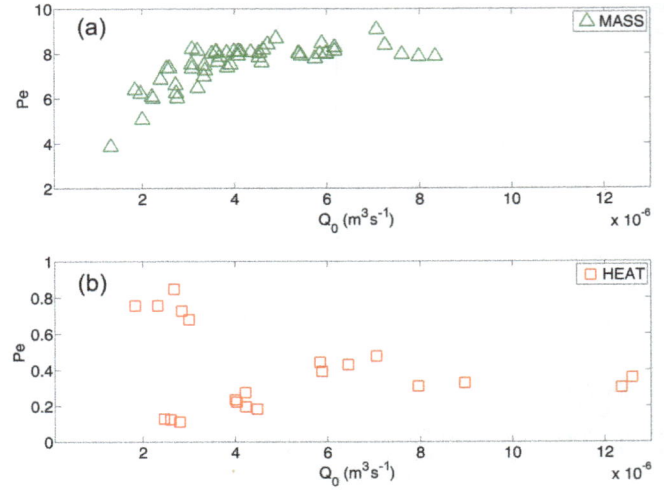

Figure 12. Peclet number as a function of the injection flow rate Q_0 ($m^3\,s^{-1}$) for both mass and heat transport.

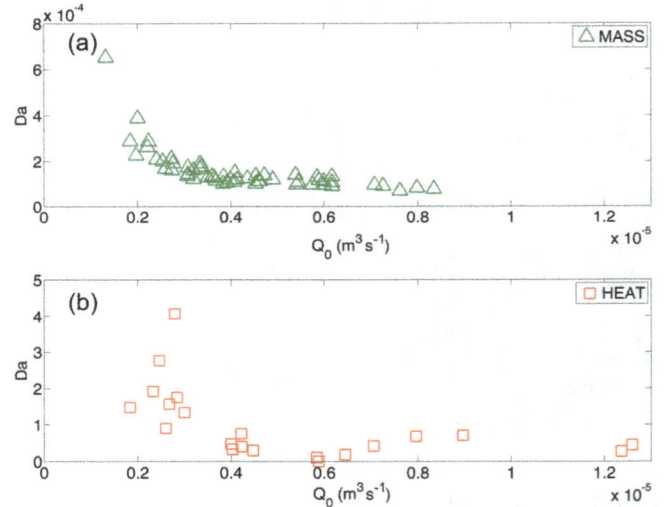

Figure 13. Da number as a function of the injection flow rate Q_0 ($m^3\,s^{-1}$) for both mass and heat transport.

ter $Q_0 = 3 \times 10^{-6}\,m^3\,s^{-1}$. Its mean value is equal to 2.714. For solute transport the S does not show a trend, and assumes a mean value equal to 2.018.

The tailing character does not exhibit a trend for either mass and heat transport. In either cases t_c is significantly higher than 1, specifically 7.70 and 30.99 for mass and heat transport, respectively.

In order to explain the transport dynamics, the trends of dimensionless numbers Pe and Da varying the injection flow rate have been investigated. Figure 12 shows the Pe as a function of Q_0 for both mass and heat experiments. As concerns mass experiments, Pe increases as Q_0 increases, assuming a constant value for high values ($Pe = 7.5$) of Q_0. For heat transport a different behavior is observed, P_e showing a con-

stant trend and being always lower than one. Even if the injection flow rate is relatively high, thermal dispersion is the dominating mechanism in heat transfer.

Figure 12 reports Da as a function of Q_0. For mass transport Da assumes very low values, on the order of magnitude of 10^{-4}.

The convective transport scale is very low with respect to the exchange transport scale; thus, the mass transport in each single fracture can be represented with the classical advection dispersion model.

As regards heat transport, Da assumes values around the unit showing a downward trend as injection flow rate increases, switching from higher to lower values than the unit.

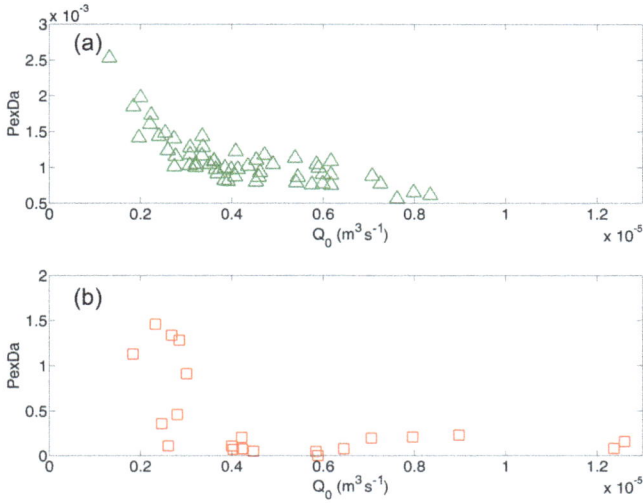

Figure 14. $Pe \times Da$ number as a function of the injection flow rate Q_0 (m^3 s^{-1}) for both mass and heat transport.

As injection flow rate increases, the convective transport timescale reduces more rapidly than the exchange timescale.

These arguments can be explained because the relationships between Q_0 and u_f show a change in slope when Da becomes lower than the unit. In other words, when Da is higher than the unit, the exchange between fracture and matrix dominates on the convective transport, giving rise to a more enhanced delay on heat transport; conversely, when Da is lower than one, convective transport dominates on fracture–matrix interactions and the delay effect is reduced.

Furthermore this effect is evident also in the trend observed in the graph $S - Q_0$ (Fig. 10). For values of Da lower than the unit, a change in slope is evident; the skewness coefficient increases more slowly. Thus for $Da > 1$ the early arrival and the tail effect of BTC increase more rapidly than for $Da < 1$.

Note that even if Da presents a downward trend as Q_0 increases, when the latter exceeds Q_{crit} a weak growth trend for Da is detected, which however assumes values lower than the unit.

Figure 14 shows the dimensionless group $Pe \times Da$ varying the injection flow rate. Regarding mass transport, $Pe \times Da$ is on the order of magnitude of 10^{-3}, confirming the fact that the fracture–matrix interaction can be neglected relative to the investigated range of injection flow rates. For heat transport, $Pe \times Da$ assumes values just below the unit, with a downward trend as Q_0 increases. t_d and t_e have the same order of magnitude.

In order to find the optimal conditions for heat transfer in the analyzed fractured medium, the thermal power exchanged per unit temperature difference $\dot{Q}$ (ML^2 T^{-1} K^{-1}) for each injection flow rate in quasi-steady-state conditions can be estimated. The thermal power exchanged can be writ-

ten as

$$\dot{Q} = \rho C_w Q_0 \left(T_{\mathrm{inj}} - T_{\mathrm{out}}\right). \tag{49}$$

The outlet temperature T_{out} can be evaluated as a function of the f_{URF} using the following expression:

$$T_{\mathrm{out}} = T_0 + \left(T_{\mathrm{inj}} - T_0\right) \int\limits_0^\infty f_{\mathrm{URF}}\left(t\right) \mathrm{d}t. \tag{50}$$

Substituting Eq. (50) into Eq. (49), the thermal power exchanged per unit temperature difference is

$$\frac{\dot{Q}}{\left(T_{\mathrm{inj}} - T_0\right)} = \left(1 - \int\limits_0^\infty f_{\mathrm{URF}}\left(t\right) \mathrm{d}t\right) \rho C_W Q_0. \tag{51}$$

Figure 15 shows the similarities between the relationship $\dot{Q} / \left(T_{\mathrm{inj}} - T_0\right) - Q_0$ (Fig. 15a) and $Da - Q_0$ (Fig. 14b). Higher Da values correspond to higher values of $\dot{Q} / \left(T_{\mathrm{inj}} - T_0\right)$. The thermal power exchanged increases as the Damköhler number increases, as shown in Fig. 15c. These results highlight that for the observed case study the optimal condition for thermal exchange in the fractured medium is obtained when the exchange timescale is lower than the convective transport scale, or rather when the dynamics of fracture–matrix exchange are dominant on the convective ones.

Moreover, in a similar way to Da, $\dot{Q} / \left(T_{\mathrm{inj}} - T_0\right)$ shows a weak growth trend when Q_0 exceeds Q_{crit}. This means that the nonlinear flow regime improves the fracture–matrix thermal exchange; however, at high values of injection flow rates, convective and dispersion timescales are less than the exchange timescale. Nevertheless, these results have been observed in a small range of Da numbers close to the unit. In order to generalize these results, a larger range of Da numbers should be investigated.

In order to estimate the effective thermal conductivity coefficient k_e, the principle of conservation of heat energy can be applied to the whole fractured medium. Neglecting the heat stored in the fractures, the difference between the heat measured at the inlet and at the outlet must be equal to the heat diffused into the matrix:

$$\rho C_W Q_0 \left(T_{\mathrm{inj}} - T_{\mathrm{out}}\right) = \int\limits_{A_f} k_e \left.\frac{\mathrm{d}T_m}{\mathrm{d}z}\right|_{z=wf/2} \mathrm{d}A_f, \tag{52}$$

where A_f is the whole surface area of the whole active fracture network and the gradient of T_m can be evaluated according to Eq. (19) using temperature instead of concentration as a variable. Then the average effective thermal conductivity $\overline{k}_e$ can be obtained as

$$\overline{k}_e = \frac{\rho_w C_w Q_0 \left(T_{\mathrm{inj}} - T_{\mathrm{out}}\right)}{\int\limits_{A_f} \left.\frac{\mathrm{d}T}{\mathrm{d}z}\right|_{z=wf/2} \mathrm{d}A_f}. \tag{53}$$

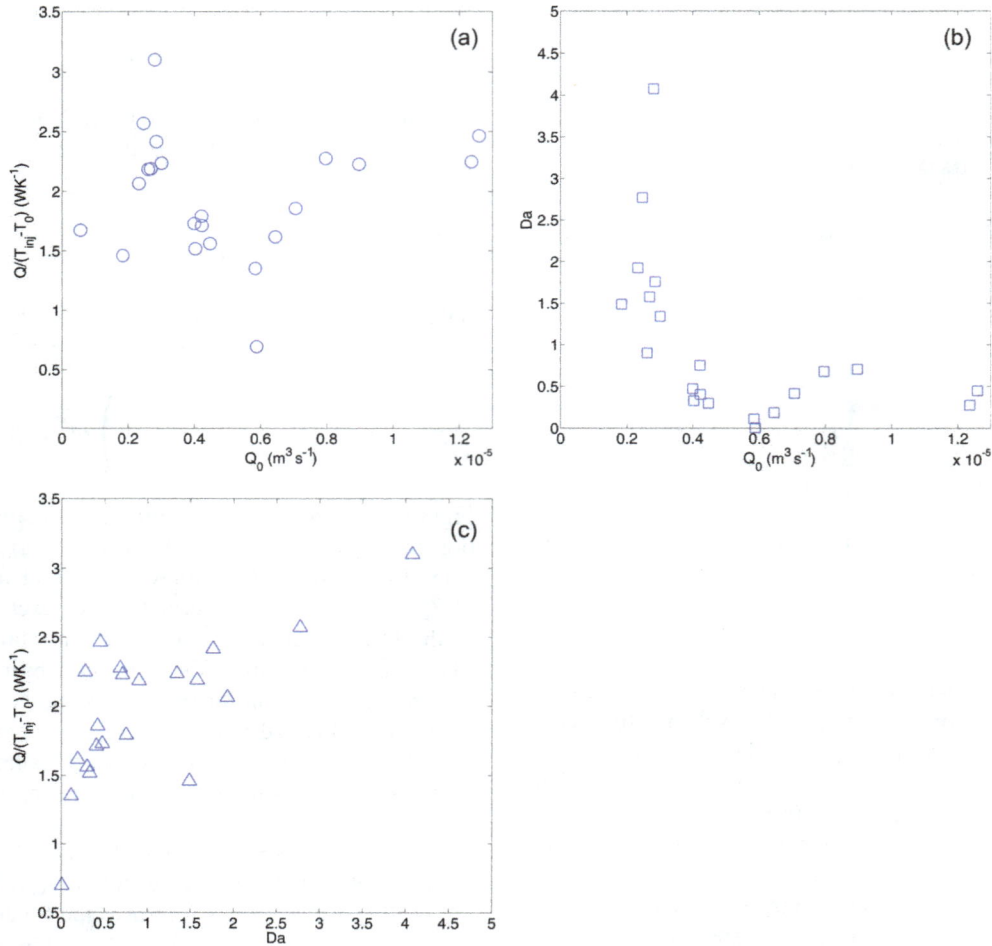

Figure 15. Heat power exchanged per difference temperature unit $\dot{Q}/(T_{\text{inj}} - T_0)$ as a function of the injection flow rate Q_0 (m^3 s^{-1}) **(a)**, Damköhler number Da as a function of the injection flow rate **(b)**, and power exchanged per difference temperature unit as a function of the Damköhler number **(c)**.

The average effective thermal conductivity has been estimated for each injection flow rate (Fig. 16) and assumes a mean value equal to $\overline{k}_e = 0.1183\,\text{Wm}^{-1}\,\text{K}^{-1}$. The estimated $\overline{k}_e$ is 1 order of magnitude lower than the thermal conductivity coefficient reported in the literature (Robertson, 1988). Fractured media have a lower capacity for diffusion, as opposed to Tang's model, which has unlimited capacity. There is a solid thermal resistance in the fluid to solid heat transfer processes which depends on the rock–fracture size ratio.

This result is coherent with previous analyses on heat transfer carried out on the same rock sample (Pastore et al., 2015). In this study Pastore et al. (2015) found that the ENM model failed to model the behavior of heat transport, in correspondence to parallel branches where the hypothesis of Tang's solution of a single fracture embedded in a porous medium having unlimited capacity cannot be considered valid. In parallel branches the observed BTCs are char-

acterized by less retardation of heat propagation as opposed to the simulated BTCs.

5 Conclusions

Aquifers offer a possibility of exploiting geothermal energy by withdrawing the heat from groundwater by means of a heat pump and subsequently supplying the water back into the aquifer through an injection well. In order to optimize the efficiency of the heat transfer system and minimize the environmental impacts, it is necessary to study the behavior of convective heat transport especially in fractured media, where flow and heat transport processes are not well known.

Laboratory experiments on the observation of mass and heat transport in a fractured rock sample have been carried out in order to analyze the contribution of thermal dispersion in heat propagation processes, the contribution of nonlinear flow dynamics to the enhancement of thermal matrix diffu-

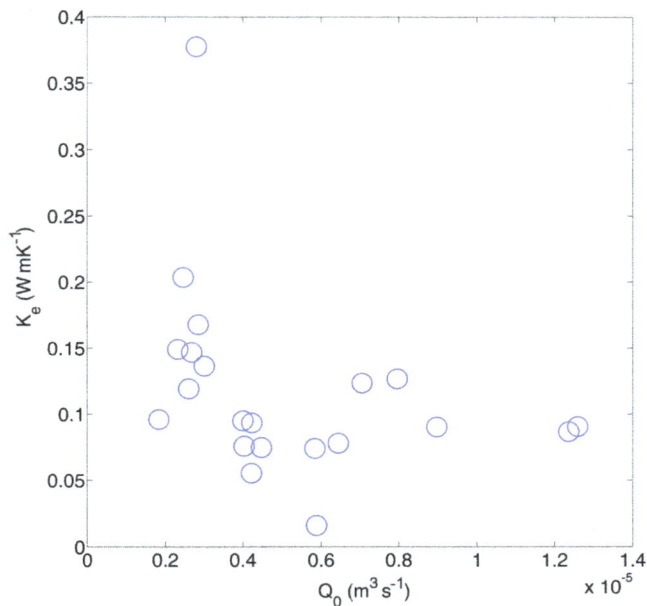

Figure 16. Effective thermal conductivity k_e $(Wm^{-1}K^{-1})$ as a function of the injection flow rate Q_0 $(m^3 s^{-1})$.

sion, and finally the optimal heat recovery and heat dissipation strategies.

The parameters that control mass and heat transport have been estimated using the ENM model based on Tang's solution.

Heat transport shows a very different behavior compared to mass transport. The estimated transport parameters show differences of several orders of magnitude. Convective thermal velocity is lower than solute velocity, whereas thermal dispersion is higher than solute dispersion, mass transfer rate assumes a very low value, suggesting that fracture–matrix mass exchange can be neglected. Non-Fickian behavior of observed solute BTCs is mainly due to the presence of the secondary path and the nonlinear flow regime. Contrarily, heat transfer rate is comparable with convective thermal velocity giving rise to a retardation effect on heat propagation in the fracture network.

The discrepancies detected in transport parameters are moreover observable through the time moment and tail character analysis which demonstrate that the dual porosity behavior is more evident in the thermal BTCs than in the solute BTCs.

The dimensionless analysis carried out on the transport parameters proves that, as the injection flow rate increases, thermal convection timescale decreases more rapidly than the thermal exchange timescale, explaining the reason why the relationship $Q_0 - u_f$ shows a change in slope for Da lower than the unit.

Thermal dispersion dominates heat transport dynamics and the Peclet number, and the product between the Peclet number and the Damköhler number is almost always less than the unit.

The optimal conditions for thermal exchange in a fracture network have been investigated. The power exchanged increases in a potential way as Da increases in the observed range.

The Explicit Network Model is an efficient computation methodology to represent flow, mass and heat transport in fractured media, as 2-D and/or 3-D problems are reduced to resolving a network of 1-D pipe elements. Unfortunately, in field case studies, it is difficult to obtain full knowledge of the geometry and parameters such as the orientations and aperture distributions of the fractures needed by the ENM, even by means of field investigation methods. However, in real case studies the ENM can be coupled with continuum models in order to represent greater discontinuities with respect to the scale of study, which generally gives rise to preferential pathways for flow, mass and heat transport.

A method to represent the topology of the fracture network is represented by multifractal analysis as discussed in Tijera at al. (2009) and Tarquis at al. (2014).

This study has permitted one to detect the key parameters to design devices for heat recovery and heat dissipation that exploit the convective heat transport in fractured media.

Heat storage and transfer in fractured geological systems is affected by the spatial layout of the discontinuities.

Specifically, the rock–fracture size ratio which determines the matrix block size is a crucial element in determining matrix diffusion on the fracture–matrix surface.

The estimation of the average effective thermal conductivity coefficient shows that it is not efficient to store thermal energy in rocks with high fracture density because the fractures are surrounded by a matrix with more limited capacity for diffusion giving rise to an increase in solid thermal resistance. In fact, if the fractures in the reservoir have a high density and are well connected, such that the matrix blocks are small, the optimal conditions for thermal exchange are not reached, as the matrix blocks have a limited capability to store heat.

On the other hand, isolated permeable fractures will tend to lead to more distribution of heat throughout the matrix.

Therefore, subsurface reservoir formations with large porous matrix blocks will be the optimal geological formations to be exploited for geothermal power development.

The study could help to improve the efficiency and optimization of industrial and environmental systems, and may provide a better understanding of geological processes involving transient heat transfer in the subsurface.

Future developments of the current study will be carrying out investigations and experiments aimed at further deepening of the quantitative understanding of how fracture arrangement and matrix interactions affect the efficiency of storing and dissipating thermal energy in aquifers. This could be achieved by means of using different formations with different fracture density and matrix porosity.

Competing interests. The authors declare that they have no conflict
of interest.

Acknowledgements. Research founded with the regional program
in support of the smart specialization and social and environmental
sustainability – FutureInResearch.

Edited by: J. M. Redondo

References

Abate, J. and Ward, W.: A unified Framework for numerically in-
verting laplace transforms, INFORMS J. Comp., 18, 408–421,
2006.

Anderson, M. P.: Heat as a ground water tracer, Ground Water, 43,
951–968, doi:10.1111/j.1745-6584.2005.00052.x, 2005.

Auradou, H., Deazerm, G., Boschan, A., Hulin, J., and Koplik, J.:
Flow channeling in a single fracture induced by shear displacement, Geothermics, 35, 575–588, 2006.

Bear, J.: Dynamics of Fluids in Porous Media, Environmental Science Series, Elsevier, Amsterdam, 764, SD-008, 1972.

Becker, M. W. and Shapiro, A. M.: Interpreting tracer breakthrough
tailing from different forced gradient tracer experiment config-
urations in fractured bedrock, Water Resour. Res., 39, 1024,
doi:10.1029/2001WR001190, 2003.

Bodin, J., Porel, G., Delay, F., Ubertosi, F., Bernard, S., and de
Dreuzy, J. R.: Simulation and analysis of solute transport in 2D
fracture/pipe networks, The SOLFRAC program, J. Contam. Hydrol., 89, 1–28, 2007.

Bravo, H. R., Jiang, F., and Hunt, R. J.: Using groundwater temperature data to constrain parameter estimation in a groundwater
flow model of a wetland system, Water Resour. Res., 38, 1153,
doi:10.1029/2000WR000172, 2002.

Bredehoeft, J. and Papadopulos, I. S.: Rates of vertical groundwater movement estimated from the Earth's thermal profile, Water
Resour. Res., 1, 325–328, 1965.

Cherubini, C.: A modeling approach for the study of contamina-
tion in a fractured aquifer, Geotech. Geol. Eng., 26, 519–533,
doi:10.1007/s10706-008-9186-3, 2008.

Cherubini, C. and Pastore, N.: Modeling contaminant propagation
in a fractured and karstic aquifer, Fresen. Environ. Bull., 19,
1788–1794, 2010.

Cherubini, C. and Pastore, N.: Critical stress scenarios for a coastal
aquifer in southeastern Italy, Nat. Hazards Earth Syst. Sci., 11,
1381–1393, doi:10.5194/nhess-11-1381-2011, 2011.

Cherubini, C., Pastore, N., and Francani, V.: Different approaches
for the characterization of a fractured karst aquifer, WSEAS
Trans. Fluid Mech., 1, 29–35, 2008.

Cherubini, C., Giasi, C. I., and Pastore, N.: Application of Mod-
elling for Optimal Localisation of Environmental Monitoring
Sensors, Proceedings of the Advances in sensor and Interfaces,
Trani, Italy, 222–227, 2009.

Cherubini, C., Hsieh, P. A., and Tiedeman, C. R.: Modeling the effect of heterogeneity on forced-gradient flow tracer tests in heterogeneous aquifers, I Congreso Internacional de Hidrologia de

Lianuras Azul, Buenos Aires, Argentina, 21–24 September, 809–
816, 2010.

Cherubini, C., Giasi, C. I., and Pastore, N.: Bench scale laboratory tests to analyze non-linear flow in fractured media, Hydrol. Earth Syst. Sci., 16, 2511–2522, doi:10.5194/hess-16-2511-
2012, 2012.

Cherubini, C., Giasi, C. I., and Pastore, N.: Evidence of non-Darcy
flow and non-Fickian transport in fractured media at laboratory
scale, Hydrol. Earth Syst. Sci., 17, 2599–2611, doi:10.5194/hess-
17-2599-2013, 2013a.

Cherubini, C., Giasi, C. I., and Pastore, N.: Laboratory tests to analyze solute transport behavior in fractured media, Rendiconti
Online Società Geologica Italiana, 24, 55–57, 2013b.

Cherubini, C., Giasi, C. I., and Pastore, N.: A laboratory physical
model to analyse flow and transport processes in fractured rock
sample at bench scale level, J. Eng. Geol. Environ., 1, 19–32,
2013c.

Cherubini, C., Giasi, C. I., and Pastore, N.: Fluid flow modeling of a
coastal fractured karstic aquifer by means of a lumped parameter
approach, Environ. Earth Sci., 70, 2055–2060, 2013d.

Cherubini, C., Giasi, C. I., and Pastore, N.: On the reliability of analytical models to predict solute transport in a fracture network,
Hydrol. Earth Syst. Sci., 18, 2359–2374, doi:10.5194/hess-18-
2359-2014, 2014.

Constantz, J., Cox, M. H., and Su, G. W.: Comparison of Heat and
Bromide as Ground Water Tracers Near Streams, Ground Water,
41, 647–656, 2003.

de Marsily, G.: Quantitative Hydrogeology: Groundwater Hydrology for Engineers, Academic Press, Orlando, Florida, 440 pp.,
1986.

Domenico, P. A. and Palciauskas, V. V.: Theoretical analysis of
forced convective heat transfer in regional ground-water flow,
Geol. Soc. Am. Bull., 84, 3803–3814, 1973.

Fahien, R. W.: Fundamental of Transport Phenomena, McGraw-
Hill, New York, 614 pp., 1983.

Ferguson, G., Beltrami, H., and Woodbury, A. D.: Perturbation of
ground surface temperature reconstruction by groundwater flow,
Geophys. Res. Lett., 33, L13708, doi:10.1029/2006GL026634,
2006.

Forchheimer, P.: Wasserbewegung durch Boden, Z. Ver. Dtsch. Ing.,
45, 1781–1788, 1901.

Geiger, S. and Emmanuel, S.: Non-fourier thermal transport in
fractured geological media, Water Resour. Res., 46, W07504,
doi:10.1029/2009WR008671, 2010.

Gisladottir, V. R., Roubinet, D., and Tartakovsky, D. M.: Particle
Methods for Heat Transfer in Fractured Media, Transport Porous.
Med., 115, 311–326, doi:10.1007/s11242-016-0755-2, 2016.

Green, D., Perry, R., and Babcock, R.: Longitudinal dispersion of
thermal energy through porous media with a flowing fluid, Aiche
J., 10, 645–651, 1964.

Hao, Y., Fu, P., and Carrigan, C. R.: Application of a dual-
continuum model for simulation of fluid flow and heat transfer
in fractured geothermal reservoir, Proceedings, 38th Workshop
on Geothermal Reservoir Engineering Stanford University, Stanford, California, 11–13 February 2013, SGP-TR-198, 2013.

Hatch, C. E., Fisher, A. T., Revenaugh, J. S., Constantz, J.,
and Ruehl, C.: Quantifying surface water-groundwater interactions using time series analysis of streambed thermal

records: Method development, Water Resour. Res., 42, W10410, doi:10.1029/2005WR004787, 2006.

Hawkins, A. J. and Becker, M. W.: Measurement of the Spatial Distribution of Heat Exchange in a Geothermal Analog Bedrock Site Using Fiber Optic Distributed Temperature Sensing, PROCEEDINGS, 37th Workshop on Geothermal Reservoir Engineering Stanford University, Stanford, California, 30 January–1 February 2012 SGP-TR-194, 2012.

Hopmans, J. W., Simunek, J., and Bristow, K. L.: Indirect estimation of soil thermal properties and water flux using heat pulse probe measurements: Geometry and dispersion effects, Water Resour. Res., 38, 1006, doi:10.1029/2000WR000071, 2002.

Ingebritsen, S. E. and Sanford, W. E.: Groundwater in Geologic Processes, 341 pp., Cambridge Univ. Press, Cambridge, UK, 1998.

Keery, J., Binley, A., Crook, N., and Smith, J. W. N.: Temporal and spatial variability of groundwater-surface water fluxes: Development and application of an analytical method using temperature time series, J. Hydrol., 336, 1–16, 2007.

Klepikova, M. V., Le Borgne, T., Bour, O., Dentz, M., and Hochreutener, R.: Heat as a tracer for understanding transport processes in fractured media: Theory and field assessment from multiscale thermal push-pull tracer tests, Water Resour. Res., 52, 5442–5457, 2016.

Kocabas, I.: Geothermal reservoir characterization via thermal injection backflow and interwell tracer testing, Geothermics, 34, 27–46, 2005.

Lu, W. and Xiang, Y.: Experiments and sensitivity analyses for heat transfer in a meter-scale regularity fracture granite model with water flow, Appl. Phys. Eng. 13, 958–968, 2012.

Ma, R., Zheng, C., Zachara, J. M., and Tonkin, M.: Utility of bromide and heat tracers for aquifer characterization affected by highly transient flow conditions, Water Resour. Res., 48, W08523, doi:10.1029/2011WR011281, 2012.

Martinez, A. R., Roubinet, D., and Tartakovsky, D. M.: Analytical models of heat conduction in fractured rocks, J. Geophys. Res. Sol.-Ea., 119, 83–98, doi:10.1002/2012JB010016, 2014.

Masciopinto, C., Volpe, A., Palmiotta, D., and Cherubini, C.: A combined PHREEQC-2/parallel fracture model for the simulation of laminar/non-laminar flow and contaminant transport with reactions, J. Contam. Hydrol., 117, 94–108, 2010.

Molina-Giraldo, N., Bayer, P., and Blum, P.: Evaluating the influence of thermal dispersion on temperature plumes from geothermal systems using analytical solutions, Int. J. Therm. Sci., 50, 1223–1231, doi:10.1016/j.ijthermalsci.2011.02.004, 2011.

Molson, J. W., Frind, E. O., and Palmer, C. D.: Thermal energy storage in an unconfined aquifer 2. Model development, validation and application, Water Resour. Res., 28, 2857–2867, 1992.

Moonen, P., Sluys, L. J., and Carmeliet, J.: A continous – Discontinuous Approach to simulate heat transfer in fractured media, Transport Porous Med., 89, 399–419, 2011.

Natarajan, N. and Kumar, G. S.: Thermal transport in a coupled sinusoidal fracture-matrix system, Int. J. Eng. Sci. Technol., 2, 2645–2650, 2010.

Neuville, A., Toussaint, R., and Schmittbuhl, J.: Fracture roughness and thermal exchange: a case study at Soultz-sous-Forêts, C. R. Geosci., 342, 616–625, 2010.

Niswonger, R. G. and Prudic, D. E.: Modeling heat as a tracer to estimate streambed seepage and hydraulic conductivity, in: Heat as a Tool for Studying the Movement of Ground Water Near

Streams, edited by: Stonestrom, D. A. and Constantz, J., 81–89, USGS Circular 1260, Reston, Virginia, USGS, 2003.

Ouyang, X. L., Xu, R. N., and Jiang, P. X.: Effective solid-to-fluid heat transfer coefficient in egs reservoirs, Proceedings of the 5th International Conference on porous Media and its Applications in Science and Engineering ICPM5, 22–27 June 2014, Kona, Hawaii, 2014.

Papadopulos, S. S. and Larson, S. P.: Aquifer storage of heated water: Part II – Numerical simulation of field results, Ground Water, 16, 242–248, 1978.

Pastore, N., Cherubini, C., Giasi, C. I., Allegretti, N. M., Redondo, J. M., and Tarquis, A. M.: Experimental study of heat transport in fractured network, Energy Proc., 76, 273–281, 2015.

Rau, G. C., Andersen, M. S., and Acworth, R. I.: Experimental investigation of the thermal dispersivity term and its significance in the heat transport equation for flow in sediments, Water Resour. Res., 48, W03511, doi:10.1029/2011WR011038, 2012.

Read, T., Bour, O., Bense, V., Le Borgne, T., Goderniaux, P., Klepikova, M. V., Hochreutener, R., Lavenant, N., and Boshero, V.: Characterizing groundwater flow and heat transport in fractured rock using fiber-optic distributed temperature sensing, Geophys. Res. Lett., 40 1–5, 2013.

Reiter, M.: Using precision temperature logs to estimate horizontal and vertical groundwater flow components, Water Resour. Res., 37, 663–674, 2001.

Robertson, E. C.: Thermal Properties of Rocks, United States Department of the Interior Geological Survey Open-File Report 88-441, Reston, Virginia, 1988.

Ronan, A. D., Prudic, D. E., Thodal, C. E., and Constantz, J.: Field study and simulation of diurnal temperature effects on infiltration and variably saturated flow beneath an ephemeral stream, Water Resour. Res., 34, 2137–2153, 1998.

Sauty, J. P., Gringarten, A. C., Fabris, H., Thiery, D., Menjoz, A., and Landel, P. A.: Sensible energy storage in aquifers 2. Field experiments and comparison with theoretical results, Water Resour. Res., 18, 253–265, 1982.

Shook, G. M.: Predicting thermal breakthrough in heterogeneous media from tracer tests, Geothermics, 30, 573–580, 2001.

Smith, L. and Chapman, D. S.: On the thermal effects of groundwater flow. 1. Regional scale systems, J. Geophys. Res., 88, 593–608, 1983.

Su, G. W., Jasperse, J., Seymour, D., and Constantz, J.: Estimation of hydraulic conductivity in an alluvial system using temperatures, Ground Water, 42, 890–901, 2004.

Tang, D. H., Frind, E. O., and Sudicky, E. A.: Contaminant transport in fractured porous media: analytical solutions for a single fractures, Water Resour. Res., 17, 555–564, 1981.

Taniguchi, M., Williamson, D. R., and Peck, A. J.: Disturbances of temperature-depth profiles due to surface climate change and subsurface water flow: 2, an effect of step increase in surface temperature caused by forest clearing in southwest-western Australia, Water Resour. Res., 35, 1519–1529, 1999.

Tarquis, A. M., Platonov, A., Matulka, A., Grau, J., Sekula, E., Diez, M., and Redondo, J. M.: Application of multifractal analysis to the study of SAR features and oil spills on the ocean surface, Nonlin. Processes Geophys., 21, 439–450, doi:10.5194/npg-21-439-2014, 2014.

Tijera, M., Cano, J., Cano, D., Bolster, D., and Redondo, J. M.: Filtered Deterministic waves and analysis of the Fractal dimension of the components of the wind velocity, Il Nuovo Cimento C, 5–6, 653–667, 2009.

Tsang, C. F. and Neretnieks, I.: Flow channeling in heterogeneous fractured rocks, Rev. Geophys., 36 257–298, 1998.

Vandenbohede, A., Louwyck, A., and Lebbe L.: Conservative solute versus Heat transport porous media during push-pull tests, Transport Porous Med., 76, 265–287, doi:10.1007/s11242-008-9246-4, 2009.

Vandenbohede, A. and Lebbe, L.: Parameter estimation based on vertical heat transport in the surficial zone, Hydrogeol. J., 18, 931–943, doi:10.1007/s10040-009-0557-5, 2010.

Whitaker, S.: Flow in porous media. I: A theoretical derivation of Darcy's law, Transport Porous Med., 1, 3–25, 1986.

Woodbury, A. D. and Smith, J. L.: On the thermal effects of three dimensional groundwater flow, J. Geophys. Res., 90, 759–767, 1985.

Wu, Y. S., Ye, M., and Sudicky, E. A.: Fracture-Flow-Enhanced Matrix Diffusion in Solute Transport Through Fractured Porous Media, Transport Porous Med., 81, 21–34, doi:10.1007/s11242-009-9383-4, 2010.

Characterization of high-intensity, long-duration continuous auroral activity (HILDCAA) events using recurrence quantification analysis

Odim Mendes[1]**, Margarete Oliveira Domingues**[2]**, Ezequiel Echer**[1]**, Rajkumar Hajra**[3]**, and Varlei Everton Menconi**[1]

[1] Space Geophysics Division (DGE/CEA), Brazilian Institute for Space Research (INPE), São José dos Campos, São Paulo, Brazil
[2] Associated Laboratory of Computation and Applied Mathematics (LAC/CTE), Brazilian Institute for Space Research (INPE), São José dos Campos, São Paulo, Brazil
[3] Laboratoire de Physique et Chimie de l'Environnement et de l'Espace (LPC2E), CNRS, Orléans 45100, France

Correspondence to: Odim Mendes (odim.mendes@inpe.br)

Abstract. Considering the magnetic reconnection and the viscous interaction as the fundamental mechanisms for transfer particles and energy into the magnetosphere, we study the dynamical characteristics of auroral electrojet (AE) index during high-intensity, long-duration continuous auroral activity (HILDCAA) events, using a long-term geomagnetic database (1975–2012), and other distinct interplanetary conditions (geomagnetically quiet intervals, co-rotating interaction regions (CIRs)/high-speed streams (HSSs) not followed by HILDCAAs, and events of AE comprised in global intense geomagnetic disturbances). It is worth noting that we also study active but non-HILDCAA intervals. Examining the geomagnetic AE index, we apply a dynamics analysis composed of the phase space, the recurrence plot (RP), and the recurrence quantification analysis (RQA) methods. As a result, the quantification finds two distinct clusterings of the dynamical behaviours occurring in the interplanetary medium: one regarding a geomagnetically quiet condition regime and the other regarding an interplanetary activity regime. Furthermore, the HILDCAAs seem unique events regarding a visible, intense manifestations of interplanetary Alfvénic waves; however, they are similar to the other kinds of conditions regarding a dynamical signature (based on RQA), because it is involved in the same complex mechanism of generating geomagnetic disturbances. Also, by characterizing the proper conditions of transitions from quiescent conditions to weaker geomagnetic disturbances inside the magnetosphere and ionosphere system, the RQA method indicates clearly the two fundamental dynamics (geomagnet-ically quiet intervals and HILDCAA events) to be evaluated with magneto-hydrodynamics simulations to understand better the critical processes related to energy and particle transfer into the magnetosphere–ionosphere system. Finally, with this work, we have also reinforced the potential applicability of the RQA method for characterizing nonlinear geomagnetic processes related to the magnetic reconnection and the viscous interaction affecting the magnetosphere.

1 Introduction

A complicated electrodynamic region populated by plasmas and ruled by the Earth's magnetic field – designated in a classical definition as magnetosphere – exists surrounding our planet (Mendes et al., 2005; Kivelson and Russell, 1995). This region is exposed to influences of the space environment and submitted to several interplanetary forcings. Initially, a summary view of the physics scenario involved is briefly described in the two following paragraphs.

In electrodynamic terms, three main solar agents ((i) electromagnetic radiation, (ii) energetic particles, and (iii) solar magnetized structures) act upon the Earth's atmosphere, which is permeated by a magnetic field created in the interior of our planet (Campbell, 2003; Hargreaves, 1992). (i) Electromagnetic radiation both heats the planet globally and ionizes the atmosphere. This ionization gives basis to a terrestrial plasma environment. (ii) Also, the incidence episodes of solar energetic particles increase the ionization in a much more localized manner. (iii) Furthermore, escaping in a continuous way from the Sun, the solar wind, superposed sometimes by coronal mass ejection structures and other peculiar solar structures (e.g. solar fast-speed streams and heliospheric current sheet), transports intrinsically the solar magnetic field to the orbit of the Earth and beyond (Kivelson and Russell, 1995). Two primary electrodynamic interactions are possible from this incidence of the magnetized solar wind plasma upon the Earth's magnetosphere. These interactions result in a transfer of energy and particles into the magnetosphere boundary. The most intense is through the magnetic reconnection process (Burch and Drake, 2009; Kivelson and Russell, 1995; Dungey, 1961), when the interplanetary magnetic field (IMF) presenting a predominantly southward orientation, in the geocentric solar magnetosphere reference system, merges into the geomagnetic field at the outer boundary and produces strong modification in a large region formed by the magnetosphere and the ionosphere – the latter is a region from about 100 to 2000 km of altitude presenting the highest quantity of ionized particles. Another competitive process is the Kelvin–Helmholtz viscous interaction (Hasegawa et al., 1997; Chen et al., 2004; Axford and Hines, 1961). Most of the time this second process is in operation when the magnetosphere acts as a closed physical system, concerning the incident frontal solar wind, due to an IMF with northward orientation. A macroscopic fluid dynamics developed by the plasma sliding at the flanks of the magnetosphere creates a kind of viscous interaction, which produces the mixing of the solar plasma inside the magnetosphere and the occurrence of ULF waves (Menk and Waters, 2013) affecting the interior regions. The former process is more efficient in energy and particle transfer than the latter one.

In a global sense, during events of solar wind transporting IMF parallel (northward) to the frontal geomagnetic field, a regime of low magnetic disturbance on the ground is noticed. However, when the IMF is strongly southward directed, anti-parallel to the geomagnetic field, intense regimes of disturbances are recorded on the ground. Nevertheless, there is a peculiar interplanetary process related to manifestations of Alfvén waves (Guarnieri et al., 2006), presenting alternation of the magnetic component orientation (in the southward–northward direction), which produces an intermediate level of geomagnetic disturbance with the typical duration of days. These nonlinear Alfvén waves are known to be the main origin of high-intensity long-duration continuous auroral electrojet (AE) activity (HILDCAA) events on the Earth (Hajra et al., 2013; Tsurutani et al., 2011b, a; Echer et al., 2011; Tsurutani et al., 1990; Tsurutani and Gonzalez, 1987). As presented in Davis and Sugiura (1966), the AE is a geomagnetic index related to the quantification of the geomagnetic disturbance produced by enhanced ionospheric electric currents flowing below and within the auroral region (https://www.ngdc.noaa.gov/stp/geomag/ae.html). The primary mechanism for these HILDCAA events is the high-speed solar wind streams (HSSs) emanating from solar coro-

nal holes accompanied by embedded Alfvén waves (Belcher and Davis, 1971; Tsurutani et al., 1994), which are characterized by significant IMF variability (see Echer et al., 2012, 2011; Tsurutani et al., 2011b, a). The sporadic magnetic reconnection (Dungey, 1961; Gonzalez and Mozer, 1974) formed between the southward component of the Alfvén waves and the Earth's magnetopause fields leads to intense substorm/convection events comprising HILDCAAs (Tsurutani et al., 1995), which are shown to last from days to weeks (Tsurutani et al., 1995, 2006; Gonzalez et al., 2006; Guarnieri, 2006; Kozyra et al., 2006; Hajra et al., 2013, 2014a). The HILDCAA events carry a large amount of solar wind kinetic energy input into the magnetosphere affecting the polar ionosphere (Gonzalez et al., 2006; Hajra et al., 2014b). More than 60 % of this energy is dissipated in the magnetosphere–ionosphere system. Another importance of these events is the accelerated relativistic electrons, known as killer electrons, in the Earth's radiation belts (Hajra et al., 2014c, 2015b, a) for their hazardous effects on orbiting spacecraft (Wrenn, 1995; Horne, 2003). The variations of AE during HILDCAAs show the nonlinear dynamics of the physical processes involved. Therefore, a dynamical characterization is of fundamental interest for a deeper insight into the electrodynamic coupling between the solar wind and the related magnetosphere.

The aim of this work is to highlight dynamical characteristics related to the HILDCAA events revealed by the AE index in the context of the electrodynamic coupling processes. With this purpose, we apply phase space analysis, the recurrence plot (RP) technique, and the recurrence quantification analysis (RQA) method (Eckmann et al., 1987; Maizel and Lenk, 1981; Trulla et al., 1996). They constitute proper tools to treat such nonlinear, non-stationary signals as in geophysics processes. Such analysis method is applied to the HILDCAA events, for the first time to our knowledge, allowing a comparison of dynamical characteristics. By applying the nonlinear tools, this work investigates AE under some distinct physical conditions of the interplanetary medium: Alfvénic fluctuations followed by HILDCAA, Alfvénic fluctuations not followed by HILDCAA (also related to co-rotating interaction regions (CIRs) and high-speed streams (HSSs)), other disturbed interplanetary conditions, and geomagnetically quiet time.

This work proceeds as follows. Section 2 describes the methods for analysis. Section 3 presents the geomagnetic database and how we apply the methodology. Section 4 shows the results and interpretations. Finally, Sect. 5 summarizes the conclusions.

2   Method of analysis

Information theory structures a branch of powerful mathematical tools to analyse nonlinear systems of signal as proposed in the seminal paper of the mathematician Claude E.

Shannon (Shannon, 1964). An analogy with the concept of entropy from physics gives basis to these tools. As reviewed and discussed in detail by Cover and Thomas (2006), the entropy H used as basis for the methods can be expressed by

$$H(X) = - \sum P(x) \log(P(x)), \quad x \in X, \tag{1}$$

where X is the set of all messages $\{x_1, \ldots, x_n\}$ that X could be, and $P(x)$ is the probability of some $x \in X$. In this work, we use quantification methods associated with this theory, precisely the method developed by Zbilut and Webber Jr. (1992) of RQA that is built from the RP, as introduced in Eckmann et al. (1987), and the proprieties of the phase space, provided in the *Cross Recurrence Plot Toolbox*.[1] Initially, these methods are used to analyse dynamical systems from a theoretical point of view. Nevertheless, since the late 1990s, they have been extended to experimental data to characterize nonlinear complex behaviour (Trulla et al., 1996; Marwan and Webber, 2015). Below we summarize the phase space, the RP, and the RQA approaches.

2.1 Phase space

A phase plot is a geometric representation of the trajectories of a dynamical system in the phase plane. It is a fundamental starting point of many approaches in nonlinear data analysis, which is based on the construction of a phase space portrait of the considered system. A review of that can be found, for instance, in N. Marwan's tutorial.[2] The state of a system can be expressed by its state variables $x_1(t)$, $x_2(t)$, $\ldots$, $x_d(t)$ – for instance, the state variables density, pressure, momentum, and magnetic field for a magneto-hydrodynamics system. The d state variables at time t establish a vector in a d-dimensional space which is called phase space. The state of a system changes in time, and, consequently, the vector in the phase space describes a trajectory representing the time evolution, i.e. the dynamics of the system. Accordingly, the appearance of the trajectory retains information about the system. Therefore, the phase space is formed by coordinates that represent each significant variable of the system to specify an instantaneous state (Marwan, 2003).

In practice, observations of a real process do not unveil all state variables, or they are not known, or they cannot be measured. Nevertheless, due to the couplings between the system components, we can reconstruct a phase space trajectory from a single observation by a time delay embedding (Takens, 1981). It yields to the so-called Takens' embedding theorem, which states that a reconstruction of the phase space trajectory $x(t)$ from a time series u_k, with a cadence Δt, allows us to present a proper dynamics of a system. In order to

do that, an embedding dimension m and a time delay τ must be identified, related to the following reconstruction:

$$x(i) = x_i = (u_i, u_i i + \tau, \ldots, u_{i+(m-1)\tau}), \tag{2}$$

where $t = i \Delta t$. Here, m is found by using the false nearest neighbour method and τ by the mutual information method (Kennel et al., 1992; Marwan and Webber, 2015). The idea behind this approach is to identify the influence of increasing the embedded dimension m in the number of neighbours along a trajectory of the system.

2.2 Recurrence plot

The RP is based on Poincaré's recurrence theorem from 1890, as discussed in Schulman (1978). It states that a dynamic system returns to a state arbitrarily close to the initial state after a particular time. Mathematically the RP is obtained by the square matrix

$$\mathbf{R}_{i,j} = \Theta(\epsilon_i - \| x_i - x_j \|), \tag{3}$$

where ϵ_i is a predefined cut-off distance, $\| . \|$ is the norm (in our case, the Euclidean norm), and $\Theta(x)$ is the Heaviside function (Eckmann et al., 1987). The binary values 0 and 1 in this matrix are represented by white and black creating visual patterns.

The characteristic typology (related to macro patterns) and texture (related to micro details) presented in the RP are the key points of the interpretation. However, the visual interpretation of RPs requires some training experience, usually done from standard systems or data libraries. For instance, as described in Marwan et al. (2007) and on the RP and RQA website http://www.recurrence-plot.tk:

i. Stationary processes are associated to homogeneous distribution of points in RP.

ii. Periodic processes present cycle patterns where the distance between periodic patterns corresponds to the period.

iii. Long diagonal lines with different distances to each other reveal a quasi-periodic process.

iv. Non-stationary processes can present interruption on the lines; they can also indicate some rare state, or RP fading to the upper left and lower right corners indicating also trend or drifts.

v. Single isolated points demonstrate heavy fluctuation in the process – in particular, if only isolated points occur, an uncorrelated or anti-correlated random process is represented.

vi. Evolutionary processes are illustrated by diagonal lines – then the evolution of states is similar at different times. However, if it has parallel lines related to the main diagonal, the system is deterministic (or even chaotic, if they

---

[1] Cross Recurrence Plot Toolbox 5.21 (R31b) by the Interdisciplinary Center for Dynamics of Complex Systems, University of Potsdam (http://tocsy.pik-potsdam.de/CRPtoolbox/.).

[2] http://www.agnld.uni-potsdam.de/~marwan/matlab-tutorials/html/phasespace.html#13.

occur beside single lines), and if the diagonal lines are orthogonal to the main diagonal, or the time is reversed or the choice of embedding is insufficient.

vii. Long bowed line structures express evolution states that are similar at different epochs although they have different velocity (the dynamics of the system could be changing).

viii. Vertical and horizontal lines/clusters are evidence that a state has no or slow change for some time, which points to a laminar state.

The establishment of quantifiers to express the characterization of the processes described in RP was a significant advance in the popularization of this tool, because it can help to express in a concise and objective way a description on the dynamics of the processes, as discussed in Marwan and Webber (2015) and references therein. Therefore, quantification from RP comes primarily from the recurrence patterns, and presents for example as point density, diagonal structures, and vertical structures in the RP. In the following text, we present four of these quantifiers to study the behaviour of physical conditions such as geomagnetically quiet intervals and HILDCAA cases.

2.3   Recurrence quantification analysis

Trulla et al. (1996) addressed the problem of quantifying the structures that appear in the RPs and used them to analyse experimental data. This approach is useful to reveal qualitative transitions in a system. The corresponding measurements capture the dynamical characters of the system as represented by the signal. Therefore, RQA provides a qualitative description of a system regarding complexity measures (Marwan et al., 2007). We refer to Marwan and Kurths (2002), and Marwan (2003) for a detailed discussion on this subject. Notably, the diagonal structures in the RP and the recurrence point density are used to measure the complexity of a physical system (Zbilut and Webber Jr., 1992; Webber Jr. and Zbilut, 1994). In the present work we restrict our analysis to four characteristic parameters described below:

1. *Recurrence rate (RR)*: This denotes the overall probability that a certain state recurs and is obtained from the RP by

$$RR = \sum_{i,j=1}^{N} \frac{R_{i,j}(\rho)}{N^2}. \qquad (4)$$

Larger values mean more recurrence.

2. *Determinism (DET)*: this represents how predictable a system is, and is expressed by the ratio of recurrence points that form diagonal lines of the RP of at least length $\ell_{\min}$ to all recurrence points, i.e.

$$DET = \frac{\sum_{\ell=\ell_{\min}}^{N} \ell P(\ell)}{\sum_{\ell=1}^{N} \ell P(\ell)}, \qquad (5)$$

where $P(\ell)$ denotes the probability to find a diagonal line of length ℓ in the RP.

3. *Laminarity (LAM)*: this measures the occurrence of laminar states and is related to intermittent regimes – namely, it is the ratio between the recurrence points forming the vertical lines and the entire set of recurrence points computed by

$$LAM = \frac{\sum_{v=v_{\min}}^{N} v P(v)}{\sum_{v=1}^{N} v P(v)}, \qquad (6)$$

where $P(v)$ denotes the probability to find a vertical line of length v in the RP. LAM does not describe the length of laminar phases. However, if this measure decreases the RP consists of more single recurrence points than vertical structures. This measurement is relatively more robust against noise in signals.

4. *Entropy (ENT)*: this reflects the complexity of the deterministic structure in the system referred to as Shannon entropy (Shannon, 1964); namely,

$$ENT = -\sum_{\ell=\ell_{\min}}^{N} p(\ell) \ln(p(\ell)), \qquad (7)$$

where $p(\ell) = P(\ell)/N_\ell$. This measure reflects the complexity of the RP concerning the diagonal lines. In this form computed from RP, the interpretation of these values differ from traditional Shannon entropy – i.e. larger values are related to low entropy compared to physics analogy (Letellier, 2006).

3   Database and methodology procedure

For the present work, we have considered an updated list of 136 HILDCAA events occurring between 1975 and 2012, compiled by Hajra et al. (2013). The events were detected from the geomagnetic AE and middle- to low-latitude disturbance Dst indices by using the four strict HILDCAA criteria (Tsurutani and Gonzalez, 1987): (i) the events have peak AE intensities greater than 1000 nT, (ii) the events last for more than 2 days, (iii) high auroral activity lasts throughout the interval, i.e. AE never drops below 200 nT for more than 2 h at a time, and (iv) the events take place outside of the main phase of a geomagnetic storm. For a better understanding, the main phase is determined by the depression in the horizontal component, from middle to low latitudes, in the geomagnetic field. This behaviour is identified and quantified using the hourly value equatorial Dst index, which represents ideally the axially symmetric disturbance magnetic field at the dipole equator on the Earth's surface. This index is derived by monitoring the equatorial ring current variations (http://wdc.kugi.kyoto-u.ac.jp/dstdir/dst2/onDstindex.

Table 1. The geomagnetically quiet intervals.

Date	$K_p \leq$	AE $\leq$	Dst $\geq$
14–18 November 2000	3^0	267 nT	−20 nT
26–30 November 2001	3^-	133 nT	−50 nT
19–25 June 2004	2^0	167 nT	0 nT
19–27 June 2006	2^0	167 nT	−9 nT
15–23 July 2006	2^0	200 nT	32 nT
1–9 December 2007	3^0	200 nT	−5 nT

html). The AE data set is provided by the OMNIweb Service (http://omniweb.gsfc.nasa.gov/) by NASA and Dst from World Data Center for Geomagnetism, Kyoto Dst index service (http://wdc.kugi.kyoto-u.ac.jp/dstdir/).

From the list, the first 16 events were eliminated due to incomplete information. Among the remaining events, 33 % were preceded by geomagnetic storm main phase (Dst < −50 nT). Thus, 80 events were analysed in this work, because we selected the events classified as pure HILDCAAs, i.e. events not preceded by any geomagnetic storm main phase.

As data sets, the high-time-resolution (1 min) AE indices were analysed to study the dynamical characterization of the HILDCAA events by the RQA method. To eliminate any marginal influences, we considered a 2280 min interval centred at the middle point of a HILDCAA event. This number of records was determined by the least interval among the events.

For a quantitative comparison of disturbance geomagnetic regimes, we also performed the same RQA during the geomagnetically quiet period listed in Table 1. The quiet days follow the criteria: $K_p \leq 3^0$, Dst $\geq$ −50 nT, and AE $\leq$ 300 nT. The planetary 3 h range K_p index was introduced by J. Bartels in 1949 and designed to be sensitive to any geomagnetic disturbance affecting the Earth (http://www.gfz-potsdam.de/en/section/earths-magnetic-field/data-products-services/kp-index/explanation/). It completes a set of indices to diagnose the level of geomagnetic disturbance in a global sense. The geomagnetic indices (Rostoker, 1972) can be obtained from the World Data Center, Kyoto, at http://wdc.kugi.kyoto-u.ac.jp/wdc/Sec3.html. In that way, different physical regimes allow us to find a distinct characterization of the signals. In our case, we investigate periods of HILDCAA events that alter a physical regime that exists during the geomagnetically quiet times.

For a more complete dynamical diagnosis, this work investigates AE index under some other different physical conditions of the interplanetary medium. Completing the earlier mentioned cases of the interplanetary Alfvénic fluctuations followed by HILDCAA (related to CIRs and HSSs), and the geomagnetically quiet time, cases of interplanetary Alfvénic fluctuations not followed by HILDCAA (also related to CIRs and HSSs) and cases of intense interplanetary conditions

(characterized by simultaneous activities in the AE, Dst and K_p indices) produced by different interplanetary causes are also analysed. Table 2 presents the CIRs/HSSs not followed by HILDCAA event. The first column shows the data set interval and the second column the 2280 min interval considered in the analysis calculations. Table 3 presents the events with AE index related to global intense geomagnetic disturbances. The first column shows the data set interval and the second column the 2280 min interval considered in the analysis calculations.

The analyses of the results allow a comparison of the dynamical characteristics of signals.

4 Results

Initially, two typical cases are shown and analysed, one from the HILDCAA events and another from the quiet time intervals. As examples for the methodology application, they help to understand the analysis and its interpretation. Figure 1 shows AE variations including a HILDCAA interval. The HILDCAA started at 17:34 UT on 30 May (day 150) and continued until 09:34 UT on 2 June (day 153) of 1986, with a total duration of about 64 h. In that figure, the double arrow horizontal line indicates the exact interval of the event. For the RQA calculation we consider the 2280 min interval centred at the middle of the HILDCAA. Two vertical dotted lines mark this interval. Figure 2 shows AE variations during a geomagnetically quiet period. The plot shows the geomagnetically quiet period from 17 to 22 July (day 198 to day 203) of 2006 (from Table 1). The region between the two vertical dotted lines shows the same 2280 min interval selected for the RQA study as in the HILDCAA case.

From the AE plots, the differences in the amplitudes between the HILDCAA interval (peak about 1200 nT) and the quiet time interval (peak about 300 nT) are remarkable, as expected. Both of them presents fluctuations in the signal intensities. The application of the RQA methodology aims to characterize the dynamical behaviour of the signals.

Figure 3 represents the phase space plots for the HILDCAA. As a value estimated by the earlier-mentioned mutual information methodology, the time delay (τ) used is 34 min. The phase space charts present snapshots of the interconnections of the records for each case. As described by the theory in Sect. 2, the geometric representation in the plot gives the trajectory of the dynamical system involved in the AE index records. Although slightly insinuated by the distribution of points, a proper representation is not achieved because the noise in the signal disturbs the identification of the trajectory. Following the same procedure, Fig. 4 gives the representation for the quiet interval shown earlier. The time delay (τ) found is also 34 min. Although the signal amplitude is quite different compared to the one of the HILDCAA event, the trajectory behaviour is similar. A question arises from the comparison – is it possible to distinguish from the dynami-

Table 2. CIRs/HSSs not followed by HILDCAA.

Data set interval	Interval considered
2008, 012–018 (Jan 12 to 17)	2008, Jan 14 (00:00)–15 (13:59)
2008, 030–036 (Jan 30 to Feb 4)	2008, Feb 2 (00:00)–3 (13:59)
2008, 058–064 (Feb 27 to Mar 3)	2008, Mar 2 (00:00)–3 (13:59)
2008, 165–171 (Jun 13 to 18)	2008, Jun 15 (00:00)–16 (13:59)
2008, 175–181 (Jun 23 to 28)	2008, Jun 26 (00:00)–27 (13:59)

Table 3. AE in global intense geomagnetic disturbances.

Event	Interval considered
2012 (Mar 9)	2012, Mar 9 (00:00)–10 (13:59)
2012 (Apr 23–24)	2012, Apr 23 (00:00)–24 (13:59)
2012 (Jun 17)	2012, Jun 17 (00:00)–18 (13:59)
2012 (Jul 15)	2012, Jun 15 (00:00)–16 (13:59)

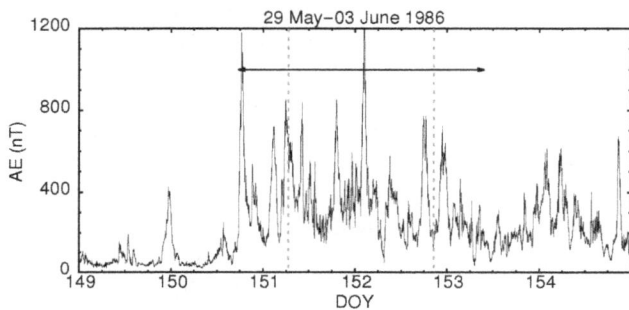

Figure 1. Geomagnetic AE index from 29 May (DOY 149) to 3 June (154) 1986 includes a HILDCAA event. The HILDCAA interval is identified by the double arrow horizontal line, and the AE interval used for the RQA is shown between the vertical dotted lines.

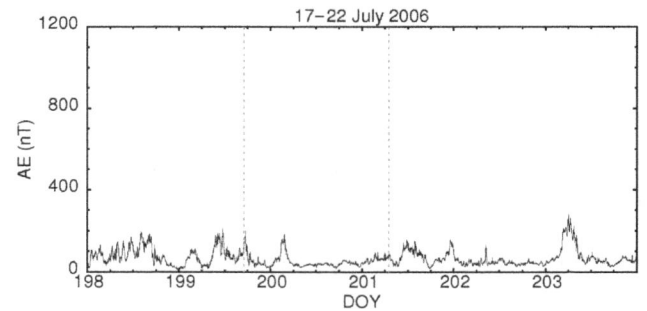

Figure 2. Geomagnetic AE index during the geomagnetically quiet period on 17 (DOY 198)–22 (203) July 2006. The AE interval used for the RQA is marked by vertical dotted lines.

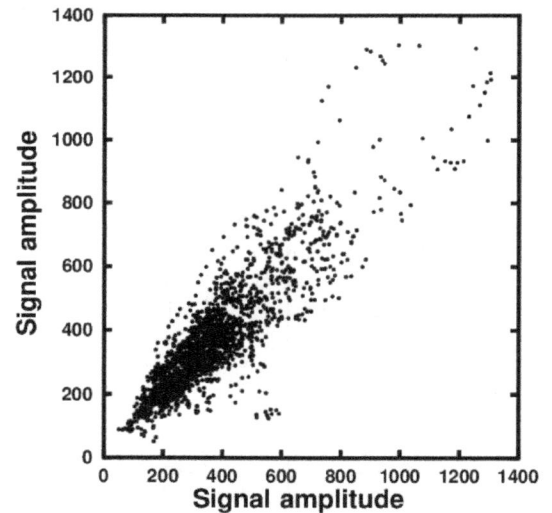

Figure 3. The phase space representation for the HILDCAA example shown in Fig. 1. The delay time is 34 min.

cal behaviour analyses the two kinds of occurrences as the AE indices point out?

To verify whether the question deserves study effort, we use the RP technique to allow a visual inspection of the signal features. Dealing with the RP theory for all the cases studied, we estimated the typical values related to these dynamical systems. The embedded dimension (m) determined by the false nearest neighbour method was found to be around 6, and following the time delay (τ) was around 34 min. The cut-off distance (ϵ) was $\approx 30\,\mathrm{nT}$ for the HILDCAAs, and $\approx 10\,\mathrm{nT}$ for the quiet intervals. For the other interplanetary conditions, the values were similar to the value of HILD-CAAs. The estimation of ϵ uses a value defined by the additive effects of the data resolution and the Gaussian noise threshold.

Related to the cases at the beginning of this section, Fig. 5 shows the RPs for the HILDCAA and Fig. 6 for the quiet interval. Here we take the embedded dimension (m) and the time delay (τ) equal to 1 for RQA calculations. These pa-

rameter choices take into account the categorization purpose of the present work, and those values do not alter our characterization process (Iwanski and Bradley, 1998; March et al., 2005; Marwan, 2011). The RPs highlight the recurrences in the signal records showing differences in the dynamical patterns between the HILDCAA interval and the quiet period. For both systems, the analyses on the large-scale patterns in the plots, designated as typology, denote that they are of the

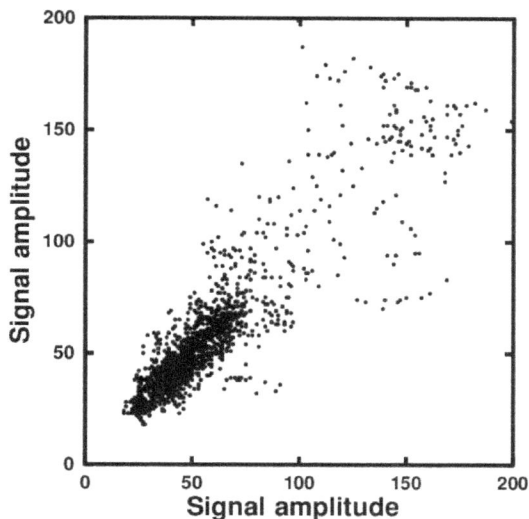

Figure 4. The phase space representation for the geomagnetically quiet period example shown, between the vertical dotted lines, in Fig. 2. The delay time is 34 min.

Table 4. RQA measures for the geomagnetically quiet interval and typical HILDCAA cases.

Case	RQA measures			
	RR	DET	LAM	ENT
Geomagnetically quiet interval	0.0203	0.357	0.518	0.719
HILDCAA period	0.0021	0.044	0.069	0.147

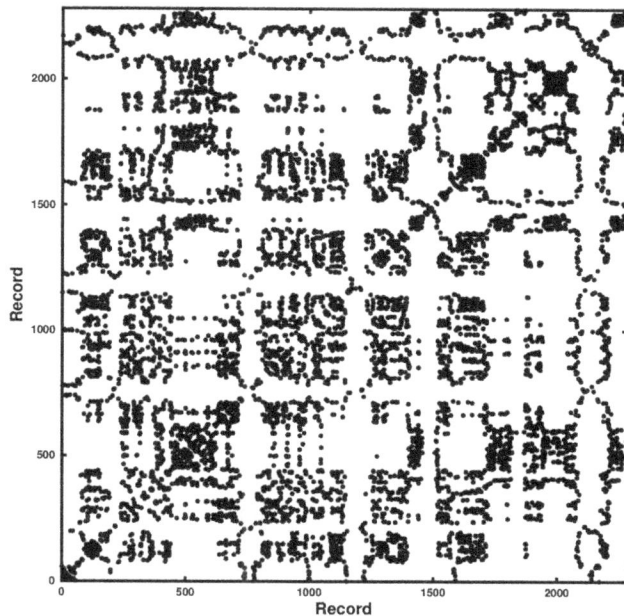

Figure 5. The RP for the HILDCAA example. The interval shown by the vertical dotted lines in Fig. 1 is used to obtain the plot.

disrupted kind – i.e. with abrupt changes in the representation of the dynamics. However, the analysis of the small-scale patterns, designated as texture, denotes a more complex dynamics in the HILDCAA event than the one in the quiet interval. To obtain an objective interpretation, we need to translate this visual appreciation to quantitative descriptors of the dynamics of the system interpreted by the AE index. As examples of this quantification, the results of the RQA dynamical parameters for the quiet and HILDCAA case examples are presented in Table 4. We verify they are about 1 order of magnitude smaller for the HILDCAA than the values for the quiet interval. Thus, we have a little evidence that encourages this kind of study.

To pursue a comprehensive answer, we apply the RQA methodology to all 80 HILDCAA events completed by the examination of other cases selected (six geomagnetically quiet intervals, five CIRs/HSSs not followed by HILDCAA, and four events of AE in global intense geomagnetic disturbances) to allow comparisons. The values of the RQA dynamical variables (RR, DET, LAM, and ENT) were obtained for each case.

Table 5 shows the minimum, maximum, mean, standard deviation, median, and mode values estimated to the HILD-

CAAs and the quiet periods. As can be seen, a difference of 1 order of magnitude for each variable exists between these cases. For minima and maxima, the differences are between half and 1 order of magnitude. The standard deviation, median, and mode are in agreement with normal distributions for the phenomena.

Finally, Fig. 7 shows the RQA dynamical parameters for all events under study. For each parameter, we normalized the values for all events concerning extreme values obtained for the parameter. The empty circles represent the HILDCAA events, and the plus signs show the quiet periods. A clear distinction between the HILDCAA events and quiet time intervals may be noted from the figure. The separation of the results for the HILDCAA event and the quiet time interval establishes a clustering of the results, which characterize two well-defined physical regimes. Further, the symbol x indicates the results for AE index in CIR/HSS events not followed by HILDCAA, and * in a whole global disturbance scenario. As also seen in the figure, parameter behaviour is similar for CIRs/HSSs causing HILDCAAs and CIRs/HSSs not causing them, and distinct from the behaviour of quiet intervals. Therefore, based on this plot, one could say that the bottom part shows the behaviour of Alfvénic solar wind intervals, CIRs and HSSs, while the top part shows the behaviour related to the slow solar wind interval. The analysis taking into account the AE in a whole global disturbance scenario regarding geomagnetic behaviour shows larger spreading values for the parameters (except by the RR parameter); nevertheless, values are also different to the one in the quiet time regime. Based on the current geophysical knowl-

Table 5. The RQA results considering two typical cases.

Value	HILDCAA period				Geomagnetically quiet interval			
	RR	DET	LAM	ENT	RR	DET	LAM	ENT
Min	0.0010	0.010	0.014	0.000	0.0115	0.251	0.397	0.574
Max	0.0056	0.086	0.139	0.273	0.0307	0.357	0.536	0.766
Mean	0.0016	0.031	0.049	0.091	0.0195	0.321	0.473	0.672
SD	0.0005	0.012	0.020	0.073	0.0065	0.046	0.058	0.075
Med	0.0015	0.028	0.046	0.104	0.0194	0.345	0.487	0.690
Mod	0.0013	0.010	0.014	0.000	0.0115	0.251	0.397	0.574

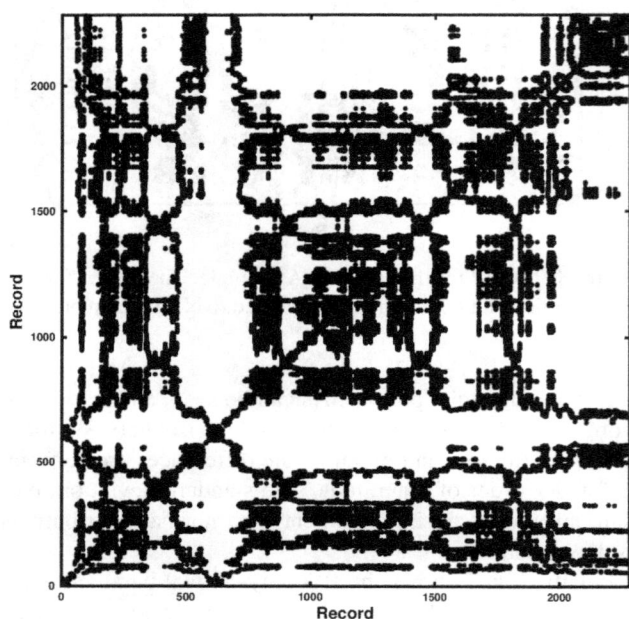

Figure 6. The RP for the geomagnetically quiet period example. The interval shown by the vertical dotted lines in Fig. 2 is used to obtain the plot.

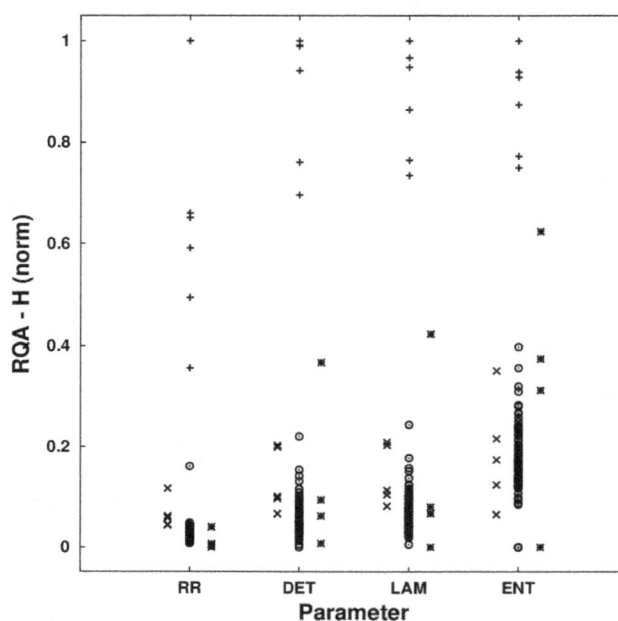

Figure 7. Normalized representation of the RQA parameters for auroral electrojet (AE) indices in HILDCAA events (○), in CIRs/HSSs not followed by HILDCAA (x), in a global geomagnetic disturbance scenario (*), and in the geomagnetically quieter intervals (+).

edge, the RQA patterns in the signals for these events help to characterize/identify the standard physical features. Examining the physics of every case in the active interplanetary regimes, one might point out that the AE signature relates to HILDCAA that is connected to long-duration, large-amplitude Alfvénic fluctuations; to CIRs/HSSs not followed by HILDCAA connected to short-term Alfvénic fluctuations and with or without a small interplanetary southward magnetic amplitude; and to events in a global geomagnetic disturbance scenario connected to small-amplitude southward interplanetary magnetic field without Alfvénic fluctuations or to a large southward interplanetary magnetic amplitude.

Thus, the RQA result comparisons lead us to achieve some interpretations.

The HILDCAAs seem unique events regarding visible, intense manifestations of interplanetary Alfvénic waves; how-

ever, they are similar to the other kinds of conditions regarding a dynamical signature (based on RQA), because the effect of HILDCAA is involved in the same complex mechanism of generating geomagnetic disturbances.

Allowing an interpretation of the geomagnetic disturbances, mainly the AE studied here, the physics scenario could be properly interpreted according to a basic view. As is well known, the fundamental mechanisms are the magnetic reconnection and viscous interaction with a transfer of energy and particles by electrodynamics interaction and generation of geomagnetic disturbance on ground. Supported by the parameter clustering behaviours shown in Fig. 7, the interpretation obtained from the RQA examination of AE index is in agreement with those fundamental mechanisms. Although describing an expected result, the quantitative study using this method indicates in a clear way categories of phenomena (showed in Fig. 7). On the one hand, during geomagnetically quiet conditions, the effective interaction is the ram pressure on the solar front side of the magnetosphere and the development of viscous interaction at flanks. On the other hand, during HILDCAA events, the two fundamental electrodynamics interactions (magnetic reconnection and viscous interaction) with a transfer of energy and particles are indeed happening. In principle, interplanetary phenomena producing both of those coupling mechanisms, as processes examined in Ma et al. (2014), concern the mechanisms related to interplanetary Alfvén waves. In this kind of occurrence, magnetic disturbances can be detected by magnetometers at the polar regions as the HILDCAA events. Although they can be clearly noticed at high latitudes, those disturbances are noticed as weak worldwide manifestations. CIR/HSS occurrences not followed by HILDCAA related to short-term Alfvénic fluctuations and with or without small southward interplanetary magnetic amplitude produce sporadic, low AE index disturbances, designated as geomagnetic substorms. Events in a whole global disturbance scenario related to large southward interplanetary magnetic amplitude produce geomagnetic storms and associated geomagnetic substorms.

Identified as distinct regimes by the RQA diagnosis, the geomagnetically quiet intervals and HILDCAA events seem the proper conditions of transitions from quiescent conditions to weaker geomagnetic disturbances inside the magnetosphere and ionosphere system. Therefore, those RQA features can be useful for other study purposes. The RQA method gives a clear indication of the dynamics to be evaluated by magneto-hydrodynamics simulations, as developed by Ma et al. (2014) or Chen et al. (2004), to understand the processes involved in a transfer of energy and particles into the magnetosphere-ionosphere system.

5 Conclusions

Obtained from a diagnosis of features of a nonlinear system analysis, a physics scenario of the auroral electrojet (AE) index is built with the aid of the recurrence quantification analysis (RQA) information extracted from the recurrence plot (RP) calculation. We performed this analysis using 80 HILDCAA events completed by the examination of other cases selected (six geomagnetically quiet intervals, five CIRs/HSSs not followed by HILDCAA, and four events of AE in global intense geomagnetic disturbances) to allow comparisons.

Some significant RQA variables (RR, DET, LAM, and ENT) quantify and characterize the dynamical signatures of the AE index related to HILDCAA occurrences and other interplanetary environment conditions.

The key findings are as follows:

- The quiet intervals as compared to HILDCAA intervals are characterized by larger values of DET, LAM, and

ENT, which means higher predictability, lower entropy, and larger laminarity of the corresponding nonlinear dynamics.

- There is distinct clustering, identified by RQA, of the dynamical behaviours recorded on the ground produced by the interplanetary medium conditions: one regarding a geomagnetically quiet condition regime and another regarding an effective disturbed interplanetary regime.

- The RQA results identify similar dynamical behaviours for HILDCAA events and the other disturbed cases.

- On the one hand, the HILDCAAs seem unique events regarding the visible, intense manifestations of Alfvénic waves; on the other hand, they are similar to the other phenomena regarding dynamical signatures (based on RQA), because they are involved in the same complex mechanism of generating geomagnetic disturbances.

- This complex mechanism is composed by the magnetic reconnection and the viscous interaction implying ground geomagnetic effects triggered by the southward interplanetary magnetic field.

- One regime of clustering is AE index organized by geomagnetically quiet conditions, related to a predominant interaction from the incidence of ram pressure on the solar front side of the magnetosphere and the development of viscous interaction at flanks, while there is a northward interplanetary magnetic field (IMF). Another regime is AE organized by disturbed interplanetary conditions, with the presence of the southward IMF.

As the geomagnetically quiet intervals and HILDCAA events characterize the proper conditions of transitions from quiescent conditions to weaker geomagnetic disturbances inside the magnetosphere and ionosphere system, the RQA method gives a clear indication of the two fundamental dynamics to be evaluated with magneto-hydrodynamics simulations to understand in a better way the fundamental processes related to energy and particle transfer into the magnetosphere–ionosphere system.

With the present work, we have also demonstrated the potential applicability of the RQA method for characterizing of nonlinear geomagnetic processes related to magnetic reconnection and viscous interaction affecting the magnetosphere, mainly with the aid of magneto-hydrodynamics simulations.

Author contributions. All authors discussed the idea and the approach for the work development and took part in the preparation of the paper. OM and MOD worked also in the application of the methodology.

Competing interests. The authors declare that they have no conflict of interest.

Acknowledgements. Margarete Oliveira Domingues and Odim Mendes thank the MCTIC/FINEP (CT-INFRA grant 0112052700) and FAPESP (grant 2015/25624 − 2) for the financial support. Odim Mendes, Margarete Oliveira Domingues and Ezequiel Echer thank the Brazilian CNPq agency (grants 312246/2013-7, 306038/2015-3 and 301233/2011-0, respectively). Rajkumar Hajra thanks the FAPESP 2012/00280-0 for a postdoctoral research fellowship at INPE, and now the work supported by ANR under financial agreement ANR-15-CE31-0009-01 at LPC2E/CNRS. Varlei Everton Menconi thanks the MCTIC-PCI program (grant 455097/2013-5) by the research fellowship at INPE. The authors would like to thank the team of Interdisciplinary Center for Dynamics of Complex Systems, University of Potsdam, for the RQA tools (http://tocsy.pik-potsdam.de/), OMNIweb Service (http://omniweb.gsfc.nasa.gov/) by NASA and the World Data Center for Geomagnetism, Kyoto, Japan (http://wdc.kugi.kyoto-u.ac.jp/), where the geomagnetic indices used in this study were collected from. The authors thank Olga Verkhogfolyadova and two anonymous referees for constructive and useful suggestions leading to significant improvement of the manuscript.

Edited by: Giovanni Lapenta

References

Axford, W. I. and Hines, C. O.: A unifying theory of high-latitude geophysical phenomena and geomagnetic storms, C. J. Phys., 39, 1433–1464, https://doi.org/10.1139/p61-172, 1961.

Belcher, J. W. and Davis, L. J.: Large-amplitude Alfvèn waves in the interplanetary medium: 2, J. Geophys. Res., 76, 3534–3563, 1971.

Burch, J. L. and Drake, J. F.: Reconnecting magnetic fields, Am. Sci., 97, 392–399, 2009.

Campbell, W. H.: Introduction to geomagnetic fields, Cambridge, 2003.

Chen, Q., Otto, A., and Lee, L. C.: Tearing instability, Kelvin-Helmholtz instability, and magnetic reconnection, J. Geophys. Res., 430, 1755–1758, 2004.

Cover, T. M. and Thomas, J. A.: Elements of Information Theory, Wiley-Interscience, New Jersey, 2006.

Davis, T. N. and Sugiura, M.: Auroral electroject activity index AE and its universal time variations, J. Geophys. Res,, 71, 785, https://doi.org/10.1029/JZ071i003p00785, 1966.

Dungey, J. W.: Interplanetary magnetic field and the auroral zones, Phys. Rev. Lett., 6, 47–48, 1961.

Echer, E., Gonzalez, W. D., Tsurutani, B. T., and Kozyra, J. U.: High speed stream properties and related geomagnetic activity during the whole heliospheric interval, Sol. Phys., 274, 303–320, https://doi.org/10.1007/s11207-011-9739-0, 2011.

Echer, E., Tsurutani, B. T., and Gonzalez, W. D.: Extremely low geomagnetic activity during the recent deep solar cycle minimum, Proc. Int. Astron. Union, 7, 200–209, https://doi.org/10.1017/S174392131200484X, 2012.

Eckmann, J. P., Kamphorst, S., and Ruelle, D.: Recurrence plots of dynamical systems, Europhys. Lett., 4, 973–977, 1987.

Gonzalez, W. D. and Mozer, F. S.: A quantitative model for the potential resulting from reconnection with an arbitrary interplanetary magnetic field, J. Geophys. Res., 79, 4186–4194, 1974.

Gonzalez, W. D., Guarnieri, F. L., Clua-Gonzalez, A. L., Echer, E., Alves, M. V., Oginoo, T., and Tsurutani, B. T.: Recurrent Magnetic Storms: Corotating Solar Wind Streams, chap. Magnetospheric energetics during HILDCAAs, AGU, https://doi.org/10.1029/167GM15, 2006.

Guarnieri, F. L.: Recurrent Magnetic Storms: Corotating Solar Wind Streams, chap. The nature of auroras during high-intensity long-duration continuous AE activity (HILDCAA) events: 1998–2001, AGU, https://doi.org/10.1029/167GM19, 2006.

Guarnieri, F. L., Tsurutani, B. T., Gonzalez, W. D., Clua-Gonzalez, A. L., Grande, M., Soraas, F., and Echer, E.: ICME and CIR storms with particular emphases on HILDCAA events, in: ILWS WORKSHOP 2006, COSPAR, GOA, India, 2006.

Hajra, R., Echer, E., Tsurutani, B. T., and Gonzalez, W. D.: Solar cycle dependence of high-intensity long-duration continuous AE activity (HILDCAA) events, relativistic electron predictors?, J. Geophys. Res., 118, 5626–5638, https://doi.org/10.1002/jgra.50530, 2013.

Hajra, R., Echer, E., Tsurutani, B. T., and Gonzalez, W. D.: Superposed epoch analyses of HILDCAAs and their interplanetary drivers: solar cycle and seasonal dependences, J. Atmos. Sol. Terr. Phys., 121, 24–31, https://doi.org/10.1016/j.jastp.2014.09.012, 2014a.

Hajra, R., Echer, E., Tsurutani, B. T., and Gonzalez, W. D.: Solar wind-magnetosphere energy coupling efficiency and partitioning: HILDCAAs and preceding CIR storms during solar cycle 23,, J. Geophys. Res., 119, 2675–2690, https://doi.org/10.1002/2013JA019646, 2014b.

Hajra, R., Tsurutani, B. T., Echer, E., and Gonzalez, W. D.: Relativistic electron acceleration during high-intensity, long-duration, continuous AE activity (HILDCAA) events: solar cycle phase dependences, Geophys. Res. Lett., 41, 1876–1881, https://doi.org/10.1002/2014GL059383, 2014c.

Hajra, R., Tsurutani, B. T., Echer, E., Gonzalez, W. D., Brum, C. G., Vieira, L. E. A., and Santolik, O.: Relativistic electron acceleration during HILDCAA events: are precursor CIR magnetic storms important?, Earth, Planets Space., 67, https://doi.org/10.1186/s40623-015-0280-5, 2015a.

Hajra, R., Tsurutani, B. T., Echer, E., Gonzalez, W. D., and Santolik, O.: Relativistic (E > 0.6, > 2.0, and > 4.0 mev) electron acceleration at geosynchronous orbit during high-intensity, long-duration, continuous AE activity (HILDCAA) events, Astrophys. J., 799, https://doi.org/10.1088/0004-637X/799/1/39, 2015b.

Hargreaves, J. K.: The solar-terrestrial environment, Camnbridge, 1992.

Hasegawa, H., Fujimoto, M., Phan, T. D., Rème, H., Balogh, A., Dunlop, M. W., Hashimoto, C., and TanDokoro, R.: Transport of solar wind into Earth's magnetosphere through rolled-up Kelvin-Helmholtz vortices, Nature, 102, 151–161, 1997.

Horne, R. B.: Rationale and requirements for a european space weather programme, in: Space Weather Workshop: Looking Towards a European Space Weather Programme, pp. 139–144, European Space Agency, Nordwijk, the Netherlands, 2003.

Iwanski, J. S. and Bradley, E.: Recurrence plots of experimental data: to embedded or not to embed?, Chaos, 8, 861–871, 1998.

Kennel, M. B., Brown, R., and Abarbanel, H. D. I.: Determining embedding dimension for phase-space reconstruction using a geometrical construction, Phys. Rev. A, 45, 3403, https://doi.org/10.1103/PhysRevA.45.3403, 1992.

Kivelson, M. G. and Russell, C. T. (Eds.): Introduction to Space Physics, Cambridge, 2 edn., 1995.

Kozyra, J. U., Crowley, G., Emery, B. A., Fang, X., Maris, G., Mlynczak, M. G., Niciejewski, R. J., Palo, S. E., Paxton, L. J., Randall, C. E., Rong, P. P., Russell, J. M., Skinner, W., Solomon, S. C., Talaat, E., Wu, Q., and Yee, J. H.: Recurrent Magnetic Storms: Corotating Solar Wind Streams, chap. Response of the upper/middle atmosphere to coronal holes and powerful high-speed solar wind streams in 2003, AGU, https://doi.org/10.1029/167GM24, 2006.

Letellier, C.: Estimating the Shannon Entropy: Recurrence Plots versus Symbolic Dynamics, Phys. Rev. Lett., 96, 254102-1–254102-4, 2006.

Ma, X., Otto, A., and Delamere, P. A.: Interaction of magnetic reconnection and kelvin-helmholtz modes for large magnetic shear: 1. kelvin-helmholtz trigger, J. Geophys. Res.-Space Physics, 119, 781–797, https://doi.org/10.1002/2013JA019224, 2014.

Maizel, J. V. and Lenk, R. P.: Enhanced graphic matrix analysys of nucleic acid and protein sequences, Genetic, P. Natl. Acad. Sci. USA, 78, 7665–7669, 1981.

March, T. K., Chapman, S. C., and Dendy, R. O.: Recurrence plot statistics and the effect of embedding, Phys. D, 200, 173–184, 2005.

Marwan, N.: Encounters with neighbours. current developments of concepts based on recurrence plots and their applications, Ph.D. thesis, Institut Für Physik, Fakultät Mathematik und Naturwissenschaften, Universität Potsdam, 2003.

Marwan, N.: How to avoid potential pitfalls in recurrence plot based data analysis, Int. J. Bifurcation Chaos, 21, 1003, 2011.

Marwan, N. and Kurths, J.: Nonlinear analysis of bivariate data with cross recurrence plots, Phys. Lett. A, 302, 299–307, https://doi.org/10.1016/S0375-9601(02)01170-2, 2002.

Marwan, N. and Webber, C. L. J.: Recurrence Quantification Analysis, chap. Mathematical and computational foundations of recurrence quantifications, 3–43, Springer, 2015.

Marwan, N., Carmen, M., Romano, M. T., and Kurths, J.: Recurrence plots for the analysis of complex systems, Phys. Reports, 438, 237–329, https://doi.org/10.1016/j.physrep.2006.11.001, 2007.

Mendes, O., Domingues, M. O., and da Costa, A. M.: Introduction to planetary electrodynamics: a view of electric fields, currents and related magnetic fields, Adv. Space Res., 35, 812–828, https://doi.org/10.1016/j.asr.2005.03.139, 2005.

Menk, F. and Waters, C. L.: Magnetoseismology, Wiley-VCH, 1 edn., 2013.

Rostoker, G.: Geomagnetic indices, Rev. Geophysics, 10, 935–950, https://doi.org/10.1029/RG010i004p00935, 1972.

Schulman, L. S.: Note on the quantum recurrence theorem, Phys. Rev. A, 18, 2379–2380, 1978.

Shannon, C. E.: The Mathematical Theory of Communication, University of Illinois, Urbana, IL, University of Illinois, Urbana, IL, 1964.

Takens, F.: Dynamical Systems and Turbulence, Warwick 1980, vol. 898, chap. Detecting strange attractors in turbulence, 366–381, Springer Berlin Heidelberg, Berlin, https://doi.org/10.1007/BFb0091924, 1981.

Trulla, L. L., Giuliani, A., Zbilut, J. P., and Webber, C. L. J.: Recurrence quantification analysis of the logistic equation with transients, Phys. Lett. A, 223, 255–260, 1996.

Tsurutani, B. T. and Gonzalez, W. D.: The cause of High-Intensity, Long-Duration, Continuous AE Activity (HILDCAAs): interplanetary Alfvèn wave trains, Planet. Space Sci., 35, 405, 1987.

Tsurutani, B. T., Gould, T., Goldstein, B. E., Gonzalez, W. D., and Sugiura, M.: Interplanetary Alfvèn waves and auroral (substorm) activity: IMP 8, J. Geophys. Res., 95, 2241–2252, https://doi.org/10.1029/JA095iA03p02241, 1990.

Tsurutani, B. T., Ho, C., Smith, E. J., Neugebauer, M., Goldstein, B. E., Mok, J. S., Arballo, J. K., Balogh, A., Southwood, D. J., and Feldman, W. C.: The relationship between interplanetary discontinuities and Alfvèn waves: Ulysses observations, Geophys. Res. Lett., 21, 2267–2270, https://doi.org/10.1029/94GL02194, 1994.

Tsurutani, B. T., Gonzalez, W. D., Clua-Gonzalez, A. L., Tang, F., Arballo, J. K., and Okada, M.: Interplanetary origin of geomagnetic activity in the declining phase of the solar cycle, J. Geophys. Res., 100, 21717–21733, 1995.

Tsurutani, B. T., Gonzalez, W. D., lua-Gonzalez, A. L., Guarnieri, F. L., Gopalswamy, N., Grande, M., Kamide, Y., Kasahara, Y., Mann, I., Lu, G., McPherron, R., Soraas, F., and Vasyliunas, V.: Corotating solar wind streams and recurrent geomagnetic activity: a review, J. Geophys. Res., 111, A07S01, https://doi.org/10.1029/2005JA011273, 2006.

Tsurutani, B. T., Echer, E., and Gonzalez, W. D.: The solar and interplanetary causes of the recent minimum in geomagnetic activity (MGA23): a combination of midlatitude small coronal holes, low IMF B_Z variances, low solar wind speeds and low solar magnetic fields, Ann. Geophys., 29, 839–849, https://doi.org/10.5194/angeo-29-839-2011, 2011a.

Tsurutani, B. T., Echer, E., Guarnieri, F. L., and Gonzalez, W. D.: The properties of two solar wind high speed streams and related geomagnetic activity during the declining phase of solar cycle 23, J. Atmos. Sol. Terr. Phys., 73, 164–177, 2011b.

Webber Jr., C. L. and Zbilut, J. P.: Dynamical assessment of physiological systems and states using recurrence plot strategies, J. Appl. Physiol., 76, 965–973, 1994.

Wrenn, G. L.: Conclusive evidence for internal dielectric charging anomalies on geosynchronous communications spacecraft, J. Spacecraft Rockets, 32, 514–520, 1995.

Zbilut, J. P. and Webber Jr., C. L.: Embeddings and delays as derived from quantification of recurrence plots, Phys. Lett. A, 171, 199–203, 1992.

3

A matrix clustering method to explore patterns of land-cover transitions in satellite-derived maps of the Brazilian Amazon

Finn Müller-Hansen[1,2], **Manoel F. Cardoso**[3], **Eloi L. Dalla-Nora**[3], **Jonathan F. Donges**[1,4], **Jobst Heitzig**[1], **Jürgen Kurths**[1,2], **and Kirsten Thonicke**[1]

[1]Potsdam Institute for Climate Impact Research, Telegrafenberg A31, 14473 Potsdam, Germany
[2]Department of Physics, Humboldt University Berlin, Newtonstraße 15, 12489 Berlin, Germany
[3]Center for Earth System Science, National Institute for Space Research, Rodovia Presidente Dutra 40, 12630-000 Cachoeira Paulista, São Paulo, Brazil
[4]Stockholm Resilience Center, Stockholm University, Kräftriket 2B, 114 19 Stockholm, Sweden

Correspondence to: Finn Müller-Hansen (mhansen@pik-potsdam.de)

Abstract. Changes in land-use systems in tropical regions, including deforestation, are a key challenge for global sustainability because of their huge impacts on green-house gas emissions, local climate and biodiversity. However, the dynamics of land-use and land-cover change in regions of frontier expansion such as the Brazilian Amazon are not yet well understood because of the complex interplay of ecological and socioeconomic drivers. In this paper, we combine Markov chain analysis and complex network methods to identify regimes of land-cover dynamics from land-cover maps (TerraClass) derived from high-resolution (30 m) satellite imagery. We estimate regional transition probabilities between different land-cover types and use clustering analysis and community detection algorithms on similarity networks to explore patterns of dominant land-cover transitions. We find that land-cover transition probabilities in the Brazilian Amazon are heterogeneous in space, and adjacent subregions tend to be assigned to the same clusters. When focusing on transitions from single land-cover types, we uncover patterns that reflect major regional differences in land-cover dynamics. Our method is able to summarize regional patterns and thus complements studies performed at the local scale.

1 Introduction

Land-use/cover change does not only affect local ecosystems and climate but has global consequences for the Earth system (Foley et al., 2005). Land use emits about 25 % of annual greenhouse gases to the atmosphere worldwide. Particularly in tropical regions, increasing demand for food, fiber and biofuels drives land conversion from forest biomes to agriculturally used areas (Lambin and Meyfroidt, 2011). In order to analyze the causes of tropical deforestation, it is thus crucial to understand the dynamics of land-cover changes that occur after deforestation, compare them between regions and connect them to socioeconomic and political drivers. Furthermore, this could help to better understand the effects of land-use intensification that can potentially reverse deforestation trends, as hypothesized in forest transition theory (Meyfroidt and Lambin, 2011).

The Brazilian Amazon is one of the world's key regions with highly dynamic land-use change and is subject to multiple pressures (Laurance and Williamson, 2001; Keller et al., 2009; Davidson et al., 2012). Economic activities such as unsustainable logging and agricultural expansion of cattle ranching and soybean cultivation lead to a fragmentation of the landscape resulting in biodiversity loss (Laurance et al., 2002). Global climate change may decrease precipitation and increase forest fires (Chen et al., 2011). All these pressures are increasing the risk of destabilizing the ecosystem and crossing a tipping point with irreversible consequences (Lenton et al., 2008; Nepstad et al., 2008; Staal et al., 2015).

In the 1970s and 1980s, deforestation was mostly driven by large infrastructure and settlement programs, but more recent years saw mainly market drivers pushing the deforestation frontier further, while government programs tried to contain it (Fearnside, 2005). Since 2005, deforestation rates in the Brazilian Amazon have been reduced enormously. In recent years, the rates are fluctuating around $6000\,\mathrm{km}^2$ per year, which is a reduction of about 80 % compared to the peak of deforestation activities in 2004 (INPE, 2017). The changes are explained by new monitoring programs, public policies and supply chain interventions (Nepstad et al., 2014; Dalla-Nora et al., 2014; Gibbs et al., 2015). However, there are warnings that deforestation may increase again (Fearnside, 2015; Aguiar et al., 2016).

In order to understand deforestation rates, it is crucial to take subsequent land uses and their dynamics into account. This paper focuses on developing methods to detect patterns of land-cover dynamics using data from remote sensing and identifying large-scale differences between subregions of the Brazilian Amazon as a sample region. To do so, we draw on the theory of Markov chains that has been used in the context of land-system science to describe and analyze land-cover dynamics (Bell and Hinojosa, 1977; Baker, 1989). Markov chains are stochastic systems that are described by transition probabilities between discrete states, here referring to a specific land-use or land-cover type. An ensemble of such chains describes a collection of land patches that undergo stochastic transitions between land-cover classes. Because simple Markov models do not take spatial correlations into account, they often form only one part of hybrid land-cover models that introduce stochasticity into the model (see, e.g., Brown et al., 2000; Subedi et al., 2013). For example, Fearnside (1996) applied a Markov analysis to estimate greenhouse gas emissions from land-use change in the Brazilian Amazon and found that carbon storage in the land system decreases as it approaches an equilibrium.

In the past, most studies using Markov analysis focused on small regions due to limited data availability. Modern geographic information systems (GISs) enable the detection of land-cover changes at an unprecedented scale using satellite images (Lu et al., 2004). Automated algorithms allow the classification of land use and land cover of vast regions. Furthermore, it is possible to compare the land-use dynamics between different subregions and find differences and similarities based on consistent data sets. For example, Levers et al. (2015) combined different sources of land-use indicators and used self-organizing maps to identify archetypical land uses and regions with similar land-use change in Europe.

In this study, we use Markov transition probability matrices as a descriptor of aggregate land-cover dynamics estimated from high-resolution land-cover data for three time slices of land cover over 6 years in the Brazilian Amazon. To our knowledge, Markov analysis has so far not been applied to investigate interregional heterogeneity of land-cover dynamics. This paper explores this idea by comparing transition matrices from different subregions in the Brazilian Amazon to identify patterns of similar land-cover dynamics drawing on large data sets derived from satellite imagery. While previous studies mostly worked with predefined regions to

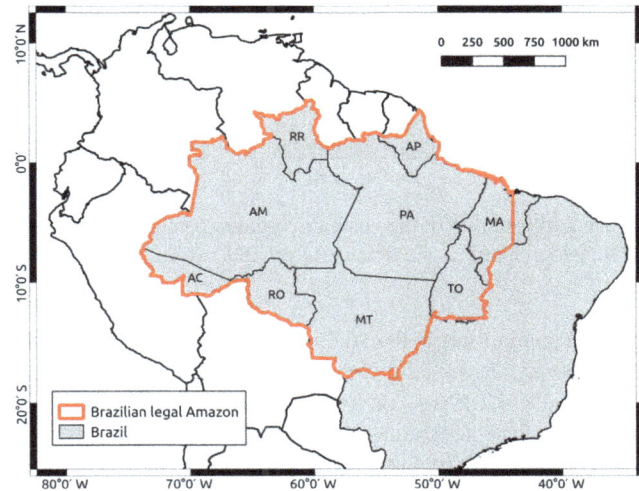

Figure 1. Map of the Brazilian legal Amazon and its nine federal states: Acre (AC), Amapá (AP), Amazonas (AM), Maranhão (MA), Mato Grosso (MT), Pará (PA), Rondônia (RO), Roraima (RR) and Tocantins (TO).

compare land-cover dynamics, we develop methods to identify regions with similar land-cover dynamics, which allows a large-scale analysis of land-cover change patterns. With this methodology, we approach the hypothesis that different land-cover dynamics can be identified by the characteristics of their transition matrix and a partition of subregions, for example, into remote, frontier and consolidated areas, can be detected from the data.

The paper is structured as follows. In the subsequent Sects. 2 and 3, we present the details of the proposed method and describe the data that we apply it to. Section 4 gives results from the analysis and discusses them, pointing to possible interpretations but also restrictions of the method. Section 5 concludes with an outlook on how the method could be applied to further analyses.

2 Data: land-cover maps of the Brazilian Amazon

In this study, we use land-cover maps of the Brazilian legal Amazon (cf. Fig. 1) produced by the TerraClass project (INPE and EMBRAPA, 2017) for the years 2008, 2010 and 2012. The land-cover maps are derived from high-resolution Landsat-5 thematic mapper (TM) and MODIS imagery using a mix of techniques including supervised learning and classification by spectral properties of different land-cover types and their annual variations (for details, see Almeida et al., 2016; Coutinho et al., 2013). The maps consist of polygons that represent patches of land attributed to 1 of 16 specific land-cover types (see Table S1 in the Supplement). The maps are based on the PRODES project that distinguishes between forest, patches not belonging to the rain forest biome (mainly savanna), hydrography (i.e., lakes and rivers) and de-

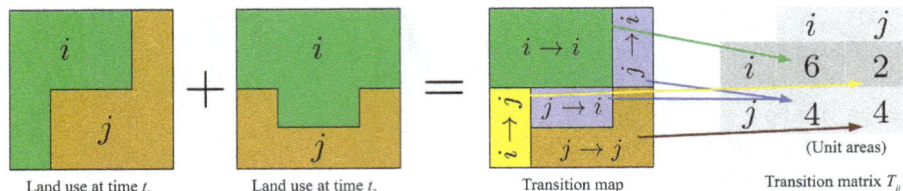

Figure 2. Illustration of the geometric union operation that combines the information of two land-cover maps into a transition map and how the transition matrices are obtained from this map.

forested patches larger than 6.25 ha (INPE, 2017). TerraClass further specifies the land cover of formerly deforested areas according to 12 types, including different kinds of pasture land, secondary vegetation and annual crops. Coutinho et al. (2013) evaluated the accuracy of land-cover detection using the method described in Congalton and Green (2009). Considering a very small sample of the data set, they found up to 58 % commission and up to 34 % omission errors. Almeida et al. (2016) found that the dominant land cover on previously deforested land is pasture (62 % as of 2008) followed by secondary vegetation (21 %). Annual crops only covered about 5 % of the total deforested areas.

This paper focuses on relevant transitions between major land-cover classes occurring in different subregions of the Brazilian Amazon. Therefore, we first exclude patches that could not be classified, e.g., due to cloud cover. Second, we discard land-cover types that do not change by definition, i.e., lakes and rivers and patches not belonging to the rain forest biome. Third, we aggregate similar land-cover types into six new classes. These classes combine different types of less intensively used pasture as well as types that only make up small fractions of the Amazon like mining and urban patches (see Table S1) and group land-cover types between which high confusion errors exist, thus decreasing them. In a final step of the data preparation, we assign patches to N different subregions. Depending on the scale of spatial aggregation of our analysis, the subregions either correspond to the legal municipalities of the Brazilian Amazon ($N = 770$, as of 2007) or to the mesoregions ($N = 30$) as defined by the Instituto Brasileiro de Geografia e Estatística (Brazilian Institute of Geography and Statistics, IBGE, 2016).

3 A method to explore patterns of land-cover transitions

In order to compare land-cover dynamics between different subregions of the Amazon, we proceed in two steps. First, we calculate the area in a given region that undergoes a transition from one land-cover type to another between two reference years (including the lumping of several land-cover types into one class) and normalize the obtained matrices. Second, we compare the transition matrices between subregions by means of cluster analysis and network methods. In this section, we describe the steps of the method in detail.

3.1 Extraction and normalization of transition matrices

Markov chains are stochastic systems, in which the probability distribution of the next time step only depends on the current state of the system; hence, the system has no memory. A subregion can be thought of as consisting of a number of land patches that undergo transitions between land-cover classes. Markov analysis then describes how the set of patches may change over time. Although the Markov property, i.e., that the transition probability only depends on the present state of the system, can be shown to hold approximately for land-use systems (Robinson, 1978), the transition rates are generally not constant over time, which means the system is not stationary. This is not surprising because of the various climatological and socioeconomic drivers and political decisions influencing land-cover dynamics (Walker, 2004). Even though Markov chain analysis may oversimplify land-cover dynamics because it does not take the underlying processes explicitly into account and may therefore not be suitable to project future land-cover change, it serves here as a first approximation in obtaining a general understanding of the land-cover dynamics observed in the data.

We obtain the transition matrices of subregions by calculating the areas in a given subregion that undergo a transition from a land-cover class i to another class j. The transition matrix of one subregion $\mathbf{T}(t)$ is an $n \times n$ matrix with elements $T_{ij}(t)$, $i, j \in \{1, \ldots, n\}$, where n is the number of land-cover classes. The transition matrix depends on time, indicating the nonstationarity of the Markov process. In the following, however, we omit the time dependence for ease of notation. With the aggregation described above, the number of land-cover classes n is 6. We estimate $\mathbf{T}$ from the data by first projecting the coordinates of the patches (in the data given in the South American Datum (SAD69) coordinate system) to the South America Albers Equal Area Conic projection. Second, we compute the geometric union with GIS software combining the information contained in the two land-cover maps of the reference years into one data set. Finally, we sum up the area of all patches in one subregion that undergo the same transition. Figure 2 illustrates the creation of the transition matrix $\mathbf{T}$ from the data.

To estimate transition probabilities, we have to normalize the transition matrices. Thereby, we also make subregions of different total area comparable. We normalize the rows of the transition matrices to 1, which allows us to focus on relative changes in single land-cover classes

$$p_{ij} = \frac{T_{ij}}{\sum_k T_{ik}} \text{ for } i, j : 1 \ldots n. \tag{1}$$

The normalization does not work if one land-cover class i does not figure in the data of one subregion, as $\sum_k T_{ik}$ would be equal to zero. In such cases, we set the diagonal element $T_{ii} = 1$ and all other elements of the ith row to zero, implying that we handle the land-cover class in the particular subregion as if no change occurs.

In statistical terms, $\mathbf{p} = (p_{ij})$ is a stochastic matrix (compare Norris, 1997) with the properties $p_{ij} \geq 0$ and $\sum_j p_{ij} = 1$ for $i = 1 \ldots n$. It corresponds to the maximum likelihood estimation of the transition probability matrix of a first-order Markov chain where land-cover classes correspond to the states of the Markov chain and the rows of $\mathbf{p}$ specify the transition probabilities between the states (Anderson and Goodman, 1957).

Figure 3a presents a visualization of the Markov chain and the calculated transition probabilities estimated for the whole Brazilian Amazon. The figure shows that there are transitions between almost all aggregated classes, but they occur with very different probabilities. After deforestation, about two-thirds of the areas are used as pasture, whereas the rest is mostly classified as secondary vegetation. Furthermore, transitions occur frequently between pasture partly covered with woody vegetation (dirty pasture) and clean pasture. The former makes also frequent transitions to secondary vegetation. Finally, there are considerable transitions from and to the "other" class, in which we aggregated the following minor land-cover types from the original TerraClass classification: "mosaic of uses", "urban area", "mining", "reforestation" and "others".

Alternatively to the Markov analysis, one could normalize the sum of the transition matrix elements $\mathbf{T}_{ij}$ to 1. Such a normalization would keep the information on the initial distribution of land-cover classes in one subregion but would not allow to analyze relative changes in individual land-cover classes.

The transition probability matrix $\mathbf{p}$, representing the dynamics of an underlying Markov chain process, includes information on the patches that undergo changes and the patches that remain in their land-cover class. To only consider changes, we set the diagonal elements to zero before normalizing the rows of $\mathbf{T}$ to 1:

$$q_{ij} = \begin{cases} \frac{\mathbf{T}_{ij}}{\sum_{k \neq i} T_{ik}} & \text{for } i \neq j \\ 0 & \text{for } i = j. \end{cases} \tag{2}$$

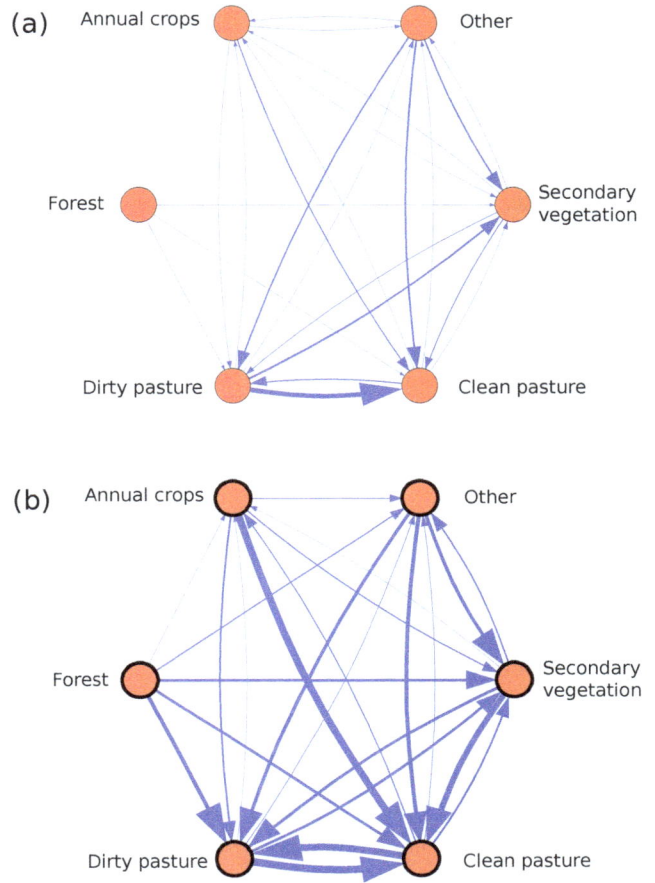

Figure 3. Illustration of the normalized transition matrices between simplified classes derived for the whole Brazilian Amazon from the TerraClass data set (changes between 2010 and 2012): **(a)** Markov transition matrix $\mathbf{p}$ (self-loops omitted) and **(b)** conditional transition matrix $\mathbf{q}$. The strengths of the arrows are scaled with the transition probabilities except for those representing small values. Arrows with very small values (below 0.005) are not shown. The values are given in Tables S2 and S3.

Hence, $\mathbf{q} = (q_{ij})$ estimates the probability to make a transition from a single land-cover class i conditional on there being a transition to a different land-cover class j. Figure 3b shows a visualization of this conditional transition matrix for the whole Brazilian Amazon. For land-use classes that have a high proportion of patches remaining in the same class, this figure allows inspecting the relative shares of transitioning patches more easily.

The normalized matrices $\mathbf{p}$ and $\mathbf{q}$ describe the transitions between all land-cover classes. In the following, we are particularly interested in comparing transition probabilities from a single land-cover class to all others, formally represented by the rows of the normalized matrices. If we only focus on the rows, we solve the above-mentioned problem of missing land-cover classes in a subregion by simply discarding the respective subregions from the analysis. To increase the robustness, we also discard subregions having less than $1\,km^2$ of the considered land-cover class.

As described above, we estimated the normalized transition matrices **p** and **q** for all mesoregions and municipalities separately. This spatial segmentation was chosen because it makes the analysis compatible with other data (e.g., socioeconomic data sets provided by the IBGE). Additionally, the areas of the municipalities reflect to some degree the population density and therefore potential land-use activities. In principle, a segmentation into regular grid cells could provide complementary information and insights. However, to keep the presentation clear, we focus here on mesoregions and municipalities.

In general, the lower the spatial aggregation, i.e., the smaller the size of the subregions, the higher the variability in space and in time. We can observe this when comparing the mesoregion and municipality maps and transitions between different times. Figure 4 shows two exemplary components of the matrices **q** calculated for each municipality. The two maps highlight these subregions in darker colors in which the transition probability from clean pasture to secondary vegetation and vice versa is high compared to transitions to other land covers. In Fig. 4a, we can observe that transitions from clean pasture to secondary vegetation are infrequent compared to other transitions except in the central north and the southwest. Figure 4b suggests that along a horizontal band from the west to the east and in the north (state of Roraima) the transition probability from secondary vegetation to clean pasture is higher than in the other parts of the Brazilian Amazon. The maps in Fig. 4 and similar maps for all other possible transitions contain the information that we aim to aggregate using clustering analysis. The next section therefore describes this second step of our method.

3.2 Construction of similarity networks and clustering analysis of land-cover transitions

Clustering methods are a basic technique described in the machine learning and data mining literature (Jain and Dubes, 1988; Gan et al., 2007). In recent years, the basic problem of clustering nodes in complex networks has also gained a lot of interest in complex systems science (Fortunato, 2010). In this paper, we choose a combination of established and more recent clustering methods to compare and test the robustness of our results. The chosen established methods are hierarchical clustering and the k means algorithm. The other methods are based on complex networks that we construct from a difference measure. To partition the network, we apply two different community detection algorithms, the fast greedy and Louvain algorithms (Clauset et al., 2004; Blondel et al., 2008).

The first method applies hierarchical clustering that merges data points or clusters based on their distance in the abstract data space. In the context of this analysis, a data point x is either a full normalized transition matrix (flat-

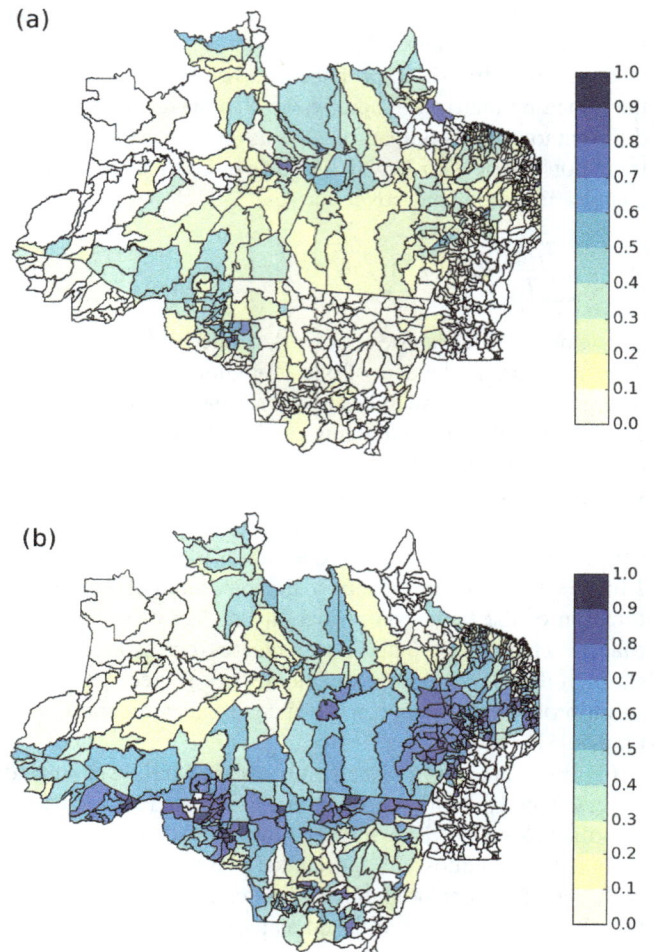

Figure 4. Map of two selected components of the conditional transition matrices **q** for each municipality of the Brazilian legal Amazon. Colors indicate the shares of areas that make a transition from **(a)** clean pasture to secondary vegetation and **(b)** secondary vegetation to clean pasture.

tened, such that $x \in \mathbb{R}^{n^2}$) or a single row of such a matrix ($x \in \mathbb{R}^n$). Each data point corresponds to an individual subregion. We choose to calculate the distance between two data points x and y by the ℓ_1 norm, also called Manhattan distance, $d(x, y) = \sum_i \text{abs}(x_i - y_i)$. This distance is easy to interpret in the context of probabilities and compared to the euclidean metric does not punish outliers of a cluster as much. The distances between two clusters or one cluster and one data point are calculated using the complete linkage algorithm that takes the maximal distance between the points of two clusters. This algorithm identifies compact clusters with small diameters (Jain and Dubes, 1988). Hierarchical clustering produces a dendrogram of cluster partitions. The clusters are obtained by cutting the dendrogram at a certain level determining the number of clusters.

The second method uses the k means algorithm. The algorithm works in an iterative manner: it associates data points to centroids and adjusts the position of the centroids by minimizing the within-cluster sum of squared distances. The k means algorithm inherently requires the choice of the euclidean metric to calculate distances.

The network methods both require the construction of a similarity network first. In the network, each node v_α represents a subregion and nodes with similar dynamics are linked by an edge $e_{\alpha\beta}$, where the Greek character indices refer to subregions. The connectivity of the network can also be represented by an adjacency matrix $\mathbf{A} = (A_{\alpha\beta})$. To determine the similarity, we use a normalized version of the Manhattan distance as the difference measure $d(\mathbf{x}, \mathbf{y}) = \frac{1}{2k}\sum_i \mathrm{abs}(x_i - y_i)$, where k is the number of land-cover classes n if we compare whole transition matrices and $k = 1$ if we only consider transitions from single land-cover classes. The metric is 0 if and only if transition probabilities are equal and 1 if they are completely different. We set a threshold d_{th} to transform the data into a network with the adjacency matrix $\mathbf{A}$:

$$A_{\alpha\beta} = \begin{cases} 1 & \text{if} \quad d(\mathbf{x}_\alpha, \mathbf{x}_\beta) < d_{\mathrm{th}} \\ 0 & \text{else.} \end{cases} \tag{3}$$

This adjacency matrix contains all information on the similarity network. The threshold d_{th}, which determines the subregions that are connected, is chosen such that only links that are significantly different from a distribution of difference measures of random vectors or matrices are realized. In order to obtain d_{th}, we use a Monte Carlo simulation: we generate a large number (10^6) of random samples of vectors or matrices, the values of which are drawn from a uniform distribution and rows are normalized. From the computed distribution of pairwise difference measures, we use the fifth percentile to determine the threshold d_{th}.

A visualization of such a similarity network is shown in Fig. 5 for transitions from clean pasture to other land-cover types. The nodes of the network represent data points for the municipality drawn around it. Links are present between regions that have a difference measure below the significant threshold $d_{\mathrm{th}} = 0.11$, which we obtain as described above from a Monte Carlo simulation of normalized random vectors of dimension 4 (because transitions to four other classes are possible). A visual inspection of the network suggests that similar transition probabilities are detected in regions of the eastern and the southern Amazon, whereas there are less similar transitions in the northern part. The inset in Fig. 5 furthermore shows a histogram of all pairwise differences. The threshold is indicated as a red vertical line. From tests with different thresholds and different underlying data, we can conclude that the patterns observed in the similarity networks hardly depend on the exact choice of the threshold (or link density). Thus, the construction of the network is robust with respect to variations of the threshold.

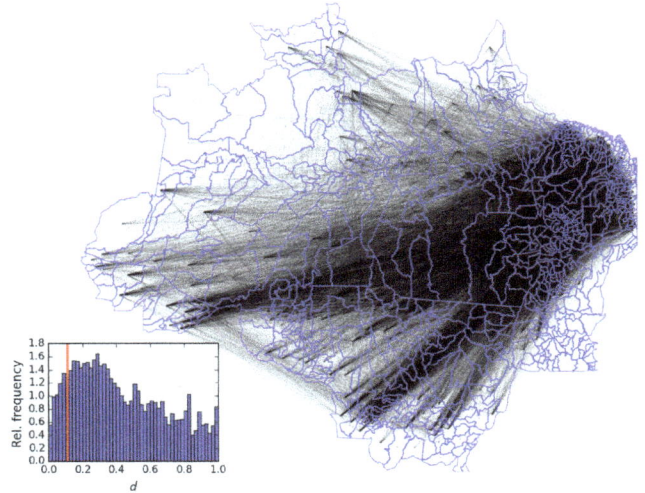

Figure 5. Illustration of a similarity network with a spatial division in municipalities for transitions from clean pasture to other land-cover classes between 2010 and 2012. Inset: histogram of difference metric values with threshold in red.

The visual inspection of similarity networks is difficult and may not be reliable. Therefore, we applied community detection algorithms to the networks to infer information about the network structure. These algorithms identify clusters of nodes on the network (in the literature the clusters are often called communities, hence the name) that have a high internal connectivity. Most of these algorithms are based on the idea of optimizing modularity Q, a network measure that compares the frequency of links inside of communities to the frequency of links between communities (Fortunato, 2010). For a network with adjacency matrix $\mathbf{A}$ and clusters C, the modularity is given by

$$Q = \frac{1}{2m}\sum_{\alpha,\beta} A_{\alpha\beta} - \frac{k_\alpha k_\beta}{2m}\delta(C_\alpha, C_\beta), \tag{4}$$

where $k_\alpha = \sum_\beta A_{\alpha\beta}$ is the degree of node α and m is the number of edges in the network. The term $\delta(C_\alpha, C_\beta)$ only gives a contribution if nodes α and β belong to the same cluster. In the following, we constrain our comparison to the fast greedy and the Louvain algorithms, which are computationally efficient and yield comparatively high modularity values. The general idea of the fast greedy algorithm as described in Clauset et al. (2004) is to subsequently join clusters such that the increase in modularity is highest after the join. This produces a dendrogram, similar to the output of the hierarchical clustering method, which can be cut at the level of highest modularity Q. In contrast, the Louvain algorithm developed in Blondel et al. (2008) proceeds in two iterative steps. It first checks subsequently if the reassignment of single nodes to other clusters leads to an improvement in modularity. In a second step, it builds a new network combining all nodes of a community found in the previous step into one node and

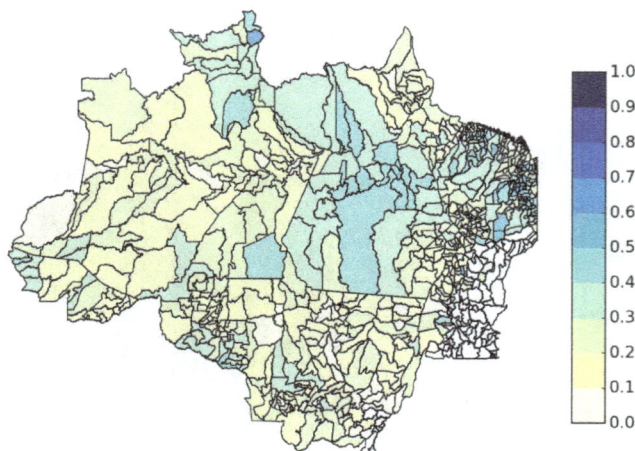

Figure 6. Relative areas that undergo changes in land-use classes between the years 2010 and 2012 (excluding primary forest).

sums up all edges between communities to form weighted new edges.

In the following, we apply these algorithms to the same heterogeneous data. A comparison between the different methods will show whether the clustering can be considered robust.

4   Spatial heterogeneity of land-cover transitions and discussion of clustering patterns

This section describes patterns of land-cover change found in the Brazilian Amazon when applying the clustering algorithms of differently normalized transition matrices or single rows of them. We show the spatial comparison of transitions between 2010 and 2012 with the threshold for the construction of the similarity networks set to $d_{th} = 0.11$ (see Sect. 3.2). Comparisons of transitions between other years are shown in the Supplement.

As explained in the methods section, we considered different normalizations of the transition matrices: the Markov matrices $\mathbf{p}$ that also contain information about patches remaining in the same land-cover class and conditional transition matrices $\mathbf{q}$ that disregard this information. First, we note that the majority of land patches do not change their class from one time step to the next. This is illustrated in Fig. 6, where the relative area of patches that make a transition to a different land-cover class is plotted (excluding primary forest), i.e., the sum of the diagonal elements of the transition matrix divided by the sum of all elements. Only in the central Amazon and in some of the smaller municipalities there are considerable fractions of up to 50 % of the area undergoing a change in land-cover class. Because we are interested in the changes, we will focus our discussion first on the conditional transitions matrices $\mathbf{q}$ and compare only single rows between the municipalities.

As an example, Fig. 7 displays the result of the clustering analysis for transitions from clean pasture to other land-cover classes. To make the clustering comparable, we fixed the number of clusters for the hierarchical and k means clustering to the one obtained from the fast greedy network clustering algorithm. As we can see from the figure, there are clearly distinguishable clusters in the south and the northwest of the Amazon colored in orange and cyan for all four different clustering algorithms. These clusters are identified independently of the chosen clustering algorithm. In the other parts of the Amazon region, the clusters vary dependent on the applied clustering algorithm. Both network community detection algorithms identify similar clusters, even though the Louvain algorithm finds seven and the fast greedy algorithm reveals five communities in the data. Also, some clustering algorithms seem to find two clusters for a group of municipalities, where other algorithms only find one (compare, e.g., the fast greedy with the k means algorithm). In addition to the two relatively stable clusters, we can observe in Fig. 7 that most clusters consist of adjacent municipalities. This suggests that neighboring municipalities have a high likelihood to exhibit similar relative land-cover changes.

In order to interpret the clusters, we analyzed the cluster centroids, i.e., the mean of all data points in a cluster weighted by the area of the considered land patches in the subregion. Figure 8 shows the cluster centroids from the hierarchical clustering. The bars indicate the shares of patches making a transition from clean pasture to another land-cover class and thus show which transitions are dominating or are absent in the cluster. They allow a straightforward interpretation of different clusters. For instance, in municipalities belonging to the orange cluster, most of the areas are converted to annual crops while only a small fraction makes the transition to dirty pasture. This is in line with a previous study by Macedo et al. (2012) who found that cropland expanded mostly into pasture in the region between 2006 and 2010. The orange cluster is located inside the Mato Grosso state, one of the biggest producers of soybeans in Brazil, which are detected as annual crops in the data. As we can see, the clusters generally differ by their relative shares of land-cover types such as dirty pasture and secondary vegetation. When comparing the cluster centroids between algorithms, these shares differ for the unstable clusters while the cluster centroids of the stable clusters are almost the same.

So far, we discussed transitions from clean pasture to other land-cover classes as one example. But our analysis has shown that the stable clusters identified in Fig. 8 can also be found when considering transitions from other land-cover classes, e.g., from secondary vegetation (see Figs. S1 and S2). However, the same patterns are not found for all transitions from single land-cover types. This is not surprising considering typical land-cover sequences (often called land-use trajectories) that follow total deforestation and are discussed in the literature (Ramankutty et al., 2007; Alves et al., 2009; de Espindola et al., 2012). According to these studies, a com-

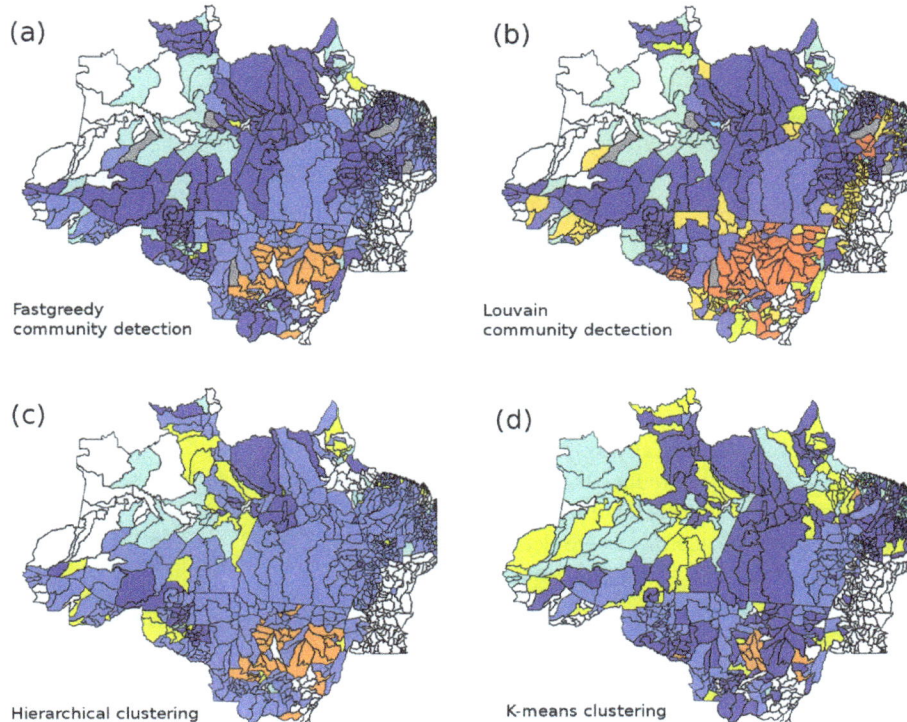

Figure 7. Comparison of network (**a, b**) and classical (**c, d**) clustering algorithms for conditional transitions from clean pasture to other land-cover classes between 2010 and 2012. Each cluster is visualized by one color. White regions lack data to estimate the transition matrix, grey regions are not connected to the similarity network. The number of clusters for the hierarchical and k means clusters was chosen to match the outcome of the fast greedy algorithm (five). The Louvain algorithm detects seven clusters.

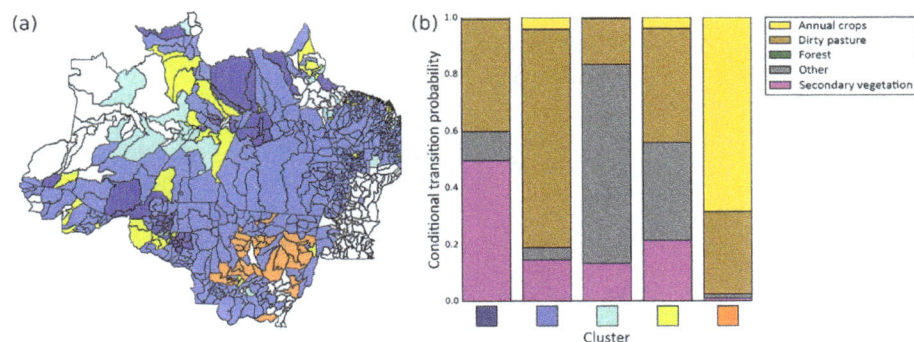

Figure 8. (**a**) Hierarchical clustering with conditionally normalized transition probabilities from clean pasture to other land-cover classes between 2010 and 2012, as in Fig. 7c. (**b**) Cluster centroids showing the conditional transition probabilities of the average over the respective cluster indicated by cluster color.

mon trajectory is that cleared forest patches are converted to pasture land or used for small-scale subsistence agriculture. After a while, as the soil degrades, the areas are often abandoned, leaving them for regrowth of secondary vegetation. Later, they may be cleared again and reused as pasture or they are converted to more intensive agricultural cropland, e.g., for soybean cultivation. These accounts are generally consistent with our results.

In addition to the clustering based on transitions from single land-cover classes, we tried to identify regions that are similar regarding the transitions between all land-cover classes. The clustering based on the full Markov matrices **p** proved to be very unreliable due to the high heterogeneity and dimensionality of the data (see Fig. S3). Furthermore, the analysis of the difference measure showed that only a small fraction of municipalities are significantly similar to each other compared to random matrices. The clustering based on the full conditional transition matrix **q** turned out to be highly dependent on the assumptions we made to fill in missing data. Thus, we can conclude that a general classification of land-cover dynamics only based on the full transition probability matrices between different land-cover types is not reliable.

This may have several reasons. First, the underlying processes of land-cover change in the Amazon are very heterogeneous in space and time and are therefore difficult to compare. Second, the areas of the municipalities may be too small for a reliable estimation of transition probabilities. For this reason, we also analyzed the transition matrices at the level of mesoregions (see Fig. S5). However, there was no reliable clustering at this spatial aggregation either. Third, the classification of land-cover types in the TerraClass data set comes with considerable errors. We tried to reduce the errors by aggregating some of the original classes. However, there is not yet an evaluation of the performance of change detection available for this data set, which makes an estimation of the errors in our analysis difficult.

The Brazilian Amazon has been broadly divided into mostly undisturbed, frontier and consolidated areas. For example, Becker (2005) distinguishes between the arch, i.e., densely populated areas in the south and the east of the legal Amazon, new frontier regions in the central Amazon and the mostly undisturbed west. Aguiar et al. (2007) used this partition to analyze interregional differences in factors potentially determining deforestation and found that the importance and combination of factors such as protected areas, distance to roads and access to markets differs between the three subregions. Although these studies focus on the 1990s and large-scale socioeconomic patterns may have changed since then, our analysis suggests that there are no clear patterns in the estimated transition probabilities which correspond to a spatial partition such as the one proposed by Becker (2005).

5 Conclusions

This paper has explored variations of a method that is able to provide important information on the dynamics of land covers, including the ability to quantify and compare land-cover transition frequencies and identify regions of similar patterns of land-cover change. We have applied different clustering techniques to find patterns in the subregional transition probabilities between land-use classes and detected patterns of subregions presenting similar transition dynamics that are consistent with other studies. In some regions, such as northern Mato Grosso where transitions from pasture to annual crops dominate, spatial patterns of relative land-use changes are consistent between different clustering methods. However, our analysis also indicates that relative land-use changes do not follow clearly distinguishable patterns that are linked to earlier socioeconomic partitions of the Brazilian Amazon.

The integration of socioeconomic data into the framework described in this paper could potentially yield insights about the underlying drivers and processes of land-cover transitions and how regionally different transition probabilities are determined. Furthermore, the analysis presented in this paper could potentially be used to parameterize models of land-cover change that track aggregate areas with different land-cover types. By controlling specific transition rates as functions of socioeconomic drivers, such models, to be developed in future research, could give rough ideas about possible future developments of land cover and thus support the planning of future land-use policies in the Amazon region.

Code availability. The input data were downloaded from the public archives indicated in Sect. 2 (INPE and EMBRAPA, IBGE). The code for the preparation of the data and the analysis as well as the processed data are stored in the long-term archive of the Potsdam Institute for Climate Impact Research. They will be made available upon request by email to the corresponding author.

Competing interests. The authors declare that they have no conflict of interest.

Acknowledgements. Finn Müller-Hansen acknowledges funding by the DFG (IRTG 1740/TRP 2011/50151-0). Jonathan F. Donges is grateful for financial support by the Stordalen Foundation (via the Planetary Boundary Research Network PB.net) and the EarthLeague's EarthDoc program. We thank Ana Cano Crespo for help with the TerraClass data and Tim Kittel, Catrin Ciemer and Silvana Tiedemann as well as the members of the ECOSTAB and COPAN flagships at PIK for fruitful discussions. The data preparation for this paper was carried out using ArcGIS with a licence provided by the German Research Centre for Geosciences (GFZ, Potsdam). The data analysis relies on the following python packages: scipy, scikit-learn, pandas, igraph, networkx, shapefile and matplotlib. We thank all the contributors and developers of these packages.

Edited by: S. Vannitsem

References

Aguiar, A. P. D., Câmara, G., and Escada, M. I. S.: Spatial statistical analysis of land-use determinants in the Brazilian Amazonia: Exploring intra-regional heterogeneity, Ecol. Model., 209, 169–188, doi:10.1016/j.ecolmodel.2007.06.019, 2007.

Aguiar, A. P. D., Vieira, I. C. G., Assis, T. O., Dalla-Nora, E. L., Toledo, P. M., Oliveira Santos-Junior, R. A., Batistella, M.,

Coelho, A. S., Savaget, E. K., Aragão, L. E. O. C., Nobre, C. A., and Ometto, J. P. H.: Land use change emission scenarios: Anticipating a forest transition process in the Brazilian Amazon, Glob. Change Biol., 22, 1821–1840, doi:10.1111/gcb.13134, 2016.

Almeida, C. A., Coutinho, A. C., Esquerdo, J. C. D. M., Adami, M., Venturieri, A., Diniz, C. G., Dessay, N., Durieux, L., and Gomes, A. R.: High spatial resolution land use and land cover mapping of the Brazilian Legal Amazon in 2008 using Landsat-5/TM and MODIS data, Acta Amazonica, 46, 291–302, doi:10.1590/1809-4392201505504, 2016.

Alves, D. S., Morton, D. C., Batistella, M., Roberts, D. A., and Souza Jr., C.: The Changing Rates and Patterns of Deforestation and Land Use in Brazilian Amazonia, in: Amazonia and Global Change, edited by: Keller, M., Bustamante, M., Gash, J., and Dias, P. S., chap. 2, 11–24, American Geophysical Union, Washington, DC, doi:10.1029/GM186, 2009.

Anderson, T. W. and Goodman, L. A.: Statistical Inference about Markov Chains, Ann. Math. Stat., 28, 89–110, available at: http://www.jstor.org/stable/2237025, 1957.

Baker, W. L.: A review of models of landscape change, Landscape Ecol., 2, 111–133, doi:10.1007/bf00137155, 1989.

Becker, B. K.: Geopolítica da Amazônia, Estudos Avançados, 19, 71–86, doi:10.1590/S0103-40142005000100005, 2005.

Bell, E. and Hinojosa, R.: Markov analysis of land use change: continuous time and stationary processes, Socio-Econ. Plan. Sci., 11, 13–17, doi:10.1016/0038-0121(77)90041-6, 1977.

Blondel, V. D., Guillaume, J.-L., Lambiotte, R., and Lefebvre, E.: Fast unfolding of communities in large networks, J. Stat. Mech.: Theory E., 10, P10008, doi:10.1088/1742-5468/2008/10/P10008, 2008.

Brown, D. G., Pijanowski, B. C., and Duh, J. D.: Modeling the relationships between land use and land cover on private lands in the Upper Midwest, USA, J. Environ. Manage., 59, 247–263, doi:10.1006/jema.2000.0369, 2000.

Chen, Y., Randerson, J. T., Morton, D. C., DeFries, R. S., Collatz, G. J., Kasibhatla, P. S., Giglio, L., Jin, Y., and Marlier, M. E.: Forecasting fire season severity in South America using sea surface temperature anomalies, Science, 334, 787–791, doi:10.1126/science.1209472, 2011.

Clauset, A., Newman, M. E. J., and Moore, C.: Finding community structure in very large networks, Phys. Rev. E, 70, 1–6, doi:10.1103/PhysRevE.70.066111, 2004.

Congalton, R. G. and Green, K.: Assessing the Accuracy of Remotely Sensed Data – Principles and Practices, 2nd edn., CRC Press, Boca Raton, FL, 2009.

Coutinho, A. C., Almeida, C., Venturieri, A., Esquerdo, J. C. D. M., and Silva, M.: Uso e cobertura da terra nas áreas desflorestadas da Amazônia Legal: TerraClass 2008, Tech. rep., EMBRAPA, Brasilia, and INPE, Belém, http://ainfo.cnptia.embrapa.br/digital/bitstream/item/87809/1/TerraClass-completo-baixa-pdf.pdf, 2013.

Dalla-Nora, E. L., de Aguiar, A. P. D., Lapola, D. M., and Woltjer, G.: Why have land use change models for the Amazon failed to capture the amount of deforestation over the last decade?, Land Use Policy, 39, 403–411, doi:10.1016/j.landusepol.2014.02.004, 2014.

Davidson, E. A., de Araújo, A. C., Artaxo, P., Balch, J. K., Brown, I. F., C. Bustamante. M. M., Coe, M. T., DeFries, R. S., Keller, M., Longo, M., Munger, J. W., Schroeder, W., Soares-Filho,

B. S., Souza, C. M., and Wofsy, S. C.: The Amazon basin in transition, Nature, 481, 321–328, doi:10.1038/nature10717, 2012.

de Espindola, G. M., de Aguiar, A. P. D., Pebesma, E., Câmara, G., and Fonseca, L.: Agricultural land use dynamics in the Brazilian Amazon based on remote sensing and census data, Appl. Geogr., 32, 240–252, doi:10.1016/j.apgeog.2011.04.003, 2012.

Fearnside, P. M.: Amazonian deforestation and global warming: carbon stocks in vegetation replacing Brazil's Amazon forest, Forest Ecol. Manag., 80, 21–34, doi:10.1016/0378-1127(95)03647-4, 1996.

Fearnside, P. M.: Deforestation in Brazilian Amazonia: History, rates, and consequences, Conservation Biology, 19, 680–688, doi:10.1111/j.1523-1739.2005.00697.x, 2005.

Fearnside, P. M.: Deforestation soars in the Amazon, Nature, 521, p. 423, doi:10.1038/521423b, 2015.

Foley, J. A., DeFries, R., Asner, G. P., Barford, C., Bonan, G., Carpenter, S. R., Chapin, F. S., Coe, M. T., Daily, G. C., Gibbs, H. K., Helkowski, J. H., Holloway, T., Howard, E. A., Kucharik, C. J., Monfreda, C., Patz, J. A., Prentice, I. C., Ramankutty, N., and Snyder, P. K.: Global Consequences of Land Use, Science, 309, 570–574, doi:10.1126/science.1111772, 2005.

Fortunato, S.: Community detection in graphs, Physics Reports, 486, 75–174, doi:10.1016/j.physrep.2009.11.002, 2010.

Gan, G., Ma, C., and Wu, J.: Data Clustering: Theory, Algorithms, and Applications, ASA-SIAM Series on Statistics and Applied Probability, SIAM, Philadelphia, ASA, Alexandria, VA, doi:10.1137/1.9780898718348, 2007.

Gibbs, B. H. K., Rausch, L., Munger, J., Schelly, I., Morton, D. C., Noojipady, P., Barreto, P., Micol, L., and Walker, N. F.: Brazil's Soy Moratorium, Science, 347, 377–378, doi:10.1126/science.aaa0181, 2015.

IBGE: Downloads Geociências, http://downloads.ibge.gov.br/downloads_geociencias.htm (last access: 21 February 2017), 2016.

INPE: Projeto PRODES – Monitoramento da Floresta Amazônica Brasileira por Satélite, http://www.obt.inpe.br/prodes/index.php, last access: 21 February 2017.

INPE and EMBRAPA: Projeto TerraClass, http://www.inpe.br/cra/projetos_pesquisas/dados_terraclass.php, last access: 21 February 2017.

Jain, A. K. and Dubes, R. C.: Algorithms for Clustering Data, Prentice Hall, Eaglewood Cliffs, NJ, 1988.

Keller, M., Bustamante, M., Gash, J., and Dias, P. S.: Amazonia and Global Change, American Geophysical Union, Washington, DC, doi:10.1029/GM186, 2009.

Lambin, E. F. and Meyfroidt, P.: Global land use change, economic globalization, and the looming land scarcity, P. Natl. Acad. Sci., 108, 3465–3472, doi:10.1073/pnas.1100480108, 2011.

Laurance, W. F. and Williamson, G. B.: Positive feedbacks among forest fragmentation, drought, and climate change in the Amazon, Conserv. Biol., 15, 1529–1535, doi:10.1046/j.1523-1739.2001.01093.x, 2001.

Laurance, W. F., Lovejoy, T. E., Vasconcelos, H. L., Bruna, E. M., Didham, R. K., Stouffer, P. C., Gascon, C., Bierregaard, R. O., Laurance, S. G., and Sampaio, E.: Ecosystem decay of Amazonian forest fragments: A 22-year investigation, Conserv. Biol., 16, 605–618, doi:10.1046/j.1523-1739.2002.01025.x, 2002.

Lenton, T. M., Held, H., Kriegler, E., Hall, J. W., Lucht, W., Rahmstorf, S., and Schellnhuber, H. J.: Tipping elements in the

Earth's climate system, P. Natl. Acad. Sci., 105, 1786–1793, doi:10.1073/pnas.0705414105, 2008.

Levers, C., Müller, D., Erb, K., Haberl, H., Jepsen, M. R., Metzger, M. J., Meyfroidt, P., Plieninger, T., Plutzar, C., Stürck, J., Verburg, P. H., Verkerk, P. J., and Kuemmerle, T.: Archetypical patterns and trajectories of land systems in Europe, Reg. Environ. Change, 1–18, doi:10.1007/s10113-015-0907-x, 2015.

Lu, D., Mausel, P., Brondízio, E., and Moran, E.: Change detection techniques, Int. J. Remote Sens., 25, 2365–2401, doi:10.1080/0143116031000139863, 2004.

Macedo, M. N., DeFries, R. S., Morton, D. C., Stickler, C. M., Galford, G. L., and Shimabukuro, Y. E.: Decoupling of deforestation and soy production in the southern Amazon during the late 2000s, P. Natl. Acad. Sci. USA, 109, 1341–1346, doi:10.1073/pnas.1111374109, 2012.

Meyfroidt, P. and Lambin, E. F.: Global Forest Transition: Prospects for an End to Deforestation, Annu. Rev. Env. Resour., 36, 343–371, doi:10.1146/annurev-environ-090710-143732, 2011.

Nepstad, D., McGrath, D., Stickler, C., Alencar, A., Azevedo, A., Swette, B., Bezerra, T., DiGiano, M., Shimada, J., Seroa da Motta, R., Armijo, E., Castello, L., Brando, P., Hansen, M. C., McGrath-Horn, M., Carvalho, O., and Hess, L.: Slowing Amazon deforestation through public policy and interventions in beef and soy supply chains, Science, 344, 1118–1123, doi:10.1126/science.1248525, 2014.

Nepstad, D. C., Stickler, C. M., Filho, B. S., and Merry, F.: Interactions among Amazon land use, forests and climate: prospects for a near-term forest tipping point, Philos. T. R. Soc. B, 363, 1737–1746, doi:10.1098/rstb.2007.0036, 2008.

Norris, J.: Markov Chains, Cambridge University Press, Cambridge, MA, doi:10.1017/CBO9780511810633, 1997.

Ramankutty, N., Gibbs, H. K., Achard, F., Defries, R., Foley, J. A., and Houghton, R. A.: Challenges to estimating carbon emissions from tropical deforestation, Glob. Change Biol., 13, 51–66, doi:10.1111/j.1365-2486.2006.01272.x, 2007.

Robinson, V. B.: Information Theory and Sequences of Land Use: an Application, Prof. Geogr., 30, 174–179, doi:10.1111/j.0033-0124.1978.00174.x, 1978.

Staal, A., Dekker, S. C., Hirota, M., and van Nes, E. H.: Synergistic effects of drought and deforestation on the resilience of the south-eastern Amazon rainforest, Ecol. Complex., 22, 65–75, doi:10.1016/j.ecocom.2015.01.003, 2015.

Subedi, P., Subedi, K., and Thapa, B.: Application of a Hybrid Cellular Automaton – Markov (CA-Markov) Model in Land-Use Change Prediction: A Case Study of Saddle Creek Drainage Basin, Florida, Appl. Ecol. Environ. Sci., 1, 126–132, doi:10.12691/aees-1-6-5, 2013.

Walker, R.: Theorizing Land-Cover and Land-Use Change: The Case of Tropical Deforestation, Int. Regional Sci. Rev., 27, 247–270, doi:10.1177/0160017604266026, 2004.

Spatial and radiometric characterization of multi-spectrum satellite images through multi-fractal analysis

Carmelo Alonso[1,2], Ana M. Tarquis[2,3], Ignacio Zúñiga[4], and Rosa M. Benito[2]

[1]Earth Observation Systems, Indra Sistemas S.A., Madrid, Spain
[2]Grupo de Sistemas Complejos, U.P.M, Madrid, Spain
[3]CEIGRAM, E.T.S.I.A.A.B., U.P.M, Madrid, Spain
[4]Dpt. Física Fundamental, Facultad de Ciencias, Universidad Nacional de Educación a Distancia (UNED), Madrid, Spain

Correspondence to: Ana M. Tarquis (anamaria.tarquis@upm.es)

Abstract. Several studies have shown that vegetation indexes can be used to estimate root zone soil moisture. Earth surface images, obtained by high-resolution satellites, presently give a lot of information on these indexes, based on the data of several wavelengths. Because of the potential capacity for systematic observations at various scales, remote sensing technology extends the possible data archives from the present time to several decades back. Because of this advantage, enormous efforts have been made by researchers and application specialists to delineate vegetation indexes from local scale to global scale by applying remote sensing imagery.

In this work, four band images have been considered, which are involved in these vegetation indexes, and were taken by satellites Ikonos-2 and Landsat-7 of the same geographic location, to study the effect of both spatial (pixel size) and radiometric (number of bits coding the image) resolution on these wavelength bands as well as two vegetation indexes: the Normalized Difference Vegetation Index (NDVI) and the Enhanced Vegetation Index (EVI).

In order to do so, a multi-fractal analysis of these multi-spectral images was applied in each of these bands and the two indexes derived. The results showed that spatial resolution has a similar scaling effect in the four bands, but radiometric resolution has a larger influence in blue and green bands than in red and near-infrared bands. The NDVI showed a higher sensitivity to the radiometric resolution than EVI. Both were equally affected by the spatial resolution.

From both factors, the spatial resolution has a major impact in the multi-fractal spectrum for all the bands and the vegetation indexes. This information should be taken in to account when vegetation indexes based on different satellite sensors are obtained.

1 Introduction

Soil moisture is a critical condition affecting interaction of land surface and atmosphere. Remotely sensed data are an important source of information and can indirectly measure soil moisture in space and time. However, the signal only penetrates the top few centimetres, and soil moisture at deeper layers must be estimated. One method to estimate soil moisture at deeper layers is through vegetation indexes. Several authors have investigated the potential of vegetation indexes to estimate root zone soil moisture. The Normalized Difference Vegetation Index (NDVI) and Enhanced Vegetation Index (EVI) have been used by several authors (Wang et al., 2007; Ben-Ze'ev et al., 2006; Deng et al., 2007) in different conditions to find significant estimations with root zone soil moisture. For the estimation of these indexes near-infrared (NIR), red and blue wavelengths are needed (Huete et al., 2014).

The images provided by the satellites show the land surface in a wide range of wavelengths (from visible to thermal infrared or microwaves) and also with a great variety of spatial resolutions (from a few kilometres to tens of centimetres). The analysis of these varied images and their synergic possibilities are a challenging problem, especially with new sensors, which have small spatial resolution and a large range of radiometric quantification. Fractal analysis offers significant potential for improvement in the measurement and analysis of spatially and radiometrically complex remote sensing data. This analysis also provides quantitative insight on the spatial complexity in the information of the landscape contained within these data.

In the general mathematical framework of fractal geometry, many analytical methods have been developed, including the following: textural homogeneity, which has been characterized using the fractal dimension (Fioravanti, 1994) (it has also been used as a spatial measure for describing the complexity of remote sensing imagery; Lam and De Cola, 1993)); and changes in the image complexity, which have been detected through the spectral range of hyperspectral images affecting the fractal dimension (Qiu et al., 1999). Similarly, De Cola (1989) and Lam (1990) have found that fractal dimension also depends on the spectral bands of Landsat-7 TM imagery.

Motivated by the fractal geometry of sets (Mandelbrot, 1983), the development of multi-fractal theory, introduced in the context of turbulence, has been applied in many areas, such as earthquake distribution analysis (Hirata and Imoto, 1991), soil pore characterization (Kravchenko et al., 1999; Tarquis et al., 2003), image analysis (Sánchez et al., 1992) or remote sensing (Tessier et al., 1993; Cheng and Agterberg, 1996; Schmitt et al., 1997; Laferrière and Gaonac'h, 1999; Cheng, 2004; Lovejoy et al., 2001b; Du and Yeo, 2002; Parrinello and Vaughan, 2002; Harvey et al., 2002; Turiel et al., 2005).

The acquisition of remotely sensed multiple spectral images is thus a unique source of data for determining the scale-invariant characteristics of the radiance fields related to many factors, such as soil and bedrock chemical composition, humidity content and surface temperature (e.g. Laferrière and Gaonac'h, 1999; Maître and Pinciroli, 1999; Lovejoy et al., 2001a, b; Harvey et al., 2002; Beaulieu and Gaonac'h, 2002; Gaonac'h et al., 2003). In one of the schemes used in the multi-fractal analysis, the satellite image is considered as a mass distribution of a statistical measure on the space domain studied, and it is analysed through a multi-fractal (MF) spectrum (Cheng, 2004; Mao-Gui Hu, 2009; Tarquis et al., 2014), which gives either geometrical or probabilistic information about the pixel distribution with the same singularity. Other techniques focus on the variations of a measure analysing the moments of the absolute differences of their values at different scales, e.g. the Generalized Structure Function and the Universal Multi-fractal model (Lovejoy et al., 2001a, 2008; Renosh et al., 2015)

The aim of this work is to characterize by MF analysis the image patterns in the wavelength range for the common bands of the satellites used, as well as both NDVI and EVI indexes. In order to investigate how the image information is affected by the sampling with different spatial and radiometric resolutions, we have also analysed images of the same

site but acquired by two different satellites: Landsat-7 and Ikonos-2.

We present a comparative analysis of multi-fractal (MF) tools applied to multi-spectral images obtained by Ikonos-2 and LANDSAT-7. Both satellites have several bands in visible and near-infrared spectral regions in common that can be used in vegetation-index estimation. However, the bands have different spatial resolution (4 m for Ikonos-2 and 30 m for LANDSAT-7), and radiometric resolution (11 bits for Ikonos-2 and 8 bits for LANDSAT-7). The bands we have chosen are red, green, blue and near infrared. For each of those bands, the MF spectrum has been calculated directly from the Hölder exponents α and the singularity spectrum $f(\alpha)$. The same calculations were applied for NDVI and EVI estimated on red, blue and near-infrared bands for each image.

2   Materials and methods

2.1   Images

As already noted, in this work we have analysed two images of the same site acquired from different satellites, Landsat-7 and Ikonos-2. Both are multi-spectral images with several bands that cover several regions of the electromagnetic spectrum in the visible and near-infrared wavelength.

Landsat-7 was put in orbit in April 1999. This satellite follows a sun-synchronous orbit at 705 km of altitude, with an equatorial crossing time of 10:00 LT in the descending node. It requires 98.8 min to circle the Earth, tracing a worldwide reference system (WRS) of just over 230 ground paths. Over at least three decades, Landsat-7 orbits over each of these paths once every 16 days in a repetitive cycle (Mika, 1997).

The main Landsat-7 sensor for Earth observation is the Enhanced Thematic Mapper Plus (ETM+). The ETM+ operates as a whiskbroom scanner and acquires data for seven spectral bands: visible (ETM+#1, from 0.45 to 0.52 μm; ETM+#2, from 0.53 to 0.61 μm; ETM+#3, from 0.63 to 0.69 μm), near infrared (ETM+#4, from 0.78 to 0.9 μm), shortwave infrared (ETM+#5, from 1.55 to 1.75 μm, and ETM+#7, from 2.09 to 2.35 μm) and thermal infrared (ETM+#6, from 10.4 to 12.5 μm). The ETM+ ground sampling distance (pixel size in the images) is 30 m for the six reflective bands and 60 m for the thermal band. The ETM+ also acquires images for a panchromatic band (ETM+#8, from 0.52 to 0.9 μm) with a 15 m ground sampling distance. The radiometric resolution of the Landsat-7 data is 8 bits per pixel or 256 grey levels for the pixel digital value.

Ikonos-2 was launched in September 1999. Its panchromatic sensor, with a resolution of 0.82 m, provided the first very high-resolution images of the Earth's surface from earth observation satellites (EOS). The Ikonos-2 orbiting altitude is approximately 681 km; it is inclined 98.1° to the equator and it provides sun-synchronous operation. The equatorial crossing time of Ikonos-2 is 10:30 LT in the descending node. The orbit provides daily access to sites within 45° of nadir (Dial et al., 2003).

The multi-spectral sensor simultaneously collects blue (IK#1, from 0.445 to 0.516 μm), green (IK#2, from 0.506 to 0.595 μm), red (IK#3, from 0.632 to 0.698 μm) and near-infrared (IK#4, from 0.757 to 0.853 μm) bands with 3.28 m resolution at nadir. Both images, panchromatic and multi-spectral, have a radiometric resolution of 11 bits per pixel or 2048 grey levels for the pixel digital value.

The Landsat-7 multi-spectral image used in this study was acquired on 6 August 2000 at 10:46 LT. and it corresponds to the scene with WRS coordinates: path and row 201 and 32, respectively. This scene is located in the central region of Spain and it covers a square surface of approximately 180 km side length, located around Madrid. Solar azimuth and elevation angles for this scene are 132.44 and 58.62° respectively.

The Ikonos-2 datum used in this study is a multi-spectral image acquired on 8 August 2000 at 11:03 LT. It covers an area of 11 km^2 located near Aranjuez, south of Madrid, in the central region of Spain. Solar azimuth and elevation angles for this scene are 139.5 and 60.79° respectively. Both images were corrected geometrically to the same cartographic projection: Universal Transverse Mercatorprojection (UTM), zone 30° N, by a co-registration process.

The analysis has been carried out on a subset that covers (approximately) the same area in both the Landsat-7 and the Ikonos-2 images, corresponding to a region located north of the town of Aranjuez. The representative elements of the land used in the selected area are irrigation crops, pastures, heaths, unirrigated land cultivations and olive groves. The Landsat-7 subset image is a square of 512×512 pixels with a resolution of 30 m covering a somewhat larger surface than the Ikonos-2 image. The Ikonos-2 image consists of a square subset with 2048×2048 pixels and 4 m resolution.

2.2 Vegetation indexes

Vegetation is one of the landscape elements that has received the most attention in the field of image analysis. Therefore, there are many parameters that can be used to obtain information on vegetation from remote sensing imagery.

One of the main parameters is made up of the vegetation indexes. These indexes allow to detect the presence of vegetation in an area and its activity, since its values are related to this activity. For this, we can use the reflectance values corresponding to the different wavelengths, interpreting these in relation to the photosynthetic activity. Of these indexes, the most commonly used is the NDVI.

The NDVI is defined by

$$\text{NDVI} = \frac{\text{NIR} - \text{R}}{\text{NIR} + \text{R}}, \tag{1}$$

where NIR is the pixel value in the near-infrared band and R the pixel value in the red band. The values of this in-

dex are within the range $(-1, 1)$ and their positive values are sensitive to the proportion of soil and vegetation in each pixel (Carlson and Ripley, 1997). Pixels with NDVI < 0.2 are considered without vegetation or bare soil. Pixels with NDVI > 0.5 are considered as fully covered by vegetation.

EVI, the other vegetation index, is defined by

$$\text{EVI} = 2.5 \frac{(\text{NIR} - \text{R})}{(L + \text{NIR} + C_1\text{R} - C_2\text{B})}, \tag{2}$$

where NIR is the pixel value in the near-infrared band, R the pixel value in the red band and B the pixel value in the blue band. L, C_1 and C_2 are constants with the values 1, 6 and 7.5 respectively. The main characteristic of this index is that it corrects some distortions caused by the light dispersion from aerosols, as well as the background soil (Huete et al., 2014).

2.3 Multi-fractal image analysis

A monofractal object can be measured by counting the number N of δ size boxes needed to cover the object. The measure depends on the box size as

$$N(\delta) \propto \delta^{-D_0}, \tag{3}$$

where

$$D_0 = \lim_{\delta \to 0} \frac{\log N(\delta)}{\log \frac{1}{\delta}} \tag{4}$$

is the fractal dimension. D_0 is calculated from slope of a log–log plot. However, many examples are found where a single scaling law cannot be applied and it is necessary to do a multi-scaling analysis.

There are several methods for implementing multi-fractal analysis. The universal multi-fractal (UM) model assumes that multi-fractals are generated from a random variable with an exponentiated extreme Levy distribution (Lavallée et al., 1991; Tessier et al., 1993). In UM analysis, the scaling exponent $K(q)$ is highly relevant. This function for the moments q of a cascade conserved process is obtained according to Schertzer and Lovejoy (1987), as follows:

$$K(q) = \begin{cases} \dfrac{C_1(q^{\alpha_L} - q)}{\alpha_L - 1} & \text{if } \alpha_L \neq 1 \\ C_1 q \log(q) & \text{if } \alpha_L = 1 \end{cases}, \tag{5}$$

where C_1 is the mean intermittency codimension and α_L is the Levy index. These are known as the UM parameters.

Other method is the moment method developed by Halsey et al. (1986) and applied to this case study. This method mainly uses three functions: $\tau(q)$, known as the mass exponent function; α, the coarse Hölder exponent; and $f(\alpha)$, multi-fractal spectrum (MFS). A measure (or field), defined in two-dimensional image embedding space ($n \times n$ pixels) and with values based on grey tones (for 8 bits, from 0 to 255), cannot be considered as a geometrical set and therefore cannot be characterized by a single fractal dimension.

To characterize the scaling property of a variable measured on the spatial domain of the studied area, it divides the image into a number of self-similar boxes. Applying disjoint covering by boxes in an "up-scaling" partitioning process we obtain the partition function $\chi(q,\delta)$ (Feder, 1989) defined as follows:

$$\chi(q,\delta) = \sum_{i=1}^{N(\delta)} \mu_i^q(\delta) = \sum_{i=1}^{N(\delta)} m_i^q, \tag{6}$$

where m is the mass of the measure, q is the mass exponent, δ is the length size of the box and $N(\delta)$ is the number of boxes in which $m_i > 0$. Based on this, the mass exponent function $\tau(q)$ shows how the moments of the measure scales with the box size:

$$\tau(q) = \lim_{\delta \to 0} \frac{\log <\chi(q,\delta)>}{\log(\delta)} = \lim_{\delta \to 0} \frac{\log <\sum_{i=1}^{N(\delta)} m_i^q>}{\log(\delta)}, \tag{7}$$

where $<>$ represents a statistical moment of the measure $\mu_i(\delta)$ defined on a group of non-overlapping boxes of the same size partitioning as the area studied.

The singularity index, α, can be determined by the Legendre transformation of the $\tau(q)$ curve (Halsey et al., 1986) as follows:

$$\alpha(q) = \frac{d\tau(q)}{dq}. \tag{8}$$

The number of cells of size δ with the same α, $N_\alpha(\delta)$, is related to the cell size as $N_\alpha(\delta) \propto \delta^{-f(\alpha)}$, where $f(\alpha)$ is a scaling exponent of the cells with common α. Parameter $f(\alpha)$ can be calculated as follows:

$$f(\alpha) = q\alpha(q) - \tau(q). \tag{9}$$

MFS, shown as plot of α versus $f(\alpha)$, quantitatively characterizes variability of the measure studied with asymmetry to the right and left indicating domination of small and large values respectively (Evertsz and Mandelbrot, 1992). There are three characteristic values obtained from MFS, the singularity $\alpha(q)$ values for $q = \{0, 1, 2\}$. The first value ($\alpha(0)$) corresponds to the maximum of MFS and it is related to the box-counting dimension of the measure support; the second value is related to information or entropy dimension ($\alpha(1)$) and the third with the correlation dimension. The entropy dimension quantifies the degree of disorder present in a distribution. According to Andraud et al. (1994) and Gouyet (1996), a $\alpha(1)$ value close to 2.0 characterizes a system uniformly distributed throughout all scales, whereas a $\alpha(1)$ close to 0 reflects a subset of the scale in which the irregularities are concentrated. These three values will be shown from each calculation of MFS.

The width of the MF spectrum (Δ) indicates overall variability (Tarquis et al., 2001, 2014) and we have split it in two sections. Section I correspond to values $\alpha(q) < \alpha(0)$ or

$q > 0$ and section II to values with $\alpha(q) > \alpha(0)$ or $q < 0$. In section I the amplitude, or semi-width, was calculated with differences $\Delta^+ = \alpha(0) - \alpha(+5)$, and in section II with $\Delta^- = \alpha(-5) - \alpha(0)$.

To study the asymmetry of the MFS we have chosen the asymmetry index (AI) estimated as follows (Xie et al., 2010):

$$AI = \frac{\Delta\alpha_L - \Delta\alpha_R}{\Delta\alpha_L + \Delta\alpha_R} \quad \begin{array}{l} \Delta\alpha_L = \alpha_0 - \alpha_{min} \\ \Delta\alpha_R = \alpha_{max} - \alpha_0 \end{array}. \tag{10}$$

In our case, α_0 is the singularity for $q = 0$ or $\alpha(0)$, α_{min} is $\alpha(+5)$ and α_{max} is $\alpha(-5)$. Therefore, we can rewrite the AI as follows:

$$AI = \frac{\Delta^+ - \Delta^-}{\Delta^+ + \Delta^-}. \tag{11}$$

Expressing AI as Eq. (11), we can see that it is a normalized index based on the amplitudes Δ^+ and Δ^-.

There are several works relating the UM model and the multi-fractal formalism based on $\tau(q)$ (Gagnon et al., 2003; Aguado et al., 2014; Morató et al., 2017, among others) through the following equations:

$$f(\alpha) = E - c(\gamma); \alpha = E - \gamma, \tag{12}$$
$$\tau(q) = E(q-1) - K(q), \tag{13}$$

where E is the Euclidean dimension in which the measure is embedded, in this case $E = 2$, and $c(\gamma)$ is the codimension of the singularity of the density of the multi-fractal measure γ.

3 Results and discussion

3.1 Radiometric influence in the multi-fractal spectrum

To study the influence of radiometric resolution on Ikonos-2 image information complexity, the original pixel code (11 bits) has been transformed to 8 bits through a rescaling based on minimum and maximum values between 0 and 255, with the aim of preserving the initial histogram shape.

We first discuss the results obtained for the 2048 × 2048 pixel Ikonos-2 image shown in Fig. 1, in bands combination of false colour (IK#4, IK#3, IK#2 band combination in RGB visualization). In Fig. 2 IK#1, IK#2, IK#3 and IK#4 band histograms are shown. In the right column are histograms with the original radiometric resolution and in the left column the corresponding histograms are rescaled to 8 bits. The histograms present a bimodal structure with a narrow peak of low-value pixels (dark grey) showing a sharp maximum and a wider peak around a second lower maximum. For bands IK#1, IK#2 and IK#3, the narrow peak maximum corresponds to vegetation, mainly irrigation crops, showing strong water absorption. This effect is particularly important in band IK#3. High-value pixels (lighter grey) correspond to ground zones with lower vegetation content. However, as

Figure 1. The Ikonos-2 image in band combinations of false colour (IK#4, IK#3 and IK#2 in RGB). The image has a size of 2048×2048 pixels, each area unit corresponding to 4×4 m. The coordinates UTM (zone 30) of the upper left and lower right pixel in the image are: ULX $= 446\,037$ m, ULY $= 4\,441\,684$ m, LRX $= 454\,229$ m and LRY $= 4\,433\,492$ m.

vegetation shows high reflectivity in the near-infrared, IK#4 band histogram shows a predominance of high-value pixels (lighter grey pixels) corresponding to dense vegetation parts. For both radiometric resolutions the shapes of the histograms are very similar, as it was our intention (see Fig. 2).

We cover the image with boxes of size $\delta = 2^{-n}$ and we change the box size from 2048 to 2 pixels, that is, $\delta = 2048/2^n$ with $n = 0, 1, 2, \ldots, 10$. For each value of the parameter q, from -5 to $+5$ with increments of 0.5, the partition function (Eq. 6) is computed and $\log \chi(q, \delta)$ versus $\log \delta$ is plotted in Fig. 3. Each graph contains 11 points and from these a range of scales are selected for the least-square linear fit reaching the maximum possible scales and with a standard error in the slope, the estimated values of $\tau(q)$, less than 0.01. Then, using Eqs. (7)–(8), $\alpha(q)$ and $f(\alpha)$ are obtained. Comparing the range of scales used in both radiometric resolutions, the bands using the original data (11 bits) showed a wider range of scales for the linear fit, up to 4 pixels, whereas in the 8-bit radiometric resolution were required up to 32 pixels (see arrows in Fig. 3).

The MF spectra $f(\alpha)$ corresponding to the four bands of multi-spectral Ikonos images are shown in Fig. 4. These differences found in the multi-scaling behaviour of each band are in agreement with previous works (Cheng, 2004; Lovejoy et al., 2008). Just by visual observation, there is a remarkable difference between the bands #3 and #4, and red and NIR,

between 8 and 11 bits respectively. Higher radiometric resolution gives a higher range of possible grey values per pixel. Note that this radiometric resolution effect is manifested in both sections of the MF spectra (for $q > 0$ and for $q < 0$).

Some characteristic parameters obtained from these MF spectra are shown in Table 1 and Table 2. As expected, in both radiometric resolutions and in each band the $\alpha(0)$ is practically 2, as the measure is defined in the entire plane and it has an Euclidean dimension of 2. With respect to the $\alpha(1)$ value, certain differences are found. Comparing the bands in 8 bits to the same ones in 11 bits, the entropy dimension was always higher. However, considering the standard errors, only IK#1 (B) and IK#2 (G) bands were significantly different, with the blue band showing the highest difference. Meanwhile, red and NIR bands are not significantly different. This shows that a more spatially uniform distribution for the bands of Ikonos-2 8 bits than in 11 bits. The same behaviour is observed in the $\alpha(2)$.

The amplitudes calculated (Δ^+ and Δ^-) in Ikonos-2 11-bit bands present opposite trends (Table 1). Note that amplitude Δ^+ decreases as band wavelength grows, whereas the other amplitude Δ^- diminishes. Observing these parameters in Ikonos-2 8-bit bands (Table 2) a different trend and behaviour are found. In this case both Δ^+ and Δ^- increase as the wavelength increases for the three visible bands, but decrease for the near-infrared band (IK#4).

The AI estimated on these MFS amplitudes on each radiometric resolution are shown in the last column of Table 1 and Table 2. Comparing the bands in 8 bits to the same ones in 11 bits, the behaviour is similar: there is a decreasing trend from IK#1 to IK#4, although the range of values is different. At a resolution of 11 bits from a positive AI $= 0.240$ at blue band goes to a negative AI $= -0.237$ at NIR band. On the other hand, at a resolution of 8 bits, an AI $= 0.092$ goes to a negative AI $= -0.347$. The more symmetric MFS are found in green and red bands at a resolution of 11 bits and in blue and green bands at a resolution of 8 bits.

Doing the same study for the vegetation indexes we found the following. The bi-log plot of the partition function ($\chi(q, \delta)$) versus δ is plotted in Fig. 5 for both vegetation indexes at both radiometric resolutions. Each graph contains 11 points as the bands from where they were estimated. The linear fit was done with the same methodology as that for the four bands. In this case only the EVI at 8 bits shows a better linear trend in a wider range of scales. However, to better compare both vegetation indexes, from both radiometric resolutions, a range achieving 32 pixels (128 m) was selected as shown by arrows in Fig. 5.

The MF spectra $f(\alpha)$ of EVI and NDVI estimated for both radiometric resolutions of Ikonos images are shown in Fig. 4. Both vegetation indexes show differences due to the transformation from 11 to 8 bits. However, NDVI shows higher differences in the MFS, mainly in the part corresponding to q negative values (right side). Even EVI presents changes; its MFS is closer at both radiometric resolutions. Compar-

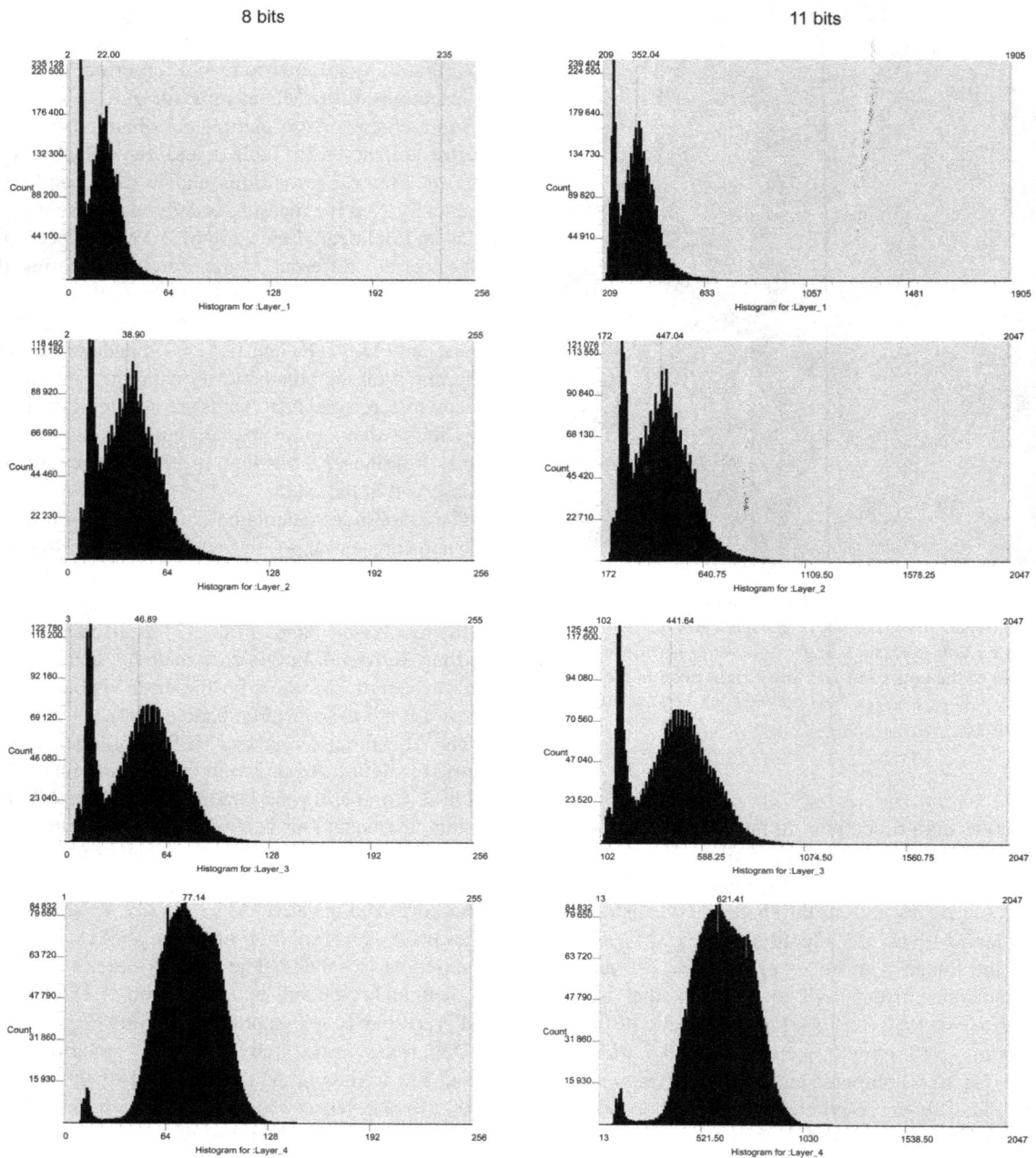

Figure 2. Histograms of the four bands of the Ikonos-2 image for the original radiometric resolution, 11 bits (right), and the minimum–maximum rescaled 8-bit radiometric resolution (left).

ing the range of $f(\alpha)$ values in the vegetation indexes to the range obtained in the four bands (left column in Fig. 4), there is a remarkable contrast. Meanwhile the NIR band of 8 bits achieves a $f(\alpha)$ value close to 0.5; EVI and NDVI achieve values closer to 0.2. These differences are higher in the 11-bit image; the red band achieves a $f(\alpha)$ value close to 0.9 and vegetation indexes again achieve values of $\sim$0.2. The same characteristic parameters obtained from the band MF spectra

were calculated for the vegetation indexes and are shown in Table 1 and Table 2.

With respect to the $\alpha(1)$ values, certain differences are found between the vegetation indexes. Comparing the NDVI in 8 bits to the same ones in 11 bits, the entropy dimension was always higher than it was found to be in the bands. However, EVI shows the contrary: entropy values of the 11-bit image are higher than the 8-bit image, although the differences are not significant. Therefore, the radiometric resolution af-

Table 1. Parameters obtained from the multi-fractal spectrum from each band of the Ikonos-2 image, and the vegetation indexes (VIs) estimated, with a pixel size of 4 m and a radiometric resolution of 11 bits. The amplitudes of α values are presented as Δ^+ and Δ^- corresponding to $\alpha(0) - \alpha(5)$ and $\alpha(-5) - \alpha(0)$ respectively. The asymmetry index (AI) corresponds to $\frac{\Delta^+ - \Delta^-}{\Delta^+ + \Delta^-}$.

	Band	q	$\alpha(q)$	Δ^+	Δ^-	AI
		0	2.001 ± 0.001			
	IK#1	1	1.938 ± 0.005	0.418	0.256	0.240
		2	1.865 ± 0.009			
		0	2.001 ± 0.001			
	IK#2	1	1.936 ± 0.005	0.377	0.313	0.093
Ikonos-2 (11 bits)		2	1.871 ± 0.007			
		0	2.001 ± 0.001			
	IK#3	1	1.937 ± 0.005	0.348	0.382	-0.047
		2	1.878 ± 0.006			
		0	2.001 ± 0.001			
	IK#4	1	1.959 ± 0.005	0.290	0.470	-0.237
		2	1.908 ± 0.009			
	VI	q	$\alpha(q)$	Δ^+	Δ^-	AI
		0	2.000 ± 0.001			
	NDVI	1	1.886 ± 0.008	0.516	1.166	-0.386
Ikonos-2 (11 bits)		2	1.779 ± 0.010			
		0	2.000 ± 0.001			
	EVI	1	1.948 ± 0.002	0.270	0.877	-0.533
		2	1.897 ± 0.004			

Table 2. Parameters obtained from the multi-fractal spectrum from each band of the Ikonos-2 image, and the vegetation indexes (VIs) estimated, with a pixel size of 4 m and a radiometric resolution of 8 bits. The amplitudes of α values are presented as Δ^+ and Δ^- corresponding to $\alpha(0) - \alpha(5)$ and $\alpha(-5) - \alpha(0)$ respectively. The asymmetry index (AI) corresponds to $\frac{\Delta^+ - \Delta^-}{\Delta^+ + \Delta^-}$.

	Band	q	$\alpha(q)$	Δ^+	Δ^-	AI
		0	2.000 ± 0.001			
	IK#1	1	1.971 ± 0.003	0.231	0.192	0.092
		2	1.930 ± 0.006			
		0	2.000 ± 0.001			
	IK#2	1	1.963 ± 0.004	0.270	0.287	-0.031
Ikonos-2 (8 bits)		2	1.914 ± 0.006			
		0	2.000 ± 0.001			
	IK#3	1	1.945 ± 0.005	0.323	0.614	-0.311
		2	1.887 ± 0.006			
		0	2.000 ± 0.001			
	IK#4	1	1.966 ± 0.004	0.248	0.512	-0.347
		2	1.923 ± 0.008			
	VI	q	$\alpha(q)$	Δ^+	Δ^-	AI
		0	2.000 ± 0.002			
	NDVI	1	1.932 ± 0.005	0.337	0.984	-0.490
Ikonos-2 (8 bits)		2	1.855 ± 0.008			
		0	2.000 ± 0.002			
	EVI	1	1.940 ± 0.004	0.300	0.874	-0.488
		2	1.873 ± 0.006			

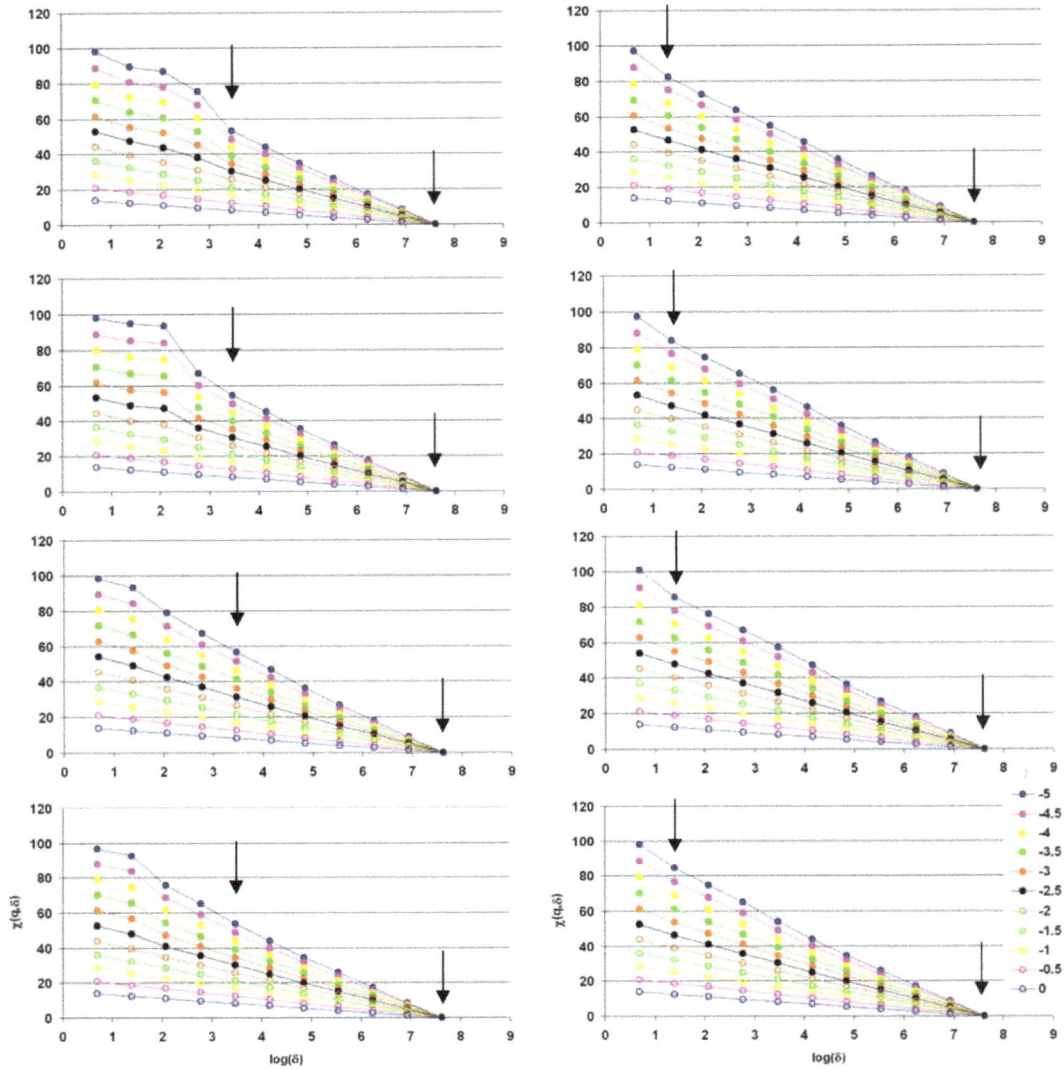

Figure 3. Bi-log plots of the partition function $\chi(q, \delta)$ versus δ for the first four bands of the Ikonos-2 satellite and for $q < 0$ values. From top to bottom we show the results for IK#1, IK#2, IK#3 and IK#4. The left column correspond to the 8-bit image and the right column to the 11-bit image. The arrows marked the range of scales used for the fit and to calculate the slope for different values of q (7 points in the left column and 10 points in the right column).

fects NDVI more than EVI. The former presents a more uniform space distribution in 8 bits than in the 11-bit image. The same behaviour is observed in $\alpha(2)$.

The amplitudes calculated (Δ^+ and Δ^-) in Ikonos-2 11-bit vegetation indexes present a similar situation (Table 1). The amplitude Δ^+ is lower than amplitude Δ^- and therefore the AI estimated is negative. This is visually perceived in Fig. 4 (right column). Observing these parameters in Ikonos-2 8 bits vegetation indexes (Table 2), similar situations are found but the values are lower. In both images (11 and 8 bits) NDVI shows higher values for both amplitudes, Δ^+ and Δ^-.

All the AI estimated for both vegetation indexes on each radiometric resolution are negative (Table 1 and Table 2), in-

dicating a high asymmetry on the right part of the MFS, as shown in Fig. 4. Comparing the AI values in 8 bits to the same ones in 11 bits, they are similar which shows that the shape of the MFS is similar as this index is a normalized index. However, the values of the amplitudes mark a higher change in NDVI than in EVI.

3.2 Spatial resolution influence in the multi-fractal spectrum

A comparison is made between Landsat, with an original pixel code of 8 bits, and the rescaled histograms from Ikonos, with an original pixel code of 11 bits. In this section, we discuss the results obtained in the MF analysis on the 512×512

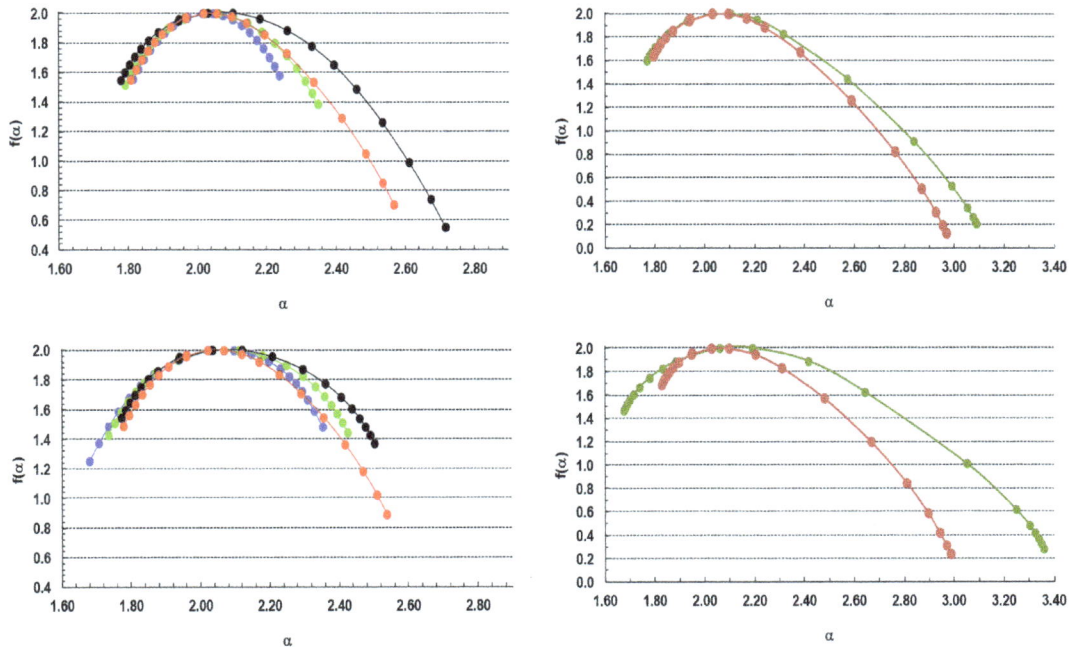

Figure 4. Multi-fractal spectrum of Ikonos-2 images for the original pixel values coded in 11 bits (lower) and the minimum–maximum rescaled to 8 bits (upper). Left column corresponds to each band analysed: IK#1 in blue colour, IK#2 in green colour, IK#3 in red colour and IK#4 in black. Right column corresponds to vegetation indexes: NDVI in green colour and EVI in brown.

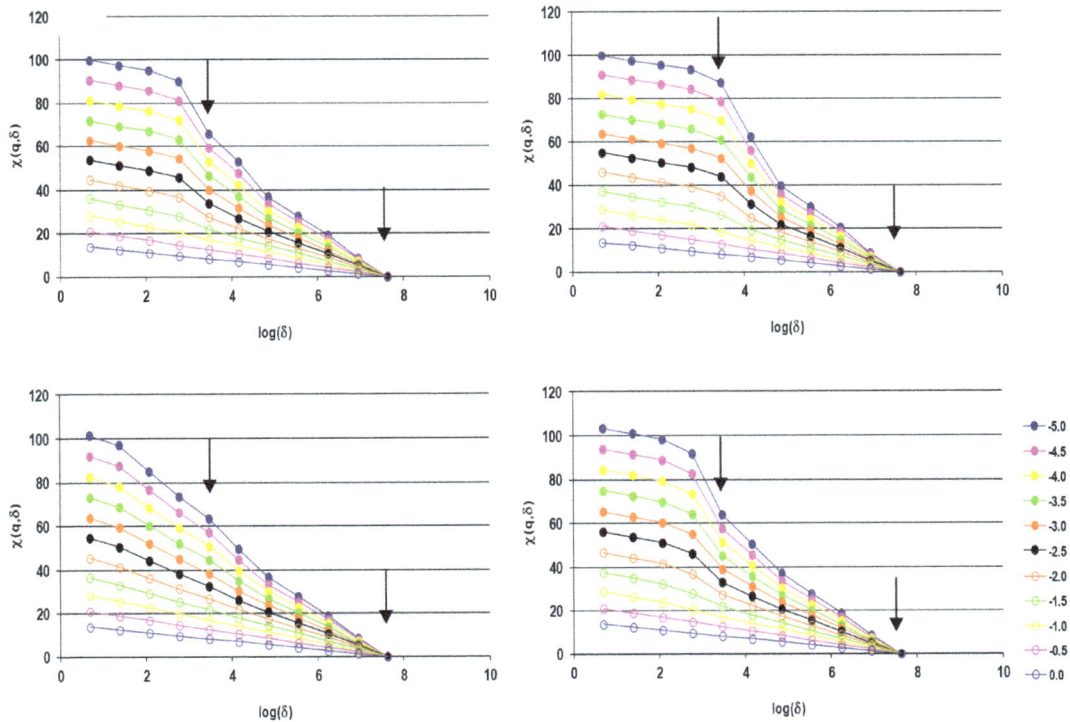

Figure 5. Bi-log plots of the partition function $\chi(q, \delta)$ versus δ for the vegetation indexes estimated from blue, red and NIR bands of the Ikonos-2 satellite and for $q < 0$ values. From top to bottom we show the results for NDVI and EVI respectively. The left column corresponds to the 8-bit image and the right column to the 11-bit image. The arrows marked the range of scales used for the fit and to calculate the slope for different values of q (7 points).

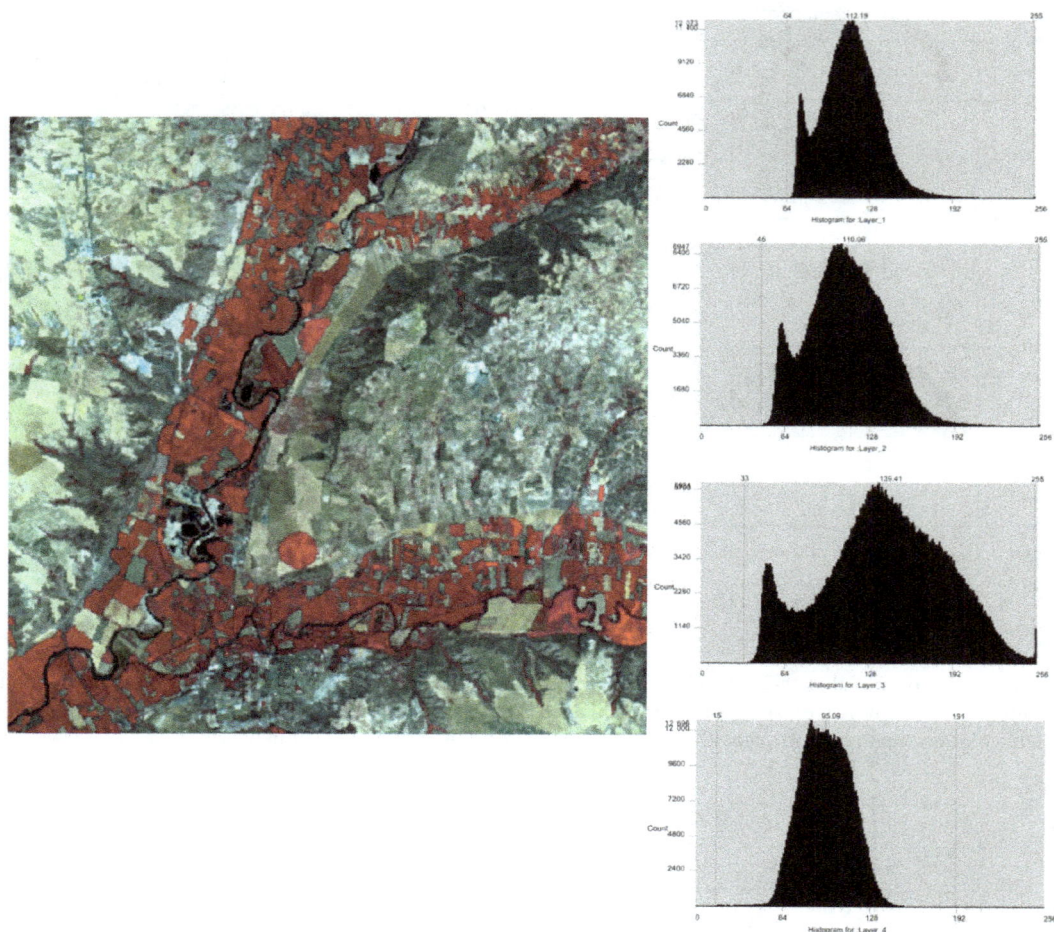

Figure 6. The Landsat-7 image and the histograms for the first four bands: blue (ETM+#1), green (ETM+#2), red (ETM+#3) and near infrared (ETM+#4). The image has a size of 512×512 pixels, each area unit corresponds to 30×30 m. The coordinates UTM (zone 30) of the upper-left and lower-right pixel in the image are ULX = 442 185 m, ULY = 4 445 568 m, LRX = 457 545 m and LRY = 4 430 208 m.

pixel Landsat-7 image shown in Fig. 6, in bands combination of false colour (ETM+#3, ETM+#2 and ETM+#1 band combination in RGB visualization).

In the right column of Fig. 6, the histograms of the Landsat-7 image for the first four bands are shown. The histograms present a bimodal structure, except for ETM+#4 (NIR), in which there is only one peak. Comparing these histograms with those obtained for Ikonos-2 8-bit image (Fig. 2), the peaks are not so abrupt and narrow. At the same time, ETM+#1, ETM+#2 and ETM+#3 bands show the absolute maximum peak at high-value pixels (light grey) and a second one at lower-value pixels (dark grey). These bands are more centred and do not show a shift to the left as the Ikonos-2 8-bit bands (IK#1, IK#2 and IK#3). In the case of the NIR band, Landsat-7 and Ikonos-2 8 bits are quite similar except for the absence of a second peak.

In the calculations, box sizes range from 512 to 2 pixels, that is, $\delta = 512/2^n$ with $n = 0, 1, \ldots, 8$. For each value of the parameters q, from -5 to $+5$ with increments of 0.5, we

compute the partition function, and the bi-log $\chi(q, \delta)$ versus δ is plotted in Fig. 7. In this case each linear fits contains only 9 points as the size of the image is 512×512 pixels. The same method was applied to select the range of scales used in the linear fit, achieving a scale of 4 pixels. Changing from pixels to metres, the scale achieved, used in Landsat-7 in the MF analysis, was ~ 120 m. In the case of the Ikonos-2 8 bits the scale was 32 pixels or 128 m, very close to Landsat-7.

The MF spectra, $f(\alpha)$, corresponding to the first four bands of multi-spectral Landsat-7 images are shown in Fig. 8. From a comparison of Figs. 4 and 8 we see that Landsat-7 image MF spectra are always located inside the corresponding Ikonos-2 MF spectra. For a given value of Hölder exponent α, the relation $f_{\text{Landsat}}(\alpha) \leq f_{\text{Ikonos}}(\alpha)$ is always satisfied. This result means that Landsat-7 images show lower complexity than Ikonos-2 8-bit images. As stated in Sect. 2.1 Ikonos-2 satellite data are coded in 11 bits in contrast with Landsat-7 8-bit-coded data. To compare both sensors, with different spatial resolution, we pass Ikonos-2 from 11 to

Table 3. Parameters obtained from the multi-fractal spectrum from each band of the Landsat-7 image, and the vegetation indexes (VIs) estimated, with a pixel size of 4 m and a radiometric resolution of 8 bits. The amplitudes of α values are presented as Δ^+ and Δ^- corresponding to $\alpha(0) - \alpha(5)$ and $\alpha(-5) - \alpha(0)$ respectively. The asymmetry index (AI) corresponds to $\frac{\Delta^+ - \Delta^-}{\Delta^+ + \Delta^-}$.

	Band	q	$\alpha(q)$	Δ^+	Δ^-	AI
		0	2.001 ± 0.001			
	ETM+#1	1	1.985 ± 0.005	0.160	0.119	0.147
		2	1.960 ± 0.010			
		0	2.003 ± 0.001			
	ETM+#2	1	1.988 ± 0.004	0.119	0.119	0.000
Landsat-7		2	1.970 ± 0.008			
		0	2.001 ± 0.001			
	ETM+#3	1	1.989 ± 0.004	0.095	0.110	-0.073
		2	1.974 ± 0.007			
		0	2.017 ± 0.001			
	ETM+#4	1	1.989 ± 0.004	0.106	0.104	0.010
		2	1.973 ± 0.008			
	VI	q	$\alpha(q)$	Δ^+	Δ^-	AI
		0	2.001 ± 0.001			
	NDVI	1	1.996 ± 0.001	0.028	0.353	-0.852
Landsat-7		2	1.992 ± 0.001			
		0	2.001 ± 0.001			
	EVI	1	1.997 ± 0.001	0.022	0.288	-0.859
		2	1.994 ± 0.001			

8 bits, observing that the latter shows more complexity than Landsat.

The MF-spectra parameters from Landsat-7 are shown in Table 3. In this section we will compare the MF spectra and the vegetation indexes of the Ikonos-2 8 bits (Table 2). The $\alpha(1)$ values from the four bands of Landsat-7 are higher than the ones presented by Ikonos-2 8 bits, indicating a more uniform space distribution. Comparing between the bands, there are not significant differences, contrary to the trend we observed among them in Ikonos-2 8 bits. The $\alpha(2)$ shows the same behaviour.

The amplitudes calculated (Δ^+ and Δ^-) in Landsat-7 bands present few variations (Table 3). The amplitude Δ^+ decreases from ETM+#1 to ETM+#3 and then presents an increase in ETM+#4 (NIR) whereas the other amplitude Δ^- remains practically constant. Observing these parameters in Ikonos-2 8-bit bands (Table 2), there are variations in value and behaviour for the four bands. In this case, both Δ^+ and Δ^- increase as the wavelength increases for the three visible bands, but decrease for the NIR band (IK#4).

The AIs estimated on these MFS amplitudes on each Landsat-7 bands are positive, except for ETM+#3 (red band). For the green band (ETM+#3) the symmetry of the MFS is complete. The band that shows certain asymmetry is the blue band (ETM+#1).

Regarding the vegetation indexes, estimated on Landsat-7 bands, we found the following. The bi-log plot of the partition function $(\chi(q,\delta))$ versus δ is plotted in Fig. 9 for both vegetation indexes. Each graph contains 9 points corresponding to the bands from which they were estimated. The linear fit was done with the same methodology as that for the four bands. EVI and NDVI show the same behaviour, and the same range of scale was selected achieving 8 pixels, as shown by arrows in Fig. 9.

The MF spectra $f(\alpha)$ of EVI and NDVI, estimated based on the Landsat-7 image, are shown in Fig. 8. Both vegetation indexes show differences mainly in the right side of the MFS (for q negative values). Comparing the range of $f(\alpha)$ values in the vegetation indexes to the range obtained in the four bands (left column in Fig. 8), there is a remarkable contrast. Meanwhile the NIR band of 8 bits achieves $f(\alpha)$ value close to 1.6, EVI and NDVI achieve values closer to 1. A similar situation was found with both images of Ikonos-2.

We are going to study the parameters obtained from the MF spectra for the vegetation indexes (Table 3). The results are quite similar to those found for the Landsat-7 bands, showing even higher values: 1.996 in NDVI and 1.997 in EVI.

The amplitude Δ^+ is quite low compared with the bands and to the vegetation indexes of Ikonos-2 8 bits. On the other hand, the amplitude Δ^- is higher than Landsat-7 bands but

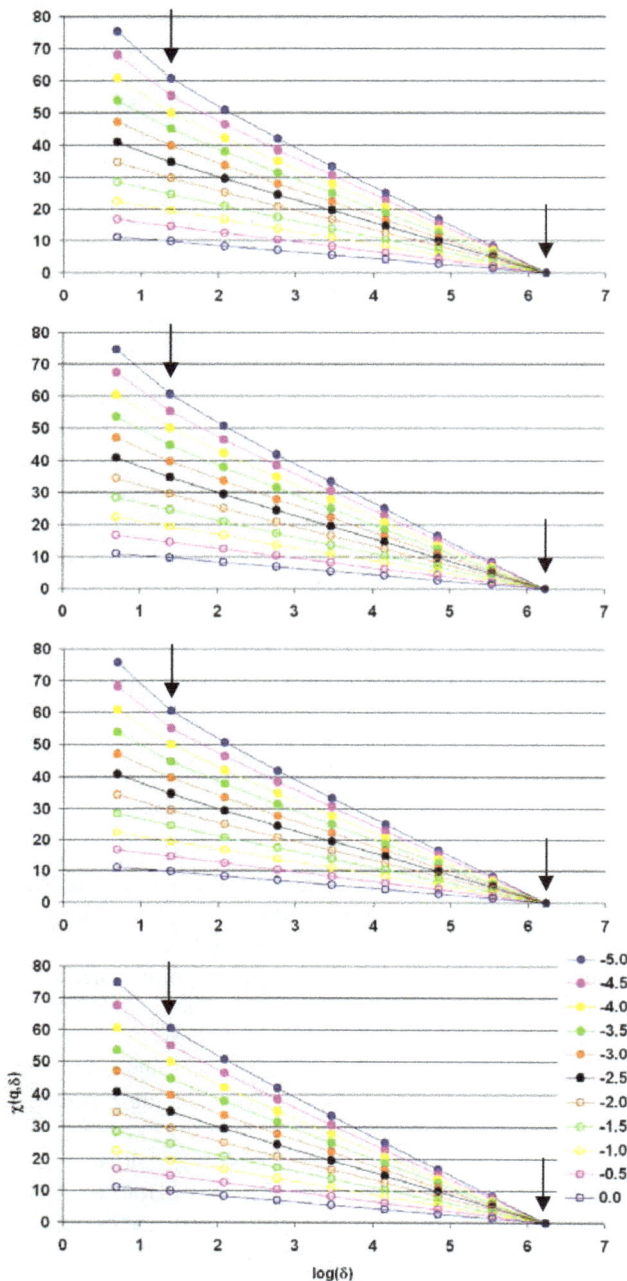

Figure 7. Bi-log plots of the partition function $\chi(q, \delta)$ versus δ for the first four bands of the Landsat-7 satellite and for $q < 0$ values. From top to bottom we show the results for ETM+#1, ETM+#2, ETM+#3 and ETM+#4. The arrows marked the range of scales used for the fit and to calculate the slope for different values of q (8 points).

by only a third of the values shown by Ikonos-2 8-bit vegetation indexes (Table 2). The AI estimated for both vegetation indexes are negative, indicating a high asymmetry on the right part of the MFS, as shown in Fig. 8. Comparing the AI values of Landsat-7 vegetation indexes with the ones of

both Ikonos-2 images, these are the highest, indicating that the most unbalanced MFS shifted totally on the right side of the spectrum.

4 Conclusions

In this work, we have used MF spectra as a successful technique for analysing common information contained in multispectral images of the site of the Earth surface acquired by two satellites, Landsat-7 and Ikonos, in four common bands in the visible (blue, green and red) and near-infrared wavelength regions used in several vegetation indexes.

The radiometric resolution has been studied by comparing MF spectra of the images acquired by Ikonos-2 coded in 11 bits and transformed into 8-bit code. The results obtained after the histogram transformation in the blue and green bands were those one would expect after the simplification applied from 11 to 8 bits, i.e. higher frequency in all the histogram bin values (see Fig. 2). In contrast, red and infrared bands showed no sensitivity at all to this transformation, keeping similar MF spectra. To our knowledge, this is the first time these differences among bands have been reported.

In order to analyse the effect of spatial resolution in each band at 4 m (Ikonos-2 with 8 bits) pixel size and 30 m (Landsat-7 with 8 bits) pixel size are compared. Obviously, the higher the spatial resolution, the higher the Hölder spectrum amplitudes in the green and blue bands are. In fact, observing the graphics of the three cases studied (Ikonos-2 11 bits, Ikonos-2 8 bits and Landsat-7 8 bits) both bands gradually reduce their $\alpha(q)$ amplitude in the negative as well as in the positive q values. However, this is not the case for red and NIR bands that present a much higher difference between Ikonos-2 8 bits and Landsat-7 curves of the MF spectra than between Ikonos-2 11 and 8 bits.

In the $q > 0$ MFS region for blue and green bands the sensitivity to both factors is very similar, the blue band ratio being slightly higher. The other two bands, red and NIR, for the same region, mainly present sensitivity to spatial resolution, showing a similar rate to blue and green bands. Observing the $q < 0$ region for blue and green, the behaviour is similar to the positive one but with a lower ratio (between 1 and 2) and, once more, the red and infrared bands show slight sensitivity to radiometric resolution. Nevertheless, in the spatial resolution the red band has a ratio similar to blue and green, and NIR shows the highest ratio (~ 8), showing the extreme influence of the lowest values contained; see histograms in Fig. 2 (Ikonos-2 8 and 11 bits) and Fig. 6 (Landsat-7).

The implications of these variations in the blue, red and NIR in the multi-scaling behaviour of two vegetation indexes, NDVI and EVI, have been also studied. The radiometric resolution showed a higher influence in the MFS of the NDVI than in EVI. This implies that the use of the blue band in the latter has a steady effect on the scaling behaviour.

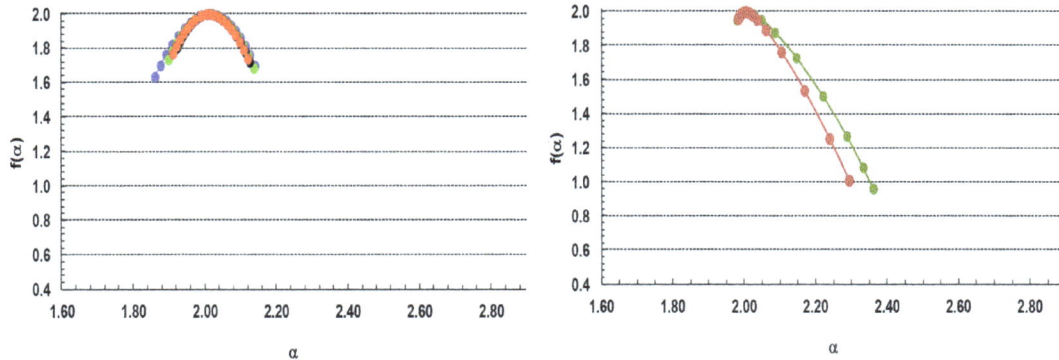

Figure 8. Multi-fractal spectrum of the Landsat-7 image for the original pixel values coded in 8 bits. Left plot corresponds to each band analysed: ETM+#1 in blue colour, ETM+#2 in green colour, ETM+#3 in red colour and ETM+#4 in black. Right plot corresponds to vegetation indexes: NDVI in green colour and EVI in brown.

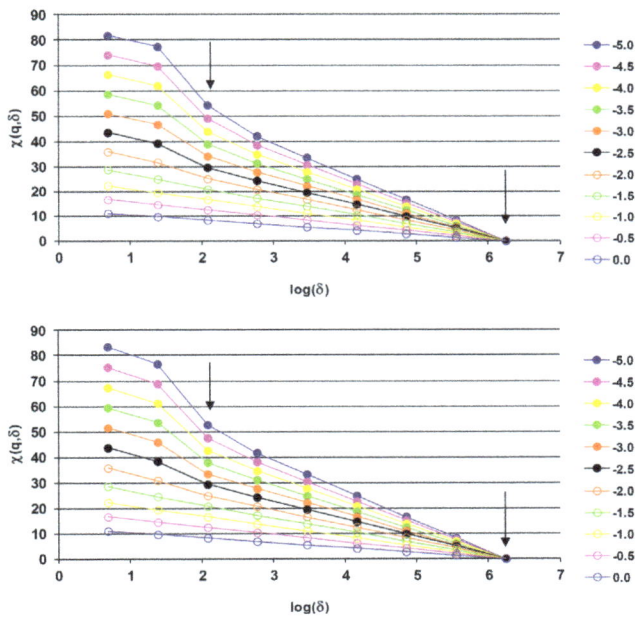

Figure 9. Bi-log plots of the partition function $\chi(q, \delta)$ versus δ for the vegetation indexes estimated from blue, red and NIR bands of the Landsat-7 satellite and for $q < 0$ values. From top to bottom we show the results for NDVI and EVI. The arrows marked the range of scales used for the fit and to calculate the slope for different values of q (7 points).

As was noted for the bands, the spatial resolution had a major impact in both vegetation indexes.

Further research will be conducted to establish a qualitative and quantitative comparison of these conclusions among several multi-fractal methodologies applied on these images.

Competing interests. The authors declare that they have no conflict of interest.

Acknowledgements. Thanks are due to the anonymous referees and the editor for their interest in and patience with this work. Discussion and comments suggested by Jose Manuel Redondo are highly appreciated. This work has been supported by the Ministerio de Economía y Competitividad (MINECO) under contract nos. MTM2012-39101 and MTM2015-63914-P.

Edited by: Asim Biswas

References

Aguado, P. L., Del Monte, J. P., Moratiel, R., and Tarquis, A. M.: Spatial Characterization of Landscapes through Multifractal Analysis of DEM, The Scientific World Journal, 2014, 563038, doi:10.1155/2014/563038, 2014.

Andraud, C., Beghdadi, A., and Lafait, J.: Entropic analysis of morphologies, Physica A, 207, 208–212, 1994.

Beaulieu, A. and Gaonac'h, H.: Scaling of differentially eroded surfaces in the drainage network of the Ethiopian plateau, Remote Sens. Environ., 82, 111–122, 2002.

Ben-Ze'ev, E., Karnieli, A., Agam, N., Kaufman, Y., and Holben, B.: Assessing Vegetation Condition In The Presence Of Biomass Burning Smoke By Applying The Aerosol-Free Vegetation Index (AFRI) On MODIS Images, Int. J. Remote Sens., 27, 3203–3221, 2006.

Carlson, T. N. and Ripley, D. A.: On the relation between NDVI, Fractional Vegetation Cover, and Leaf Area Index, Remote Sens. Environ., 62, 241–252, 1997.

Cheng, Q.: A new model for quantifying anisotropic scale invariance and for decomposition of mixing patterns, Math. Geol., 36, 345–360, 2004.

Cheng, Q. and Agterberg, F. P.: Multifractal modelling and spatial statistics, Mathematical Geology, 28, 1–16, 1996.

De Cola, L.: Fractal analysis of a classified Landsat-7 scene, Photogrammetric Engineering and Remote Sensing, 55, 601–610, 1989.

Deng, F. P., Su, G. L., and Liu, C.: Seasonal Variation Of MODIS Vegetation Indexes And Their Statistical Relationship With Climate Over The Subtropic Evergreen Forest In Zhejiang, China, IEEE Geosci. Remote Sens. Lett., 4, 236–240, 2007.

Dial, G., Bowen, H., Gerlach, F., Grodecki, J., and Oleszczuk, R.: IKONOS satellite, imagery and products, Remote Sens. Environ., 88, 23–36, 2003.

Du, G. and Yeo, T. S.: A novel multifractal estimation method and its application to remote image segmentation, IEEE T. Geosci. Remote, 40, 980–982, 2002.

Evertsz, C. J. G. and Mandelbrot, B. B.: Chaos and Fractals: New Frontiers of Science, edited by: Peitgen, H., Jurgens, H., and Saupe, D., Springer-Verlag, New York, 921, 1992.

Feder, J.: Fractals, Plenum Press, New York, 283 pp., 1988.

Fioravanti, S., Multifractals: theory and application to image texture recognition, in: Fractals in Geosciences and Remote Sensing, Proceedings of a joint JRC/EARSeL Expert meeting, Ispra, Italy, 14–15 April 1994.

Gagnon, J. S., Lovejoy, S., and Schertzer, D.: Multifractal surfaces and terrestrial topography, Europhys. Lett., 62, 801–807, 2003.

Gaonac'h, H., Lovejoy, S., and Schertzer, D.: Resolution dependence of infrared imagery of active thermal features at Kilauea volcano, Int. J. Remote Sens., 24, 2323–2324, 2003.

Gouyet, J. F.: Physics and fractal structures, Springer-Verlag: New York, 1996.

Halsey, T. C., Jensen, M. H., Kadanoff, L. P., Procaccia, I., and Shraiman, B. I.: Fractal measures and their singularities: the characterization of strange sets, Phys. Rev. A, 33, 1141–1151, 1986.

Harvey, D. C., Gaonac'h, H., Lovejoy, S., and Sxhertzer, D.: Multrifractal characterization of remotely sensed volcanic features: a case study from Kilauea volcano, Hawaii, Fractals, 10, 265–274, 2002.

Hirata, T. and Imoto, M.: Multifractal analysis of spatial distribution of micro-earthquakes in the Kanto region, Geophys. J. Int., 107, 155–162, 1991.

Huete, A., Miura, T., Yoshioka, H., Ratana, P., and Broich, M.: Indices of Vegetation Activity, in: Biophysical Applications of Satellite Remote Sensing, edited by: Hanes, J. M., Springer, 2014.

Kravchenko, A. N., Boast, C. W., and Bullock, D. G.: Multifractal analysis of soil spatial variability, Agron. J., 91, 1033–1041, 1999.

Laferrière, A. and Gaonac'h, H.: Multifractal properties of visible reflectance fields from basaltic volcanoes, J. Geophys. Res., 104, 5115–5126, 1999.

Lam, N. S.: Description and measurement of Landsat-7 TM images using fractals, Photogrammetric Engineering and Remote Sensing, 56, 187–195, 1990.

Lam, N. S. and De Cola, L.: Fractals in Geography, Prentice Hall, Englewood Cliffs, New Jersey, 1993.

Lavallée, D., Schertzer, D., and Lovejoy, S.: On the determination of the codimension function, in: Non-Linear Variability in Geophysics, edited by: Schertzer, D. and Lovejoy, S., Springer Netherlands: Dordrecht, the Netherlands, 99–109, 1991.

Lovejoy, S., Pecknold, S., and Schertzer, D.: Stratified multifractal magnetization and surface geomagnetic fields-I. Spectral analysis and modelling, Geophys. J. Int., 145, 112–126, 2001a.

Lovejoy, S., Schertzer, D, Tessier, Y., and Gaonach, H.: Multifractals and resolution-independent remote sensing algorithms: the example of ocean colour, Int. J. Remote Sensing, 22, 119–1234, 2001b.

Lovejoy, S., Tarquis, A., Gaonac'h, H., and Schertzer, D.: Single and multiscale remote sensing techniques, multifractals and MODIS derived vegetation and soil moisture, Vadose Zone J., 7, 533–546, doi:10.2136/vzj2007.0173, 2008.

Maître, H. and Pinciroli, M.: Fractal characterization of a hydrological basin using SAR satellite images, IEEE Trans. Geosci. Remote, 37, 175–181, 1999.

Mandelbrot, B. B.: The Fractal Geometry of Nature, Freeman, San Francisco, 1983.

Hu, M.-G., Wang, J.-F., and Ge, Y.: Super-Resolution Reconstruction of Remote Sensing Images Using Multifractal Analysis, Sensors (Basel), 9), 8669–8683, doi:10.3390/s91108669, 2009.

Mika, A. M.: Three decades of Landsat-7 instruments, Photogrammetric Engineering and Remote Sensing, 63, 839–852, 1997.

Morató, M. C., Castellanos, M. T., Bird, N. R., and Tarquis, A. M.: Multifractal analysis in soil properties: Spatial signal versus mass distribution, Geoderma, 287, 54–65, doi:10.1016/j.geoderma.2016.08.004, 2017.

Parrinello, T. and Vaughan, R. A.: Multifractal Analysis and feature extraction in satellite imagery. Int. J. Remote Sens., 23, 1799–1825, 2002.

Qiu, H., Lam, N. S., Quattrochi, D. A., and Gamon, J. A.: Fractal characterization of hyperspectral imagery, Photogrammetric Engineering and Remote Sensing, 65, 63–71, 1999.

Renosh, P. R., Schmitt, F. G., and Loisel, H.: Scaling analysis of ocean surface turbulent heterogeneities from satellite remote sensing: use of 2D structure functions, PLoS ONE, 10, e0126975, doi:10.1371/journal.pone.0126975, 2015.

Sánchez, A., Serna, R., Catalina, F., and Afonso, C. N.: Multifractal patterns formed by laser irradiation in GeAl thin multilayer films, Phys. Rev. B, 46, 487–490, 1992.

Schertzer, D. and Lovejoy, S.: Physical modeling and analysis of rain and clouds by anisotropic scaling multiplicative processes, J. Geophys. Res., 92, 9693–9714, 1987.

Schmitt, F., Schertzer, D., Lovejoy, S., and Marchal, P.: Multifractal analysis of satellite images: towards an automatic segmentation, in: Fractals in Engineering, Arcachon: Jules, 103–109, 1997.

Tarquis, A. M., Losada, J. C., Benito, R., and Borondo, F.: Multifractal analysis of the Tori destruction in a molecular Hamiltonian System, Phys. Rev. E., 65, 0126213, doi:10.1103/PhysRevE.65.016213, 2001.

Tarquis, A. M., Giménez, D., Saa, A., Díaz, M. C., and Gascó, J. M.: Scaling and multiscaling of soil pore systems determined by image analysis, in: Scaling methods in soil physics, edited by: Pachepsky, Y., Radcliffe, D. E., and Magdi Selim, H., CRC Press, 2003.

Tarquis, A. M., Platonov, A., Matulka, A., Grau, J., Sekula, E., Diez, M., and Redondo, J. M.: Application of multifractal analysis to the study of SAR features and oil spills on the ocean surface, Nonlin. Processes Geophys., 21, 439–450, doi:10.5194/npg-21-439-2014, 2014.

Tessier, Y., Lovejoy, S., Schertzer, D., Lavalle'e, D., and Kerman, B.: Universal multifractal indices for the ocean surface at far red wavelengths, Geophys. Res. Lett., 20, 1167–1170, 1993.

Turiel, A., Isern-Fontanet, J., García-Ladona, E., and Font, J.: Multifractal Method for the Instantaneous Evaluation of the Stream Function in Geophysics Flows, Phys. Rev. Lett., 95, 104502, doi:10.1103/PhysRevLett.95.104502, 2005.

Wang, X., Xie, H., Guan, H., and Xiaobing, Z.: Different responses of MODIS-derived NDVI to root-zone soil moisture in semi-arid and humid regions, J. Hydrol., 340, 12–24, 2007.

Xie, S., Cheng, Q., Xing, X., Bao, Z., and Chen, Z.: Geochemical multifractal distribution patterns in sediments from ordered streams, Geoderma, 160, 36–46, 2010.

5

Modeling the dynamical sinking of biogenic particles in oceanic flow

Pedro Monroy[1], **Emilio Hernández-García**[1], **Vincent Rossi**[1,2], **and Cristóbal López**[1]

[1]IFISC, Instituto de Física Interdisciplinar y Sistemas Complejos (CSIC-UIB), 07122 Palma de Mallorca, Spain
[2]Mediterranean Institute of Oceanography (UM 110, UMR 7294), CNRS, Aix Marseille Univ., Univ. Toulon, IRD, 13288, Marseille, France

Correspondence to: Pedro Monroy (pmonroy@ifisc.uib-csic.es)

Abstract. We study the problem of sinking particles in a realistic oceanic flow, with major energetic structures in the mesoscale, focussing on the range of particle sizes and densities appropriate for marine biogenic particles. Our aim is to evaluate the relevance of theoretical results of finite size particle dynamics in their applications in the oceanographic context. By using a simplified equation of motion of small particles in a mesoscale simulation of the oceanic velocity field, we estimate the influence of physical processes such as the Coriolis force and the inertia of the particles, and we conclude that they represent negligible corrections to the most important terms, which are passive motion with the velocity of the flow, and a constant added vertical velocity due to gravity. Even if within this approximation three-dimensional clustering of particles can not occur, two-dimensional cuts or projections of the evolving three-dimensional density can display inhomogeneities similar to the ones observed in sinking ocean particles.

1 Introduction

The sinking of small particles suspended in fluids is a topic of both fundamental importance and of practical implications in diverse fields ranging from rain nucleation to industrial processes (Michaelides, 1997; Falkovich and Fouxon, 2002).

In the oceans, photosynthesis by phytoplankton in surface waters uses sunlight, inorganic nutrients and carbon dioxide to produce organic matter which is then exported downward and isolated from the atmosphere (Henson et al., 2012), a process which forms the so-called biological carbon pump. The downward flux of carbon-rich biogenic particles from the marine surface due to gravitational settling, one of the key process of the biological carbon pump, is responsible (together with the solubility and the physical carbon pumps) for much of the oceans' role in the Earth's carbon cycle (Sabine et al., 2004). Although most of the organic matter is metabolized and remineralized in surface waters, a significant portion sinks into deeper horizons. It can be sequestered on various timescales spanning a few years to decades in central and intermediate waters, several centuries in deep waters and up to millions of years locked up in bottom sediments (DeVries et al., 2012). Suitable modeling of the sinking process of particulate matter is thus required to properly assess the amount of carbon sequestered in the ocean and in general to better understand global biogeochemical cycling and its influence on the Earth's climate.

This is a challenging task that involves the downward transport of particles of many different sizes and densities by turbulent ocean flows which contain an enormous range of interacting scales. In the oceanographic community, numerous studies approached this problem by considering biogenic particles transported in oceanic flow as passive particles with an added constant velocity in the vertical to account for the sinking dynamics (Siegel and Deuser, 1997; Siegel et al., 2008; Qiu et al., 2014; Roullier et al., 2014; van Sebille et al., 2015). They suggest that the sinking of particles may not be strictly vertical but oblique, meaning that the locations where the particles are formed at the surface may be distant from the location of their deposition in the seafloor sediment. Then (Siegel et al., 2008) presented the concept of statistical funnels which describe and quantify the source region of a sediment trap (subsurface collecting device of sinking particles used to get estimates of vertical fluxes). The validity of this approximation and the influence of different physical processes is however poorly discussed in these analyses.

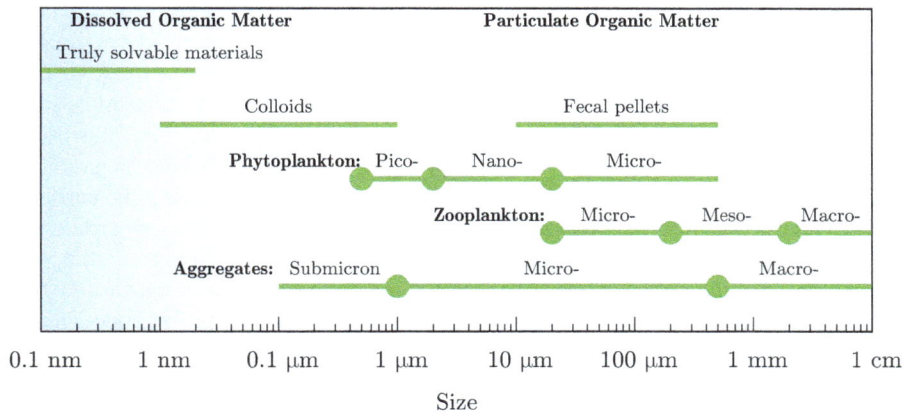

Figure 1. Size and classification of marine particles (adapted from Simon et al., 2002).

In the physical community, the framework to model sinking particles is based on the Maxey–Riley–Gatignol equation for a small spherical particle moving in an ambient flow (Maxey and Riley, 1983; Gatignol, 1983; Michaelides, 1997; Provenzale, 1999; Cartwright et al., 2010), which highlights the importance of mechanisms beyond passive transport and constant sinking velocity, such as the role of finite size, inertia and history dependence. A major outcome of these studies is that inhomogeneities and particle clustering can arise spontaneously even if the fluid velocity field is incompressible and particles do not interact (Squires and Eaton, 1991). Particle clustering and patchiness are indeed observed in the surface and subsurface of the ocean (Logan and Wilkinson, 1990; Buesseler et al., 2007; Mitchell et al., 2008).

Here we consider the theory of small but finite-sized particles driven by geophysical flows, which is, as mentioned above, conveniently based on the Maxey–Riley–Gatignol equation. In Sect. 2 we review the main characteristics of marine particles which are relevant for their sinking dynamics. In Sect. 3 we present the equations of motion describing this process, together with the approximations required to obtain them and the types of particles for which they are valid. In particular, we discuss its validity and the relevance of the different physical processes involved in the range of sizes and densities of marine biogenic particles. In Sect. 4 we use these equations to study the settling dynamics in a modeled oceanic velocity field produced by a realistic high-resolution regional simulation of the Benguela upwelling system (southwestern Africa). We estimate the relevance of physical processes such as the Coriolis force and the inertia of the particles with respect to the settling velocity. We also observe the spatial distribution of particles falling onto a plane of constant depth above the seabed and we identify clustering of particles that is interpreted with simple geometrical arguments which do not require physical phenomena beyond passive transport and constant terminal velocity. Our main results are finally summarized in the Conclusions section.

2 Characteristics of marine biogenic particles

In theory, the sinking velocities of biogenic particles depend on various intrinsic factors (such as their sizes, shapes, densities, and porosities) which can be modified along their fall by complex bio-physical processes (e.g., aggregation, ballasting, trimming by remineralization) as well as by the three-dimensional flow field (Stemmann and Boss, 2012). However, reasonable estimates of the effective sinking velocities of marine particles can be obtained by taking into account only their size and density (McDonnell and Buesseler, 2010). In our Lagrangian setting we thus consider that the two key properties of marine particles controlling their sinking dynamics are their size and density. Here we present the standard classification of marine particles according to the typical range of size and density by compiling different bibliographical sources.

2.1 Size

Because of the diversity of the shapes, the size of a particle refers to the diameter of a sphere of equivalent volume (equivalent spherical diameter; Guidi et al., 2008). The size of marine particles ranges from 1 nm (almost-dissolved colloids) to aggregates larger than 1 cm (Stemmann and Boss, 2012).

Originally, the size classification of particles was based on the minimal pore size of the nets used for their collection, which is about $\simeq 0.45$–$1.0\,\mu m$. Any material larger than $0.2\,\mu m$ (thus isolated by the filtration of seawater) is regarded as particulate organic matter, while the fraction that percolates through the filter is labeled dissolved matter. This includes colloidal and truly dissolved materials (see Fig. 1). Although this discrimination of the size continuum observed in the real ocean is somehow arbitrary, it is useful – and we will follow it – because particles smaller than 1.0 μm are not prone to sinking (Hedges, 2002).

In the following, our focus is thus on particulate matter larger than 1.0 μm (Fig. 1). Organic matter is produced in the sunlit layer of the ocean by the primary production through photosynthesis of autotrophic microbes (mainly bacteria and phytoplankton). During their lifetime growth they exude colloidal and small particles to finally form larger particles when they die. Dead phytoplankton are within the range of 1 μm (picoplankton, e.g., cyanobacteria) and a few hundred micrometers (microphytoplankton, e.g., diatoms).

Thereafter zooplankton consume live phytoplankton and inert particles and produce fecal pellets and dead bodies. Most fecal materials have enough size to sink rapidly by their own (De La Rocha and Passow, 2007). Typical sizes of such particles are 10 μm for a pellet of a copepod of 200 μm length (Jackson, 2001); krill fecal pellets are between 160 and 460 μm (McDonnell and Buesseler, 2010) and euphausiid fecal pellets span 300 μm–3 mm (Komar et al., 1981), providing the total range of 10 μm to 3 mm. Concerning the zooplankton dead bodies, they are divided into micro-, meso- and macro-, with sizes in the range 20 μm–1 cm. A detailed summary is given in Table 1.

Finally, there are the so-called organic aggregates which occur in the size range of 1 μm to 10 cm. They are typically formed in situ by physical aggregation or biological coagulation and are usually composed of numerous planktonic individuals and fecal pellets stuck together within a colloidal matrix. They are often distinguished in three size classes (Simon et al., 2002): macroscopic aggregates or macro-aggregates > 5 mm usually called marine snow; microscopic, from 1 to 500 μm, also known as micro-aggregates; and submicron particles < 1 μm (which do not sink).

2.2 Density

The density of marine particles depends on their composition, which can be divided into a mineral and an organic fraction (Maggi and Tang, 2015). The mineral or inorganic matter consists of biogenic minerals: particulate inorganic carbon (PIC), e.g., calcium carbonate produced by coccoliths with density $2700 \, \mathrm{kg \, m^{-3}}$, and biogenic silica (BSi), produced by diatoms, significantly less dense than PIC, $1950 \, \mathrm{kg \, m^{-3}}$ (Balch et al., 2010). The density of particulate organic matter (POC) ranges widely depending on its origin. For instance, the density of cytoplasm spans from 1030 to $1100 \, \mathrm{kg \, m^{-3}}$, while the one of fecal pellets ranges from 1174 to $1230 \, \mathrm{kg \, m^{-3}}$ (Komar et al., 1981). Despite this variability, it is possible to assign a range to the density of organic matter, from 1050 to $1500 \, \mathrm{kg \, m^{-3}}$.

Considering all these estimates together, the density of marine particles ranges approximately between 1050 and $2700 \, \mathrm{kg \, m^{-3}}$ (Maggi, 2013). This should be compared to standard values for seawater density in the interior of the ocean, which span roughly $1020{-}1030 \, \mathrm{kg \, m^{-3}}$. Thus most of the particle types described previously will sink. Assuming constant size and density for each particle along its downward course, we deduce that most of the particles types described previously will sink. This holds without considering biogeochemical and (dis)aggregation processes that may occur in nature, thus lowering the particle density and resulting in clustering and trapping of particles at particular isopycnals (Sozza et al., 2016). Note that we do not consider here living organisms which show vertical movements by active swimming or by controlling their buoyancy (Moore and Villareal, 1996; Azetsu-Scott and Passow, 2004).

3 Equations of motion for small spherical rigid particles

3.1 The Maxey–Riley–Gatignol equation

To describe the sedimentation of biogenic particles, we need to study the motion of single particles driven by fluid flow. A milestone to analyze the dynamics of a small spherical rigid particle of radius a subject to gravity acceleration $\mathbf{g}$ in an unsteady fluid flow $\mathbf{u}(\mathbf{r}, t)$ is given by the Maxey–Riley–Gatignol (Maxey and Riley, 1983; Gatignol, 1983; Michaelides, 1997; Cartwright et al., 2010) equation (MRG in the following):

$$
\begin{aligned}
\rho_{\mathrm{p}} \frac{d\mathbf{v}}{dt} =& \rho_{\mathrm{f}} \frac{D\mathbf{u}}{Dt} + (\rho_{\mathrm{p}} - \rho_{\mathrm{f}})\mathbf{g} - \frac{9\nu\rho_{\mathrm{f}}}{2a^2}\left(\mathbf{v} - \mathbf{u} - \frac{a^2}{6}\nabla^2\mathbf{u}\right) \\
& - \rho_{\mathrm{f}}\left(\frac{d\mathbf{v}}{dt} - \frac{D}{Dt}(\mathbf{u} + \frac{a^2}{10}\nabla^2\mathbf{u})\right) \\
& - \frac{9\rho_{\mathrm{f}}}{2a}\sqrt{\frac{\nu}{\pi}}\int_0^t \frac{\frac{d}{ds}(\mathbf{v} - \mathbf{u} - \frac{a^2}{6}\nabla^2\mathbf{u})}{\sqrt{t-s}}ds.
\end{aligned} \tag{1}
$$

The velocity of the particle is denoted by $\mathbf{v} = \mathbf{v}(t)$. The particle and fluid densities are ρ_{p} and ρ_{f}, respectively, and ν denotes the fluid kinematic viscosity. The time derivative operators $\frac{d}{dt} = \frac{\partial}{\partial t} + \mathbf{v} \cdot \nabla$ and $\frac{D}{Dt} = \frac{\partial}{\partial t} + \mathbf{u} \cdot \nabla$ denote the time rate of change following the particle itself and the time rate of change following a fluid element in the undisturbed flow field $\mathbf{u}(\mathbf{r}, t)$, respectively. This equation of motion gives the balance between the different forces acting on the particle, which correspond to the right-hand-side terms: the pressure force (the force exerted on the particle by the undisturbed flow), the buoyancy force, the drag force (Stokes drag), the added mass force resulting from the part of the fluid moving with the particle, and the history force. As will be discussed below, the validity of this equation requires several conditions, the main one being the small size of the particles. The terms with $a^2\nabla^2\mathbf{u}$ are the Faxén corrections (Faxén, 1922).

The full MRG is very complicated to manage. A further simplification is usually performed based on the single assumption of very small particles (what this exactly means

Table 1. Simplified categorization of marine biogenic particles, and their associated sizes.

Individual particles (mostly organic)	Aggregates (compounds of organic and inorganic particles)
Fecal pellets (cylindrical):	Aggregates (Simon et al., 2002):
– Krill fecal pellets: length between 400 μm and 9 mm, diameter 120 μm (McDonnell and Buesseler, 2010). ESD (160–60 μm)	– Macroscopic (marine snow): size > 500 μm.
– 10 μm, consistent with the pellet volume of a 200 μm copepod (Jackson, 2001)	– Microscopic: 1 μm < size < 500 μm.
Dead zooplankton (Stemmann and Boss, 2012):	– Submicron: size < 1 μm.
– Macrozooplankton: size > 2000 μm	
– Mesozooplankton: 200 < size < 2000 μm	
– Microzooplankton: 20 < size < 200 μm	
Dead phytoplankton (Stemmann and Boss, 2012):	
– Microphytoplankton: (size > 200 μm)	
– Nanophytoplankton: (20 < size < 200 μm)	
– Picophytoplankton: (2 < size < 20 μm)	

will be discussed later on). With this, the Faxén corrections and, as commented on below, also the history term (since $a/\sqrt{\nu} \ll 1$), can be neglected (Maxey and Riley, 1983; Michaelides, 1997; Haller and Sapsis, 2008). Note however that the history term can be relevant under some conditions, for example larger particle size (Daitche and Tél, 2011; Guseva et al., 2013, 2016; Olivieri et al., 2014). Thus we obtain the standard form of the MRG equations (Maxey and Riley, 1983):

$$\frac{d\mathbf{v}}{dt} = \beta \frac{D\mathbf{u}}{Dt} + \frac{\mathbf{u} - \mathbf{v} + \mathbf{v}_s}{\tau_p}, \qquad (2)$$

where $\beta = \frac{3\rho_f}{2\rho_p + \rho_f}$, the Stokes time is $\tau_p = \frac{a^2}{3\beta\nu}$, and $\mathbf{v}_s = (1 - \beta)\mathbf{g}\tau_p$ is the settling velocity in quiescent fluid.

Equation (2) is the starting point for most inertial particle studies (Michaelides, 1997; Balkovsky et al., 2001; Cartwright et al., 2010).

We now discuss the validity of the MRG equation Eq. (1) or rather its simplified form Eq. (2) for the range of sizes and densities of marine organisms. We do so in the context of open-ocean flows, which are typically most energetic at the mesoscale (scales of about 100 km), and where there is a strong stratification, with vertical velocities 3 or 4 orders of magnitude smaller than horizontal ones. The motion becomes more three-dimensional, and then the concepts of three-dimensional turbulence more relevant, below scales l

of some hundred of meters, with typical velocities decreasing as $l^{1/3}$ for decreasing scale and velocity gradients increasing as $l^{-2/3}$ until the Kolmogorov scale $l = \eta$ below which flow becomes smooth. Because of its direct exposure to wind, turbulence intensity is typically larger at the ocean surface, with values of turbulent energy dissipation in the range $1 \cdot 10^{-6}\,\mathrm{m^2\,s^{-3}} < \epsilon < 3 \cdot 10^{-5}\,\mathrm{m^2\,s^{-3}}$ (Jimenez, 1997), than at depth. The first condition for the validity of the MRG equation that was originally discussed by (Maxey and Riley, 1983) is that the particles have to be much smaller than the typical length scale of variation of the flow. This means that for multiscale (turbulent) flows the radius of the particle a has to be much smaller than the Kolmogorov scale η, which according to the previous values of ϵ, is typically $0.3\,\mathrm{mm} < \eta < 2\,\mathrm{mm}$ in the ocean surface (Okubo, 1971; Jimenez, 1997). Note that we only have to consider worst-case situations for assessing the validity of the different approximations. Another condition to be fulfilled is that the shear Reynolds number must be small $Re_\nabla = a^2 U / \nu L \ll 1$, where U and L are typical velocity and length scales. For a turbulent ocean with multiple scales and velocities, the most restrictive condition arises when they take the values of the Kolmogorov velocity v_η and length η, respectively, since then the velocity gradients are maxima. In this case the condition becomes $Re_\nabla = a^2/\eta^2 \ll 1$, which again is satisfied for small particles. We note that Guseva et al. (2013) found that the relative importance of the history term in Eq. (1) with respect to the drag force is of the order of a parameter which in our notation is $(Re_\nabla)^{1/2}$. This justifies neglecting the history term for small particles, although its importance increases for increasing size (Daitche and Tél, 2011; Guseva et al., 2013).

Another condition to be satisfied for the validity of the MRG equation is that the so-called Reynolds particle number, $Re_p = \frac{a|\mathbf{v}-\mathbf{u}|}{\nu}$ should fulfill $Re_p \ll 1$. Considering that gravity force dominates over other forces one has $|\mathbf{v}-\mathbf{u}| \simeq |\mathbf{v}_s| \equiv v_s$, where $\mathbf{v}_s$ is, as introduced before, the settling velocity of particles in a quiescent fluid due to Stokes drag. The Reynolds particle number is then $Re_p = \frac{av_s}{\nu}$. Note that the settling velocity depends only on the densities of particles via the parameter β. Assuming a mean density of seawater in the upper ocean as $\rho_f = 1025\,\mathrm{kg\,m^{-3}}$, the parameter β has values within the range $[0.5, 0.99]$ for the typical values of the density of marine particles previously discussed. Figure 2 shows v_s for different sizes and the regions where $Re_p > 1$ (and other parameter regions where MRG is not a good approximation) as a function of particle radius and for the limiting values of β. It reveals that Eq. (1) can not describe ocean particles larger than $300\,\mu m$ of any density, and for a limited range of densities when the particle radius exceeds approximately $100\,\mu m$. In fact, the range of application of MRG to marine particles is plotted in the blue area, which at the same time gives an estimate of the typical sinking velocities for a given particle size.

Summarizing, both the MRG and its approximation Eq. (2) are valid for marine particles with sizes within the range 1 and $200\,\mu m$. That is, it is valid for all particulate organic matter in Fig. 1 except the largest of the micro-aggregates and meso- and macro-bodies of zooplankton. The sinking velocities range from $1\,\mathrm{mm\,day^{-1}}$ to $1\,\mathrm{km\,day^{-1}}$.

3.2 The MRG equation in a rotating frame and further simplifications

We are interested in applying Eq. (2) in oceanic flows, where the particle $\mathbf{v}$ and flow $\mathbf{u}$ velocities are expressed in a frame rotating with the Earth's angular velocity $\mathbf{\Omega}$ (Elperin et al., 2002; Biferale et al., 2016; Tanga et al., 1996; Provenzale, 1999; Sapsis and Haller, 2009). Both time derivatives $\frac{d}{dt}$ and $\frac{D}{Dt}$ have to be corrected following the rule

$$\frac{d}{dt} \rightarrow \frac{d}{dt} + 2\mathbf{\Omega} \times \mathbf{v} + \mathbf{\Omega} \times (\mathbf{\Omega} \times \mathbf{r}), \tag{3}$$

$$\frac{D}{Dt} \rightarrow \frac{D}{Dt} + 2\mathbf{\Omega} \times \mathbf{u} + \mathbf{\Omega} \times (\mathbf{\Omega} \times \mathbf{r}), \tag{4}$$

where $\Omega = |\mathbf{\Omega}|$ and $\mathbf{r}$ is the particle position vector whose origin is in the rotation axis, so that Eq. (2) is now

$$\frac{d\mathbf{v}}{dt} = \beta \frac{D\mathbf{u}}{Dt} - \frac{\mathbf{v}-\mathbf{u}}{\tau_p} - 2\mathbf{\Omega} \times (\mathbf{v}-\beta\mathbf{u}) + \mathbf{v}_s'/\tau_p. \tag{5}$$

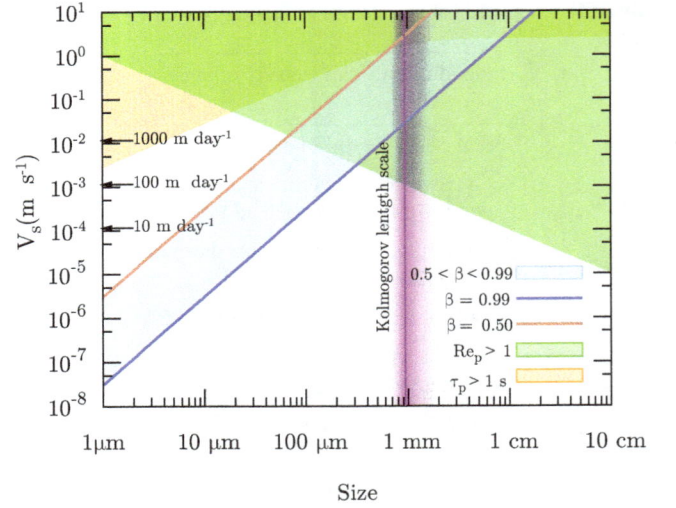

Figure 2. Sinking velocity versus particle radius for different β, which is determined by densities. The blue zone determines the values of the settling velocities at a given radius, as determined by the typical marine particle densities. The green area is determined by the condition $Re_p > 1$ for which the MRG equation is not valid. Use of the MRG equation is also unjustified for particles larger than the Kolmogorov length scale also plotted in the figure. We also show the region $\tau_p > \tau_\eta \approx 1\,\mathrm{s}$ where the additional approximation leading to Eq. (6) becomes invalid.

Two apparent forces arise in the equation, the Coriolis force $2\mathbf{\Omega} \times (\mathbf{v}-\beta\mathbf{u})$ and the centrifugal force, which is included in a modified sinking velocity $\mathbf{v}_s' = (1-\beta)(\mathbf{g}-\mathbf{\Omega} \times (\mathbf{\Omega} \times \mathbf{r}))\tau_p$. The effect of the centrifugal force is very small (on order 10^{-3} compared to gravity) and can be absorbed in a redefinition of $\mathbf{g}$. Thus, in the following we take $\mathbf{v}_s' = \mathbf{v}_s$ with the properly chosen $\mathbf{g}$.

The ratio between the particle response time and the Kolmogorov timescale is the Stokes number $St = \tau_p/\tau_\eta$, which measures the importance of a particle's inertia because of its size and density. According to the range of ϵ in the ocean mentioned before, we get $0.1\,\mathrm{s} < \tau_\eta < 5\,\mathrm{s}$, and for our range of particle sizes $10^{-6}\,s < \tau_p < 10^{-2}\,s$, so we can assume that $St \ll 1$ (see Fig. 2). This motivates us to make a second (standard) approximation (Balkovsky et al., 2001; Haller and Sapsis, 2008) of the MRG equation, expanding in powers of τ_p (note that it would be more natural to make the expansion in powers of the non-dimensional St, but we prefer to do it in τ_p to control the timescales of the problem). Assuming first the solution to Eq. (2),

$$\mathbf{v} = \mathbf{u} + \mathbf{u}_1\tau_p + \mathbf{u}_2\tau_p^2 + \ldots,$$

and using $\frac{d\mathbf{v}}{dt} = \frac{D\mathbf{u}}{Dt} + O(\tau_p)$, we get that the particle velocity at first order in τ_p is

$$\mathbf{v} = \mathbf{u} + \mathbf{v}_s + \tau_p(\beta-1)\left(\frac{D\mathbf{u}}{Dt} + 2\mathbf{\Omega} \times \mathbf{u}\right). \tag{6}$$

It is worth recalling that $\tau_p(1 - \beta) = v_s/g$, so that all dependencies on particle size and density appear in Eq. (6) through the combination of parameters defining v_s. Different combinations of size and density, taken within the ranges reported in Sect. 2, follow the same dynamics if they have the same undisturbed settling velocity v_s.

A further discussion of Eq. (6) follows. At this order only three physical processes correct the particle velocity with respect to the fluid velocity: the Stokes friction determining the settling velocity v_s, the inertial term given by $\tau_p(\beta - 1)\frac{D\mathbf{u}}{Dt}$ whose major effect is to introduce a centrifugal force pulling particles away from vortex cores (Maxey, 1987; Michaelides, 1997), and the influence of the Coriolis force $2\tau_p(\beta - 1)\mathbf{\Omega} \times \mathbf{u}$. Concerning sinking dynamics, the $\mathbf{v} = \mathbf{u} + \mathbf{v}_s$ is the most relevant approximation, and many other studies consider it, mainly in oceanographic contexts (e.g., Siegel and Deuser, 1997). Note that we can use the right-hand side of Eq. (6) with $\mathbf{u} = \mathbf{u}(\mathbf{r}, t)$ to define the particle velocity $\mathbf{v}$ as a velocity field in three-dimensional space $\mathbf{v} = \mathbf{v}(\mathbf{r}, t)$. If one uses the lowest-order approximation $\mathbf{v} \approx \mathbf{u}$ we have $\nabla \cdot \mathbf{v} = \nabla \cdot \mathbf{u} = 0$ when the fluid velocity field $\mathbf{u}$ is incompressible (which is the case for ocean flows). This means that when considering this term alone, one cannot obtain a compressible particle velocity whereas this was the main reason invoked to explain the clustering of finite-size particles (Squires and Eaton, 1991; Bec, 2003). For this reason, numerous studies (Tanga et al., 1996; Michaelides, 1997; Bec et al., 2007, 2014; Cartwright et al., 2010; Guseva et al., 2013; Gustavsson et al., 2014; Beron-Vera et al., 2015) consider the role of the additional terms. With them $\nabla \cdot \mathbf{v} = \tau_p(\beta - 1)\nabla \cdot (\frac{D\mathbf{u}}{Dt} + 2\mathbf{\Omega} \times \mathbf{u}) \neq 0$, and inertia-induced clustering may occur. In the following sections we address two main questions: (a) how relevant for the sinking dynamics are the Coriolis and centrifugal terms?; and (b) are they essential ingredients for the clustering of biogenic particles? We will study the relevance of the different terms in Eq. (6) in a realistic oceanic setting.

4 Numerical simulations

The velocity flow $\mathbf{u}$ of the Benguela region was produced by a regional simulation of a hydrostatic free-surface primitive equations model called ROMS (Regional Ocean Modelling System). The configuration used here extends from 12 to 35° S and from 4 to 19° E (blue rectangle in Fig. 3) and was forced with climatological atmospheric data (Gutknecht et al., 2013). The simulation area extends from 12 to 35° S and from 4 to 19° E (blue rectangle in Fig. 3). The velocity field data set consists of 2 years of daily averages of zonal (u), meridional (v), and vertical velocity (w) components, stored in a three-dimensional grid with a horizontal resolution of $1/12°$ and 32 vertical terrain-following levels using a stretched vertical coordinate where the layer thickness increases from the surface/bottom to the ocean interior.

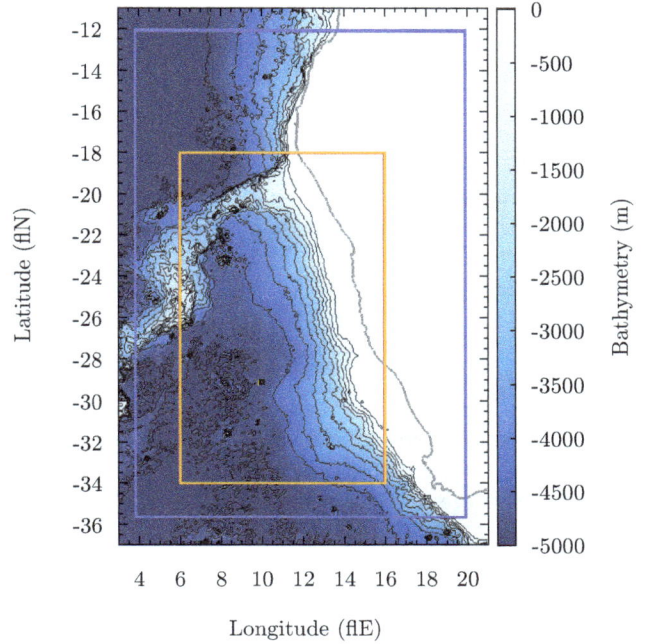

Figure 3. Map of region of study. Color corresponds to bathymetry. The blue rectangle is the region used for simulations of the ROMS model. The orange rectangle is the region for the clustering numerical experiment of Sect. 5 and the red rectangle is the release site of the sinking numerical experiments of Sect. 4. Gray represents the coastline.

In order to integrate particle trajectories from the velocity in Eq. (6), we interpolate linearly $\mathbf{u}(\mathbf{r}, t)$ from the closest space–time grid points to the actual particle locations. Given the huge disparity between the model resolution and the small particle sizes considered, it is pertinent to parameterize in some way the unresolved scales. This can be done by different approaches, from stochastic Lagrangian modeling (Brickman and Smith, 2002), to deterministic kinematic fields (Palatella et al., 2014). The first approach is adopted by adding a simple white noise to the particle velocity (Tang et al., 2012), with different intensity in the vertical and horizontal directions. Thus, we consider this noisy version of the simplified MRG:

$$\frac{d\mathbf{r}(t)}{dt} = \mathbf{v}(t), \tag{7}$$

$$\mathbf{v} = \mathbf{u} + \mathbf{v}_s + \tau_p(\beta - 1)\left(\frac{D\mathbf{u}}{Dt} + 2\mathbf{\Omega} \times \mathbf{u}\right) + \mathbf{W}. \tag{8}$$

$\mathbf{W}(t) \equiv \sqrt{2D_h}\mathbf{W}_h(t) + \sqrt{2D_v}W_z(t)$, with $(\mathbf{W}_h, W_z) = (W_x(t), W_y(t), W_z(t))$ a three-dimensional vector Gaussian white noise with zero mean and correlations $\langle W_i(t)W_j(t')\rangle = \delta_{ij}\delta(t - t')$, $i, j = x, y, z$. We consider an horizontal eddy diffusivity, D_h, depending on resolution length scale l according to the Okubo formula (Okubo, 1971; Hernandez-Carrasco et al., 2011): $D_h(l) = 2.055 \times 10^4 l^{1.55}$

($m^2\,s^{-1}$). Thus, when taking $l \sim 8\,km = 8000\,m$ (corresponding to $1/12°$), we obtain $10\,m^2\,s^{-1}$. In the vertical direction we use a constant value of $D_v = 10^{-5}\,m^2\,s^{-1}$ (Rossi et al., 2013).

In order to obtain quantitative assessment of the relative effects of the different physical terms in Eq. (8), we will compare trajectories obtained from the following expressions which only consider some of the terms of the full Eq. (8):

$$\mathbf{v}^{(0)} = \mathbf{u} + \mathbf{v}_s + \mathbf{W}, \tag{9}$$

$$\mathbf{v}^{(co)} = \mathbf{u} + \mathbf{v}_s + \tau_p(\beta - 1)2\boldsymbol{\Omega} \times \mathbf{u} + \mathbf{W}, \tag{10}$$

$$\mathbf{v}^{(in)} = \mathbf{u} + \mathbf{v}_s + \tau_p(\beta - 1)\frac{D\mathbf{u}}{Dt} + \mathbf{W}. \tag{11}$$

Besides the random noise term, the first Eq. (9) only considers the settling velocity, Eq. (10) resolves the settling velocity plus the Coriolis effect, and Eq. (11) considers the settling plus the inertial term.

For the numerical experiments we will consider a set of six values of v_s ranging from 5 to $200\,m\,day^{-1}$, with different integration times to have in all the cases a sinking to about 1000–1100 m depth. The stochastic Eq. (7) with Eqs. (8)–(11) is written in spherical coordinates and numerically integrated using a second-order Heun method with time step of 4 h (Toral and Colet, 2014). We use $R = 6371\,km$ for the Earth's radius, $g = 9.81\,m\,s^{-2}$, and the angular velocity $\boldsymbol{\Omega}$ is a vector pointing in the direction of the Earth's axis and modulus $|\boldsymbol{\Omega}| = 7.2722 \times 10^{-5}\,s^{-1}$. We take v_s and τ_p constant in each experiment because, although water density may increase with depth, this variation is at most of $10\,kg\,m^{-3}$ in the range of depths we are considering here and then the impact on v_s is below 0.1 %. We use as initial starting date 17 September 2008. The numerical experiments consist in launching $N = 6000$ particles from initial conditions randomly chosen in a square of size $1/6°$ centered at $10.0°$ E $29.12°$ S and -100.0 m depth (red rectangle in Fig. 3), and in letting them evolve for a given time t_f (stated in Table 2) following Eq. (7) with Eqs. (8)–(11). We use in each case identical initial conditions and the same sequence of random numbers for the noise terms. In this way we guarantee that any difference in particle trajectories arise from the inclusion or not of the inertial and Coriolis terms. We obtain the time-dependent positions of all the particles for each approximation to the dynamics: $\mathbf{r}_i(t)$, $\mathbf{r}_i^{(0)}(t)$, $\mathbf{r}_i^{(co)}(t)$, and $\mathbf{r}_i^{(in)}(t)$, $i = 1, \ldots, N$, following, respectively, Eqs. (8)–(11) and the corresponding final positions at $t = t_f$.

Table 2 gives the mean and the standard deviation of the depths attained by the set of particles in each numerical experiment as obtained from Eqs. (7) and (8). We find that the use of the different approximations (9)–(11) gives virtually the same results. The only differences larger than 1 cm in mean or standard deviation are the ones for the smallest unperturbed settling velocity considered, $v_s = 5\,m\,day^{-1}$, and are also reported in Table 2. The measured differences are negligible as compared with the traveled distance or even

Table 2. Mean and standard deviation of the set of depths attained, according to Eqs. (7) and (8), by the set of particles released from the red rectangle in Fig. 3 at $z = -100$ m for the different values of v_s and integration times used. The results labeled (co), (in), and (0) are obtained from the different approximations in Eqs. (9)–(11), which differ more than 1 cm from the ones obtained from Eq. (8) only in the $v_s = 5\,m\,day^{-1}$ case.

v_s (m day^{-1})	integration time t_f (days)	Mean final depth (m)	std final depth (m)
200	5	−1091.78	3.88
100	10	−1065.33	6.57
50	20	−1033.97	6.22
20	50	−1051.85	22.67
10	100	−1043.49	51.22
5	200	−1054.97	62.03
		−1054.76 (co)	62.14 (co)
		−1054.76 (in)	62.16 (in)
		−1054.72 (0)	62.14 (0)

with the model grid size. Indeed, small changes in the ROMS model configuration or in the velocity interpolation procedure would have an impact larger than this. The mean displacements in the horizontal obtained with the different approximations (not shown) also differ by less than 0.1 %. We thus conclude that the simplest approximation Eq. (9) which only considers passive transport and an added constant sinking velocity already provides a good description of the sinking process for the type of marine particles and the range of space scales and timescales considered here. Note that the depth attained by the particles is always slightly shallower than $z = -1100$ m, which is the depth that would be reached in a still fluid. It is still debated under which conditions fluid flows enhance or reduce the settling velocity (Maxey, 1987; Wang and Maxey, 1993; Ruiz et al., 2004; Bec et al., 2014).

We now perform a more stringent test going beyond the analyses of mean displacements by considering differences between individual particle trajectories. To assess the impact of the Coriolis and of the inertial effects, we compare the positions $\mathbf{r}_i^{(co)}(t)$ and $\mathbf{r}_i^{(in)}(t)$ with the simpler dynamics Eq. (9) which gives $\mathbf{r}_i^{(0)}(t)$ for each time t. To do so we compute the root mean square difference in position per particle and time, which we separate into vertical and horizontal components:

$$r_h^{(k)}(t) = \sqrt{\frac{1}{N}\sum_{i=1}^{N}\left(\mathbf{x}_i^{(0)}(t) - \mathbf{x}_i^{(k)}(t)\right)^2}, \tag{12}$$

$$r_v^{(k)}(t) = \sqrt{\frac{1}{N}\sum_{i=1}^{N}\left(z_i^{(0)}(t) - z_i^{(k)}(t)\right)^2}, \tag{13}$$

with $\mathbf{x}_i = (x_i, y_i)$, the horizontal position vectors, and the superindex (k) takes the values (co) or (in).

Figure 4 shows the influence of the Coriolis term in the horizontal component for each sinking velocity as a function of time. We observe an exponential growth in a wide range of times, which reveals the chaotic behavior of each of the compared trajectories. The value of the exponent $0.08\,\text{days}^{-1}$ is in agreement with the order of magnitude of the Lyapunov exponent calculated using the same ROMS velocity model and region (Bettencourt et al., 2012). Similar exponential growth with the same growth rate were observed for the inertial terms and the vertical components (not shown), although the absolute magnitude of these mean root square differences was much smaller.

The horizontal and vertical differences $r_{h,v}^{(\text{co})}$ at the final integration time t_f (i.e., the time at which the particles reach an approximate depth of $1000\,\text{m}$ for each value of v_s) are displayed in Fig. 5, both as a function of v_s and of t_f. Similarly, the values of $r_{h,v}^{(\text{co})}$ are presented in Fig. 6. The behavior can be understood as resulting from two factors: on the one hand smaller v_s requires larger t_f to reach the final depth, and larger integration time t_f allows for accumulation of larger differences between trajectories. On the other hand the Coriolis and inertial terms in Eqs. (10)–(11) are proportional to $\tau_p(\beta - 1) = v_\text{s}/g$ so that their magnitude decreases for smaller v_s. The combination of these two competing effects shapes the curves in Figs. 5 and 6, which for the vertical-difference case turn out to be non-monotonic in v_s or t_f.

In all cases, the differences (both in vertical and horizontal) between the simple dynamics (Eq. 9) and the corrected ones in Eqs. (10) and (11) are negligible when compared with typical particle displacements, or even with model grid sizes. For example, we imposed in our simulations a vertical displacement close to $1000\,\text{m}$, whereas the mean root square difference with respect to simple sinking is below $1\,\text{m}$ for the Coriolis case (Fig. 5) and below $1\,\text{cm}$ for the inertial case (Fig. 6). In the horizontal direction, displacements during those times are of the order of hundreds of km, whereas the corrections introduced by the Coriolis and inertial terms are in the worst cases of the order of a few kilometers or of tens of meters, respectively. In particular, the most important impact (horizontal differences of tens of kilometers) is attributed to the Coriolis term for particles sinking at $5\,\text{m}\,\text{day}^{-1}$ (Fig. 5). It is worth noting that although the small value of Rossby numbers $\simeq 0.01$ for mesoscale processes might indicate a strong influence of the Coriolis force in Eq. (8), its influence on particle dynamics becomes negligible because it is multiplied by τ_p or, equivalently, the Stokes number, which is significantly small for biogenic particles. Nevertheless, the Rossby number coincides with the ratio of inertial term to Coriolis term in Eq. (8) and its value $\simeq 0.01$ explains the difference of 2 orders of magnitude among the corrections arising from the inertial force and from Coriolis. The trajectories of the full dynamics ruled by Eq. (8) are nearly identical to the ones under the approximation which keeps only the sinking term and Coriolis, so that the correspond-

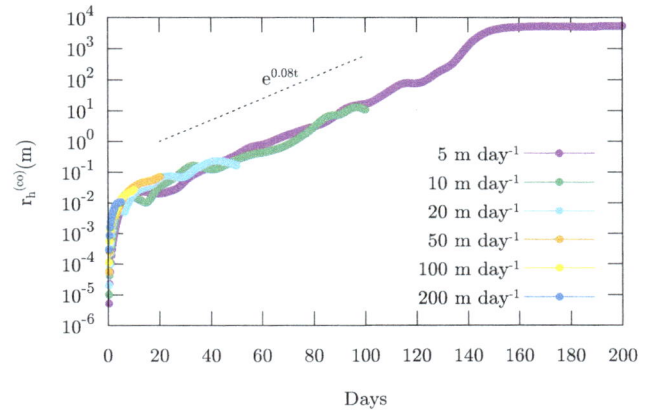

Figure 4. Root mean square difference per particle, as a function of time, between horizontal particle positions computed with Eq. (9) and with Eq. (10), i.e., with and without the Coriolis term. The different colors correspond to distinct values of the unperturbed sinking velocity. The dashed line is an exponential with slope $0.08\,\text{day}^{-1}$.

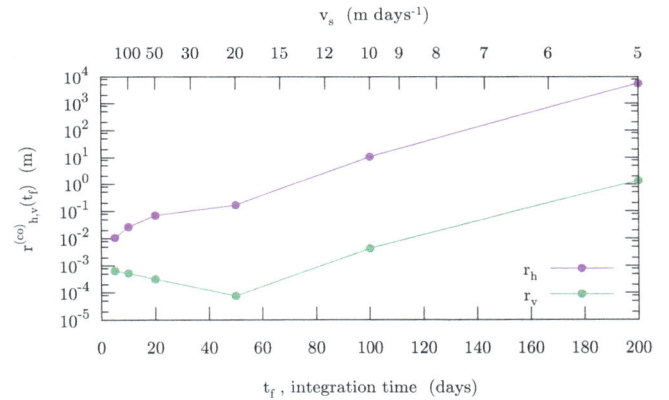

Figure 5. Root mean square difference per particle between final positions (at times t_f stated in Table 2) computed with and without the Coriolis term (Eqs. 10 and 9, respectively). Data are presented as a function of the unperturbed sinking velocity v_s used (upper horizontal scale) and of the final integration time t_f (lower horizontal scale). Upper violet line, the horizontal difference $r_h^{(\text{co})}(t_\text{f})$; lower green line, the vertical difference $r_v^{(\text{co})}(t_\text{f})$.

ing comparison to $\mathbf{r}_i^{(0)}$ gives a figure essentially identical to Fig. 5 (not shown).

In summary, for the range of sizes and densities of the marine particles considered here, the sinking dynamics is essentially given by the velocity $\mathbf{v} = \mathbf{u} + \mathbf{v}_s$, which has been the one used in some oceanographic studies (Siegel and Deuser, 1997; Siegel et al., 2008; Roullier et al., 2014). Note however that a new question arises: what is then the reason for the observed clustering of falling particles (Logan and Wilkinson, 1990; Buesseler et al., 2007; Mitchell et al., 2008)? The argument of the non-inertial dynamics of the particles does not

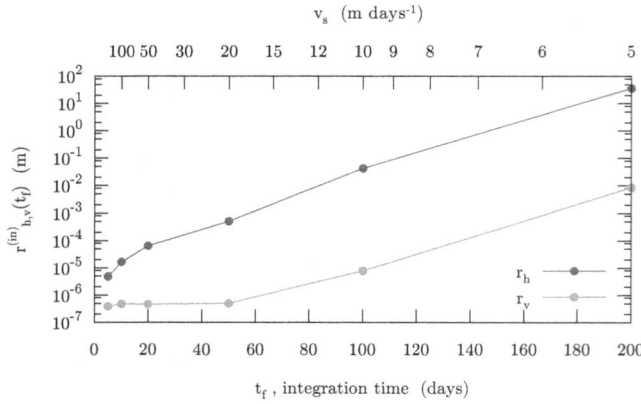

Figure 6. Root mean square difference per particle between final positions (at times t_f stated in Table 2) computed with and without the inertial term (Eqs. 11 and 9, respectively). Data are presented as a function of the unperturbed sinking velocity v_s used (upper horizontal scale) and of the final integration time t_f (lower horizontal scale). Upper violet line, the horizontal difference $r_h^{(in)}(t_f)$; lower green line, the vertical difference $r_v^{(in)}(t_f)$.

serve since $\nabla \cdot \mathbf{v} = \nabla \cdot \mathbf{u} = 0$. A possible response is explored in the next section.

5 Geometric clustering of particles

Compressibility of the particle-velocity field, i.e., $\nabla \cdot \mathbf{v} \neq 0$, which can arise from inertial effects even when the corresponding fluid-velocity field is incompressible, $\nabla \cdot \mathbf{u} = 0$, has been identified as one of the mechanisms leading to preferential clustering of particles in flows (Squires and Eaton, 1991; Balkovsky et al., 2001). This is so because $\rho(t)$, the particle density at time t at the location $\mathbf{r} = \mathbf{r}(\mathbf{r}_0, t)$ of a particle that started at $\mathbf{r}_0$ at time zero, satisfies $\rho(t) = \rho(0)\delta^{-1}$, where δ is a dilation factor equal to the determinant of the Jacobian $|\frac{\partial \mathbf{r}}{\partial \mathbf{r}_0}|$, which satisfies

$$\frac{1}{\delta}\frac{D\delta}{Dt} = \nabla \cdot \mathbf{v} \qquad (14)$$

or, using $\delta(0) = 1$:

$$\delta(t_f) = e^{\int_0^{t_f} dt \nabla \cdot \mathbf{v}}. \qquad (15)$$

Thus, particles will accumulate (i.e., higher $\rho(t_f)$) in final deep locations receiving particles whose trajectories have predominantly travelled through regions with $\nabla \cdot \mathbf{v} < 0$. We have seen however that to a good approximation $\nabla \cdot \mathbf{v} \approx \nabla \cdot \mathbf{u} = 0$ since inertial effects can be neglected for the type of marine particles we consider here, and then the three-dimensional particle-velocity field is incompressible.

We now reproduce numerically a typical situation in which clustering of marine particles is observed. We release particles uniformly in an horizontal layer close to the surface, we

let them sink within the oceanic flow and we finally observe the distribution of the locations where they touch another horizontal deeper layer. The domain chosen is the rectangle 12 to 35° S and 4 to 19° E (orange rectangle in Fig. 3). We divide the domain horizontally into squares of side $1/25°$, then initialize 1000 particles at random positions in each of them on 20 August 2008 at depth $z = -100$ m (i.e., the bottom of the euphotic layer, starting point of our biogenic particles), and then integrate each trajectory until it reaches -1000 m depth. We use Eq. (9) for the velocity, with $v_s = 50$ m day^{-1}. In order to avoid any small fluctuating compressibility arising from the noise term, we put $\mathbf{W} = \mathbf{0}$, but we have checked that the result in the presence of noise is virtually indistinguishable (not shown). At the bottom layer ($z = -1000$ m) we count how many particles arrive to each of the $1/25°$ boxes and display the result in Fig. 7a. Despite $\nabla \cdot \mathbf{v} = 0$ we see clear preferential clustering of particles in some regions related to eddies and filaments. We note that our horizontal boxes have a latitude-dependent area so that distributing particles at random in them produces a latitude-dependent initial density which could lead to some final inhomogeneities. We have checked however that for the range of displacements of the particles, this effect is everywhere smaller than 5% and thus can not be responsible for the large clustering observed in Fig. 7a. Nevertheless, this effect will be taken into account later.

We explain the observed particle clustering by considering the field displayed in Fig. 7a as a projection in two dimensions of a density field (the cloud of sinking particles) which evolves in three dimensions. Even if the three-dimensional divergence is zero, and then an homogeneous three-dimensional density will remain homogeneous, a two-dimensional cut or projection can be strongly inhomogeneous. This mechanism has been proposed to explain clustering and inhomogeneities in the ocean surface (Huntley et al., 2015; Jacobs et al., 2016), but we show here that it is also relevant for the crossing of a horizontal layer by a set of falling particles.

A simple way to confirm that this clustering arises from the two-dimensionality of the measurement is to estimate the changes in the horizontal density of evolving particle layers as if they were produced just by the horizontal part of the velocity field. This is only correct if an initially horizontal particle layer remains always horizontal during the sinking process, which is not true. But, given the huge differences in the values of the horizontal and vertical velocities in the ocean, we expect this approximation to capture the essential physics and provide a qualitative explanation of the observed clustering. We expect the approximation to become better for increasing v_s, because of the shorter sinking time during which vertical deformations could develop. Thus we compute the two-dimensional version of the dilation field, $\delta_h(\mathbf{x}, t_f)$, at each horizontal location $\mathbf{x}$ in the deep layer at

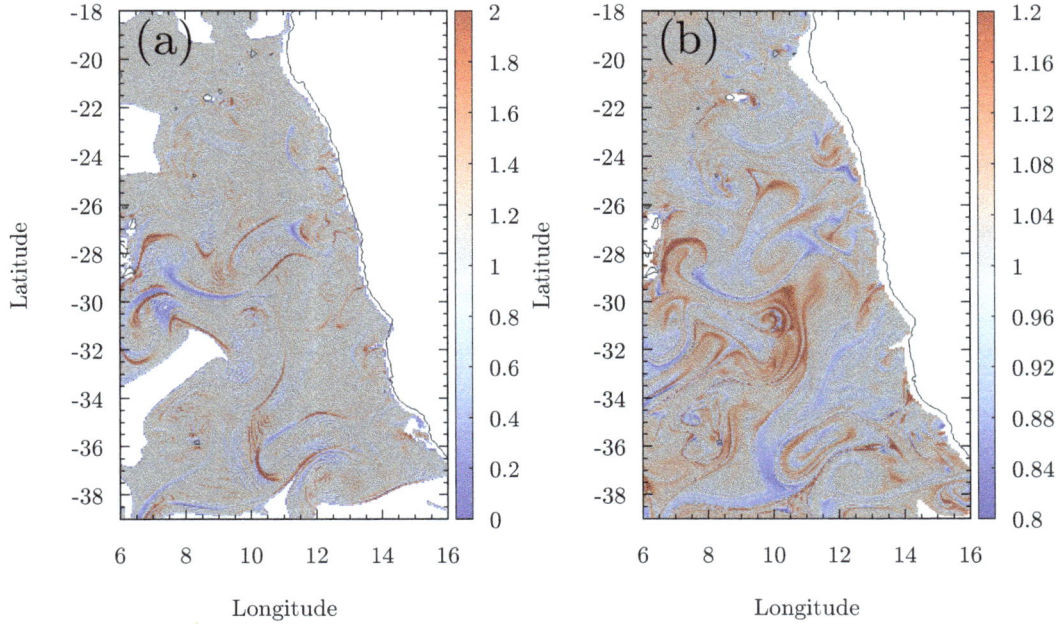

Figure 7. Results of the clustering numerical experiments of Sect. 5. **(a)** N_f/N_0, the number of particles N_f arriving to an horizontal box of size $1/25°$ in the horizontal layer at $z = -1000$ m, normalized by the number of particles $N_0 = 1000$ released from the upper $z = -100$ m layer. **(b)** The corrected dilation factor $\delta(\mathbf{x}, t_f)^{-1} \cos(\theta_f)/\cos(\theta_0)$ mapped on the final $z = -1000$ m layer. It gives the ratio between horizontal densities at the final and initial locations, corrected with the latitudinal dependence of the horizontal boxes used in panel **(a)**, to give an estimation of the local particle number ratio between lower and upper layer. The black thin line represents the coastline; white oceanic areas indicate in **(a)** regions which do not receive any falling particles; in **(b)** they are regions from which the backward integration ends up outside the domain.

$z = -1000$ m:

$$\delta_h(\mathbf{x}, t_f) = e^{\int_0^{t_f} dt \nabla_h \cdot \mathbf{v}} \qquad (16)$$

with the horizontal divergence

$$\nabla_h \cdot \mathbf{v} \equiv \frac{\partial v_x}{\partial x} + \frac{\partial v_y}{\partial y} = \frac{\partial u}{\partial x} + \frac{\partial v}{\partial y} = -\frac{\partial w}{\partial z}, \qquad (17)$$

where in the second equality we have used Eq. (9) from which $\nabla_h \cdot \mathbf{v} = \nabla_h \cdot \mathbf{u}$ and the third one is a consequence of $\nabla \cdot \mathbf{u} = 0$. In order to get the values of δ_h on a uniform grid on the $-1000m$ depth layer at the arrival date t_f of the particles in the previous simulation, we integrate backwards in time trajectories from grid points separated $1/50°$ at $z = -1000$ m until they reach -100 m. The starting date (t_f) of the backwards integration was 7 September 2008, i.e., 18 days after the release date used in the previous clustering experiment. This value correspond to the average duration time of trajectories in that experiment. Then δ_h was computed integrating in time the values of $\nabla_h \cdot \mathbf{v}$ along every trajectory using Eq. (16).

Figure 7b displays the quantity $\delta(\mathbf{x}, t_f)^{-1} \cos(\theta_f)/\cos(\theta_0)$, which gives the ratio between densities in the upper and lower layers, corrected with the angular factors controlling the area of the horizontal boxes so that this can be compared with the ratio between the particle numbers displayed

in Fig. 7a. θ_f is the latitude of point $\mathbf{x}$, and θ_0 is the latitude of the corresponding trajectory in the upper $z = -100$ m layer. As stated before, the latitudinal corrections by the cosine terms are always smaller than a 5 %. Although there is no perfect quantitative agreement, there is clear correspondence between the main clustered structures in panels (a) and (b) of Fig. 7, confirming that they originate from the horizontal dynamics in an incompressible three-dimensional velocity field. We have checked in specific cases that locations with larger differences between Fig. 7a and b correspond to places with large dispersion in the arrival times to the bottom layer, indicating deviations from the horizontality assumption.

6 Conclusions

We have studied the problem of sinking particles in a realistic oceanic flow, focussing on the range of sizes and densities appropriate for marine biogenic particles. Starting from a modeling approach in terms of the MRG Eq. (1), our conclusion is that the simplest approximation given by Eq. (9) in which particles move passively with the fluid flow with an added constant settling velocity in the vertical direction is an accurate framework to describe the sinking process in the types of flows and particles considered. A re-assessment of these assumptions may be required if more complex processes (such as aggregation/disaggregation) are included and when super-high resolution (submesoscale and below) mimicking the real ocean becomes available.

Corrections arising from the Coriolis force turn out to be about 100 times larger than the ones coming from inertial effects, in agreement with the results in Sapsis and Haller (2009) or in Beron-Vera et al. (2015), but both of them are negligible when compared to the effects of passive transport by the fluid velocity plus the added gravity term, except for very slowly sinking particles at high latitudes.

If the fluid flow field $u(r, t)$ has vanishing divergence, then the same is true for the particle velocity field defined by the approximation in Eq. (9). Then, no three-dimensional clustering can occur within this approximation. Nevertheless, we have shown that two-dimensional cuts or projections of evolving three-dimensional particle clouds display horizontal clustering.

Competing interests. The authors declare that they have no conflict of interest.

Acknowledgements. We acknowledge support from the LAOP project, CTM2015-66407-P (AEI/FEDER, EU), from the Office of Naval Research, grant no. N00014-16-1-2492, and from a Juan de la Cierva Incorporación fellowship (IJCI-2014-22343) granted to Vincent Rossi.

Edited by: Vicente Perez-Munuzuri

References

Azetsu-Scott, K. and Passow, U.: Ascending marine particles: Significance of transparent exopolymer particles (TEP) in the upper ocean, Limnol. Oceanogr., 49, 741–748, https://doi.org/10.4319/lo.2004.49.3.0741, 2004.

Balch, W. M., Bowler, B. C., Drapeau, D. T., Poulton, A. J., and Holligan, P. M.: Biominerals and the vertical flux of particulate organic carbon from the surface ocean, Geophysical Research Letters, 37, L22605, https://doi.org/10.1029/2010GL044640, 2010.

Balkovsky, E., Falkovich, G., and Fouxon, A.: Intermittent Distribution of Inertial Particles in Turbulent Flows, Phys. Rev. Lett., 86, 2790–2793, https://doi.org/10.1103/PhysRevLett.86.2790, 2001.

Bec, J.: Fractal clustering of inertial particles in random flows, Phys. Fluids, 15, 81–84, https://doi.org/10.1063/1.1612500, 2003.

Bec, J., Biferale, L., Cencini, M., Lanotte, A., Musacchio, S., and Toschi, F.: Heavy Particle Concentration in Turbulence at Dissipative and Inertial Scales, Phys. Rev. Lett., 98, 084502, https://doi.org/10.1103/PhysRevLett.98.084502, 2007.

Bec, J., Homann, H., and Ray, S. S.: Gravity-Driven Enhancement of Heavy Particle Clustering in Turbulent Flow, Phys. Rev. Lett., 112, 184501, https://doi.org/10.1103/PhysRevLett.112.184501, 2014.

Beron-Vera, F. J., Olascoaga, M. J., Haller, G., Farazmand, M., Triñanes, J., and Wang, Y.: Dissipative inertial transport patterns near coherent Lagrangian eddies in the ocean, Chaos, 25, 087412, https://doi.org/10.1063/1.4928693, 2015.

Bettencourt, J. H., López, C., and Hernández-García, E.: Oceanic three-dimensional Lagrangian coherent structures: A study of a mesoscale eddy in the Benguela upwelling region, Ocean Modell., 51, 73–83, https://doi.org/10.1016/j.ocemod.2012.04.004, 2012.

Biferale, L., Bonaccorso, F., Mazzitelli, I. M., van Hinsberg, M. A. T., Lanotte, A. S., Mussachio, S., Perlekar, P., and Toschi, F.: Coherent structures and extreme events in rotating multiphase flows, Phys. Rev. X, 6, 041036, 2016.

Brickman, D. and Smith, P. C.: Lagrangian Stochastic modeling in coastal oceanography, J. Atmos. Ocean. Technol., 19, 83–99, 2002.

Buesseler, K. O., Antia, A. N., Chen, M., Fowler, S. W., Gardner, W. D., Gustafsson, O., Harada, K., Michaels, A. F., Rutgers van der Loef, M., Sarin, M., Steinberg, D. K., and Trull, T.: An assessment of the use of sediment traps for estimating upper ocean particle fluxes, J. Mar. Res., 65, 345–416, https://doi.org/10.1357/002224007781567621, 2007.

Cartwright, J. H. E., Feudel, U., Károlyi, G., de Moura, A., Piro, O., and Tél, T.: Dynamics of Finite-Size Particles in Chaotic Fluid Flows, pp. 51–87, Springer Berlin Heidelberg, Berlin, Heidelberg, https://doi.org/10.1007/978-3-642-04629-2_4, 2010.

Daitche, A. and Tél, T.: Memory Effects are Relevant for Chaotic Advection of Inertial Particles, Phys. Rev. Lett., 107, 244501, https://doi.org/10.1103/PhysRevLett.107.244501, 2011.

De La Rocha, C. L. and Passow, U.: Factors influencing the sinking of POC and the efficiency of the biological carbon pump, Deep-Sea Res. Pt. II, 54, 639–658, https://doi.org/10.1016/j.dsr2.2007.01.004, 2007.

DeVries, T. F. P. and Deutsch, C.: The sequestration efficiency of the biological pump, Geophys. Res. Lett, 39, L13601, 2012.

Elperin, t., Kleeorin, N., and Rogachevskii, I.: Dynamics of particles advected by fast rotating turbulent fluid flows: fluctuations and large-scale structures, Phys. Rev. Lett., 81, 2898–2901, 2002.

Falkovich, G. and Fouxon, I., and Stepanov, M. G.: Acceleration of rain initiation by cloud turbulence, Nature, 419, 151–154, 2002.

Faxén, H.: Der Widerstand gegen die Bewegung einer starren Kugel in einer zähen Flüssigkeit, die zwischen zwei parallelen ebenen Wänden eingeschlossen ist, Ann. Phys., 373, 89–119, https://doi.org/10.1002/andp.19223731003, 1922.

Gatignol, R.: The Faxén formulae for a rigid particle in an unsteady non-uniform Stokes flow, J. Mec. Theor. Appl., 990, 143–160, 1983.

Guidi, L., Jackson, G., Stemmann, L., Miquel, J., Picheral, M., and Gorsky, G.: Relationship between particle size distribution and flux in the mesopelagic zone, Deep-Sea Res. Pt. I, 55, 1364–1374, 2008.

Guseva, K., Feudel, U., and Tél, T.: Influence of the history force on inertial particle advection: Gravitational effects and horizontal diffusion, Phys. Rev. E, 88, 042909, https://doi.org/10.1103/PhysRevE.88.042909, 2013.

Guseva, K., Feudel, U., Daitche, A., and Tél, T.: History effects in the sedimentation of light aerosols in turbulence: The case of marine snow, Phys. Rev. Fluids, 1, 074203, https://doi.org/10.1103/PhysRevFluids.1.074203, 2016.

Gustavsson, K., Vajedi, S., and Mehlig, B.: Clustering of particles falling in a turbulent flow, Phys. Rev. Lett., 112, 214–501, 2014.

Gutknecht, E., Dadou, I., Le Vu, B., Cambon, G., Sudre, J., Garçon, V., Machu, E., Rixen, T., Kock, A., Flohr, A., Paulmier, A., and Lavik, G.: Coupled physical/biogeochemical modeling including O_2-dependent processes in the Eastern Boundary Upwelling Systems: application in the Benguela, Biogeosciences, 10, 3559–3591, https://doi.org/10.5194/bg-10-3559-2013, 2013.

Haller, G. and Sapsis, T.: Where do inertial particles go in fluid flows?, Physica D: Nonlinear Phenomena, 237, 573–583, https://doi.org/10.1016/j.physd.2007.09.027, 2008.

Hedges, J. I.: Biogeochemistry of Marine Dissolved Organic Matter, Elsevier, https://doi.org/10.1016/B978-012323841-2/50003-8, 2002.

Henson, S., Sanders, R., and Madsen, E.: Global patterns in efficiency of particulate organic carbon export and transfer to the deep ocean, Global Biogeochem. Cy., 26, GB1028, https://doi.org/10.1029/2011GB004099, 2012.

Hernandez-Carrasco, I., López, C., Hernández-García, E., and Turiel, A.: How reliable are Finite-Size Lyapunov Exponents for the assessment of ocean dynamics?, Ocean Modell., 36, 208–218, https://doi.org/10.1016/j.ocemod.2010.12.006, 2011.

Huntley, H. S., Lipphardt, B. L., Jacobs, G., and Kirwan, A. D.: Clusters, deformation, and dilation: Diagnostics for material accumulation regions, J. Geophys. Res.-Oceans, 120, 6622–6636, https://doi.org/10.1002/2015JC011036, 2015.

Jackson, G. A.: Effect of coagulation on a model planktonic food web, Deep-Sea Res. Pt. I, 48, 95–123, https://doi.org/10.1016/S0967-0637(00)00040-6, 2001.

Jacobs, G. A., Huntley, H. S., Kirwan, A. D., Lipphardt, B. L., Campbell, T., Smith, T., Edwards, K., and Bartels, B.: Ocean processes underlying surface clustering, J. Geophys. Res.-Oceans, 121, 180–197, https://doi.org/10.1002/2015JC011140, 2016.

Jimenez, J.: Ocean turbulence at milimiter scales, Sci. Mar., 61, 47–56, 1997.

Komar, P. D., Morse, A. P., Small, L. F., and Fowler, S. W.: An analysis of sinking rates of natural copepod and euphausiid fecal pellets, Limnol. Oceanogr., 26, 172–180, https://doi.org/10.4319/lo.1981.26.1.0172, 1981.

Logan, B. E. and Wilkinson, D. B.: Fractal geometry of marine snow and other biological aggregates, Limnol. Oceanogr., 35, https://doi.org/10.4319/lo.1990.35.1.0130, 1990.

Maggi, F.: The settling velocity of mineral, biomineral, and biological particles and aggregates in water, J. Geophys. Res.-Oceans, 118, 2118–2132, https://doi.org/10.1002/jgrc.20086, 2013.

Maggi, F. and Tang, F. H.: Analysis of the effect of organic matter content on the architecture and sinking of sediment aggregates, Mar. Geol., 363, 102–111, https://doi.org/10.1016/j.margeo.2015.01.017, 2015.

Maxey, M. R.: The gravitational settling of aerosol particles in homogeneous turbulence and random flow fields, J. Fluid Mech., 174, 441–465, https://doi.org/10.1017/S0022112087000193, 1987.

Maxey, M. R. and Riley, J. J.: Equation of motion for a small rigid sphere in a nonuniform flow, Phys. Fluids, 26, 883–889, https://doi.org/10.1063/1.864230, 1983.

McDonnell, A. M. P. and Buesseler, K. O.: Variability in the average sinking velocity of marine particles, Limnol. Oceanogr., 55, 2085–2096, https://doi.org/10.4319/lo.2010.55.5.2085, 2010.

Michaelides, E. E.: Hydrodynamic Force and Heat/Mass Transfer From Particles, Bubbles, and Drops, J. Fluids Eng., 125, 209–238, https://doi.org/10.1115/1.1537258, 1997.

Mitchell, J., H., Y., Seuront, L., Wolk, F., and Li, H.: Phytoplankton patch patterns: Seascape anatomy in a turbulent ocean, J. Mar. Sys., 69, 247–253, 2008.

Monroy, P., Hernández-García, E., Rossi, V., and López, C.: Modeling the dynamical sinking of biogenic particles in oceanic flow, https://doi.org/10.20350/digitalCSIC/8504, 2017.

Moore, J. K. and Villareal, T. A. : Size-ascent rate relationships in positively buoyant marine diatoms, Limnol. Oceanogr., 41, 1514–1520, https://doi.org/10.4319/lo.1996.41.7.1514, 1996.

Okubo, A.: Oceanic diffusion diagram, Deep-Sea Res., 18, 789–802, 1971.

Olivieri, S., Picano, F., Sardina, G., Iudicone, D., and Brandt, L.: The effect of the Basset history force on particle clustering in homogeneous and isotropic turbulence, Phys. Fluids, 26, 041704, 2014.

Palatella, L., Bignami, F., Falcini, F., Lacorata, G., Lanotte, S. A., and Santoleri, R.: Lagrangian simulations and the interannual variability of anchovy egg and larva dispersal in the Sicily channel, J. Geophys. Res.-Oceans, 119, 1306–1323, https://doi.org/10.1002/2013JC009384, 2014.

Provenzale, A.: Transport by coherent barotropic vortices, Annu. Rev. Fluid Mech., 31, 55–93, 1999.

Qiu, Z., Doglioli, A., and Carlotti, F.: Using a Lagrangian model to estimate source regions of particles in sediment traps, Sci.China-Earth Sci., 57, 2447–2456, https://doi.org/10.1007/s11430-014-4880-x, 2014.

Rossi, V., Van Sebille, E., Sen Gupta, E., Garçon, V., and England, M.: Multi-decadal projections of the surface and interior pathways of the Fukushima Cesium-137 radioactive plume, Deep-Sea Res. Pt. I, 80, 37–46, 2013.

Roullier, F., Berline, L., Guidi, L., Durrieu De Madron, X., Picheral, M., Sciandra, A., Pesant, S., and Stemmann, L.: Particle size distribution and estimated carbon flux across the Arabian Sea oxygen minimum zone, Biogeosciences, 11, 4541–4557, https://doi.org/10.5194/bg-11-4541-2014, 2014.

Ruiz, J., Macias, D., and Peters, F.: Turbulence increases the average settling velocity of phytoplankton cells, P. Natl. Acad. Sci. USA, 101, 17720–17724, 2004.

Sabine, C., Feely, R., Gruber, N., Key, R., Lee, K., Bullister, J., Wanninkhof, R., Wong, C., Wallace, D., Tilbrook, B., Millero, F., Peng, T., Kozyr, A., Ono, T., and Rios, A.: The Oceanic Sink for Anthropogenic CO2, Science, 305, 367–371, 2004.

Sapsis, T. and Haller, G.: Inertial Particle Dynamics in a Hurricane, J. Atmos. Sci., 66, 2481–2492, https://doi.org/10.1175/2009JAS2865.1, 2009.

Siegel, D., Fields, E., and Buesseler, K. O.: A bottom-up view of the biological pump: Modeling source funnels above ocean sediment traps, Deep-Sea Res. Pt. I, 55, 108–127, 2008.

Siegel, D. A. and Deuser, W. G.: Trajectories of sinking particles in the Sargasso Sea: modeling of statistical funnels above deep-ocean sediment traps, Deep-Sea Res. Pt. I, 44, 1519–1541, 1997.

Simon, M., Grossart, H., Schweitzer, B., and Ploug, H.: Microbial ecology of organic aggregates in aquatic ecosystems, Aquat. Microb. Ecol., 28, 175–211, https://doi.org/10.3354/ame028175, 2002.

Sozza, A., De Lillo, F., Musacchio, S., and Boffetta, G.: Larga-scale confinament and small-scale clustering of floating particles in stratified turbulence, Phys. Rev. Fluids, 1, 05240, https://doi.org/10.1103/PhysRevFluids.1.052401, 2016.

Squires, K. D. and Eaton, J. K.: Preferential concentration of particles by turbulence, Phys. Fluids A, 3, 1169–1178, 1991.

Stemmann, L. and Boss, E.: Plankton and particle size and packaging: from determining optical properties to driving the biological pump, Annu. Rev. Mar. Sci., 4, 263–90, https://doi.org/10.1146/annurev-marine-120710-100853, 2012.

Tang, W., Knutson, B., Mahalov, A., and Dimitrova, R.: The geometry of inertial particle mixing in urban flows, from deterministic and random displacement models, Phys. Fluids, 24, https://doi.org/10.1063/1.4729453, 2012.

Tanga, P., Babiano, A., Dubrulle, B., and Provenzale, A.: Forming Planetesimals in Vortices, Icarus, 121, 158–170, https://doi.org/10.1006/icar.1996.0076, 1996.

Toral, R. and Colet, P.: Stochastic numerical methods: an introduction for students and scientists, John Wiley & Sons, 2014.

van Sebille, E., Scussolini, P., Durgadoo, J., Peeters, F., Biastoch, A., Weijer, W., Turney, C. S. M., Paris, C. B., and Zahn, R.: Ocean currents generate large footprints in marine palaeoclimate proxies, Nat. Commun., 6, 6521, https://doi.org/10.1038/ncomms7521, 2015.

Wang, L. and Maxey, M. R.: Settling velocity and concentration distribution of heavy particles in homogeneous isotropic turbulence, J. Fluid Mech., 256, 27-68, 1993.

6

Statistical analysis of Lagrangian transport of subtropical waters in the Japan Sea based on AVISO altimetry data

Sergey V. Prants, Maxim V. Budyansky, and Michael Yu. Uleysky

Laboratory of Nonlinear Dynamical Systems, Pacific Oceanological Institute of the Russian Academy of Sciences, 43 Baltiyskaya st., 690041 Vladivostok, Russia

Correspondence to: Sergey V. Prants (prants@poi.dvo.ru)

Abstract. Northward near-surface Lagrangian transport of subtropical waters in the Japan Sea frontal zone is simulated and analysed based on altimeter data for the period from 2 January 1993 to 15 June 2015. Computing different Lagrangian indicators for a large number of synthetic tracers launched weekly for 21 years in the southern part of the Sea, we find preferred transport pathways across the Subpolar Front. This cross-frontal transport is statistically shown to be meridionally inhomogeneous with "gates" and "barriers" whose locations are determined by the local advection velocity field. The gates "open" due to suitable dispositions of mesoscale eddies facilitating propagation of subtropical waters to the north. It is documented for the western, central and eastern gates with the help of different kinds of Lagrangian maps and verified by some tracks of available drifters. The transport through the gates occurs by a portion-like manner, i.e. subtropical tracers pass the gates in specific places and during specific time intervals. There are some "forbidden" zones in the frontal area where the northward transport has not been observed during all the observation period. They exist due to long-term peculiarities of the advection velocity field.

1 Introduction

The Japan Sea (JS) is a mid-latitude marginal sea with dimensions of 1600×900 km, the maximal depth of 3.72 km and the mean depth of about 1.5 km. It spans regimes from subarctic to subtropical and is characterised by many of the same phenomena found in the deep ocean: fronts, eddies, currents and streamers, deep water formation, convection and subduction. It communicates with the Pacific Ocean at the south and east through the Tsushima/Korean and Tsugaru straits, respectively. In the north it is connected with the Okhotsk Sea through the Soya (La Perouse) and Tatarsky straits. All the four channels are shallow with depths not exceeding 135 m. Bathymetry of the JS and its geographic and oceanographic features are shown in Fig. S1 in the Supplement.

Warm and saline Pacific waters enter the Tsushima Strait and split into three currents. Figure 1 with the AVISO velocity field, averaged for the period from 2 January 1993 to 15 June 2015, reflects the main known features of mesoscale near-surface circulation in the JS (Lee and Niiler, 2005; Danchenkov et al., 2006; Talley et al., 2006; Yoon and Kim, 2009; Kim and Yoon, 2010; Lee and Niiler, 2010; Ito et al., 2014). The Nearshore Branch of the Tsushima Current flows northward along the western coast of the Honshu Island (Japan). Its Offshore Branch with a meander-like path flows into the Yamato Basin. The East Korean Warm Current flows northward along the eastern coast of Korea to meet the North Korean Cold Current which is a prolongation of the Liman Cold Current flowing southward along the Siberian coast down to Vladivostok. One of the major large-scale features in the northern JS is a cyclonic gyre over the Japan Basin and the Tatarsky Strait. Some well-known persistent mesoscale eddy-like features are also indicated in Fig. 1. In the Ulleung Basin there are the warm Ulleung anticyclonic circulation (Chang et al., 2004; Mitchell et al., 2005; Shin, 2009; Lee and Niiler, 2010) with the centre at about 37° N, 130.5° E and a cyclonic circulation around 36.7° N, 132° E often called the cold Dok Eddy (Lee and Niiler, 2010). The flow over bottom topography around the Oki Spur in the southeastern part of the Sea generates the anticyclonic Oki Eddy (37.5° N, 134.2° E) (Isoda, 1994). In the western part of the Sea meandering of the East Korean Warm Current produces an anticyclonic circulation called as the anticyclonic Wonsan Eddy (39° N, 129° E) (Lee and Niiler, 2005).

The confluence of northward warm subtropical waters with southward cold subarctic ones forms one of the most remarkable features in the JS – the distinct Subpolar Front that extends across the basin near 40° N (Park et al., 2004; Talley et al., 2006). It is a boundary of physical and chemical properties such as temperature, salinity, dissolved oxygen and nutrients. Like many other hydrological fronts, the Subpolar Front is a highly productive zone with favourable fishery conditions. It is not a continuous curve crossing the basin with a maximal thermal gradient. It is rather a vast area between 38 and 41° N extending across the basin from the Korea coast to the Japanese islands. Understanding transport pathways of subtropical water in the JS is relevant to a number of applications. Physical properties (temperature and salinity), chemical properties, pollutants and biota (phytoplankton, zooplankton, larvae, etc.) are transported and mixed by currents and eddies. Transport of heat to the north is crucial for climatic applications. The ability to simulate transport adequately would be useful to deal with the aftermath of accidents at sea such as discharges of radionuclides, pollutants and oil spills. It is also crucial, for instance, for understanding transport pathways for species invasions.

Since the last decades in the twentieth century, invasions of heat-loving fish (conger eel, tuna, moonfish and triggerfish) and some tropical and subtropical marine organisms (turtles, sharks and others) have been observed in the northern part of the JS, near the coast of Russia (Ivankova and Samuilov, 1979). It is natural to assume that such invasions could be caused by intrusions of subtropical waters in the northern part of the sea across the Subpolar Front. They may be also one of the reasons for a prolongation of the warm period in the fall in Primorye province in Russia since the 1990s (Nikitin et al., 2002). From the oceanographic point of view, this transport of subtropical waters contradicts long-held beliefs on circulation in the JS. It is believed that the Subpolar Front is a transport barrier for propagation of subtropical waters to the north, at least in the western and central parts of the front area (see e.g. Danchenkov et al., 2006). In this paper we use altimetry data to simulate and analyse the northward near-surface transport of subtropical waters across the frontal area from 2 January 1993 to 15 June 2015.

The paper is organised as follows. Section 2 introduces briefly the altimetry data and simulation methods we use. Northward transport of subtropical waters across the frontal area is studied statistically in Sect. 3 for a long period of time. We compute, document and discuss preferred transport pathways and meridional distributions of artificial tracers launched in the southern part of the sea. Supplement data can be found in the online version.

2 Data and methods

Geostrophic velocities were obtained from the AVISO database (http://aviso.altimetry.fr) archived daily on a $1/4° \times 1/4°$ grid from 2 January 1993 to 15 June 2015. Our Lagrangian approach is based on solving equations of motion for a large number of passive synthetic particles (tracers) advected by the AVISO velocity field

$$\frac{d\lambda}{dt} = u(\lambda, \varphi, t), \qquad \frac{d\varphi}{dt} = v(\lambda, \varphi, t), \qquad (1)$$

where u and v are angular zonal and meridional velocities, φ and λ are latitude and longitude, respectively. Bicubical spatial interpolation and third-order Lagrangian polynomials in time are used to provide numerical results. Lagrangian trajectories are computed by integrating Eq. (1) with a fourth-order Runge–Kutta scheme with an integration step of $1/1000$ day. The merged TOPEX/POSEIDON and ERS-1/2 altimeter data sets have been shown by Choi et al. (2004) to be appropriate to study mesoscale surface circulation in the JS because of their comparatively small temporal and spatial sampling intervals. In particular, they have been shown to correlate well (0.95) with tide gauge data in the western JS (Choi et al., 2004).

We study northward transport of tracers in the central part of the JS basin between 37 and 42° N. With this aim 10^5 tracers have been launched weekly from 2 January 1993 to 15 June 2013 at the latitude 37° N from 129 to 138° E. Trajectory of each tracer has been computed for 2 years after its launch date. We fixed the location and the moment of time where and when each tracer crossed a given latitude in the central JS between 37 and 43° N. We take into account the first crossing only, because we are interested not in a net transport but in the northward transport. We stop to compute trajectories of those tracers which get into an AVISO cell with at least two corners situated at the land.

Each water parcel can be attributed to temperature, salinity, density and other properties which characterise this volume as it moves. In addition, each water parcel can be attributed to more specific characteristics which are trajectory functions called "Lagrangian indicators". They are, for example, a distance passed by a fluid particle, its displacement from an original position, its travel time and others. The Lagrangian indicators contain information about the origin, history and fate of the corresponding water masses. Lagrangian maps are plots of Lagrangian indicators versus particle initial positions. A studied area is seeded with a large number of tracers whose trajectories are computed for a given period of time to get the field of a specific Lagrangian indicator whose values are coded by colour and represented as a map in geographic coordinates.

To simulate and analyse transport across the frontal area, we solve successively a few tasks which are numbered in the text in accordance with the following diagrams and Lagrangian maps.

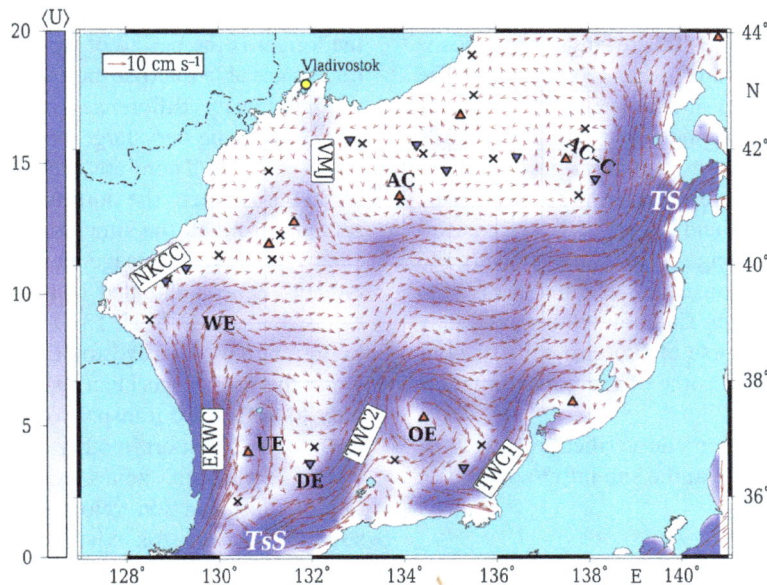

Figure 1. The AVISO velocity field averaged for the period from 2 January 1993 to 15 June 2015. Elliptic and hyperbolic stagnation points with zero mean velocity are indicated by triangles and crosses, respectively. Abbreviations: TsS (Tsushima or Korean Strait), TS (Tsugaru Strait), EKWC (East Korean Warm Current), NKCC (North Korean Cold Current), TWC1 and TWC2 (the first and second branches of the Tsushima Warm Current, UE (Ulleung Eddy), DE (Dok Eddy), OE (Oki Eddy), WE (Wonsan Eddy), AC-C (vortex pair near the eastern gate), AC (anticyclonic eddy over the Japan Basin), VMJ (Vladivostok meridional jet).

1. A meridional distribution of the number of tracers, N, crossing fixed latitudes, λ_f, in the central JS with a space step 0.1°. The corresponding data are represented as a density map which shows by colour the density of tracks of the particles crossed all the latitudes in the central JS from 2 January 1993 to 15 June 2015. Tracking maps show where the subtropical tracers, which crossed eventually the fixed zonal line through fixed meridional "gates", wandered for the whole integration period. They also can be represented as a $N(\lambda_f)$ distribution which shows how many tracers reached a fixed zonal line at the longitude λ_f for the whole period of integration.

2. Fixing initial longitudes λ_0 of launched tracers along the material line 37° N, we compute those final longitudes λ_f at which they cross a fixed zonal line for the whole period of integration. The results are represented as λ_0–λ_f plots.

3. The T–λ_f plots show when and at which longitudes the tracers, launched at 37° N, crossed the latitudes 40 and 42° N for the whole period of integration.

4. In order to document and visualise intrusions of subtropical waters into subarctic ones, we compute backward-in-time Lagrangian maps (Prants, 2015). A subbasin in the sea is seeded at a fixed date with a large number of tracers whose trajectories are computed backward in time for a given period of time. We

use three kinds of Lagrangian map in this paper. Such maps have been shown to be useful in studying large-scale transport and mixing in various basins, from bays (Prants et al., 2013) and seas (Prants et al., 2011a, 2013) to the ocean scale (Prants et al., 2011b; Prants, 2013), in quantifying propagation of radionuclides in the Northern Pacific after the accident at the Fukushima Nuclear Power Plant (Prants et al., 2011b, 2014a; Prants, 2014; Budyansky et al., 2015) and in finding potential fishing grounds (Prants et al., 2014b, c).

In order to track those subtropical waters which were able to cross the Subpolar Front and reach northern latitudes, we colour the tracers that reached the line 37° N in the past and compute how much time it took. In order to know where this or that tracer came from for a given period of time, we compute the drift maps with boundaries. The waters that entered a given area through its southern boundary are shown by one colour, and waters that came through the northern boundary are shown by another colour. The drift maps show in greyscale the finite-time displacement of tracers, D, that is a distance between final, (λ_f, φ_f), and initial, (λ_0, φ_0), positions of advected particles on the Earth sphere with the radius R_E:

$$D \equiv R_E \arccos[\sin\varphi_0 \sin\varphi_f + \cos\varphi_0 \cos\varphi_f \cos(\lambda_f - \lambda_0)]. \qquad (2)$$

"Instantaneous" stagnation elliptic and hyperbolic points are indicated by triangles and crosses, respectively. They are points with zero velocity which are computed daily. Up(down)ward orientation of one of the triangle's top means anticyclonic (cyclonic) rotations of water around them. The triangles are coloured as (blue), marking elliptic points for anticyclones (cyclones). The elliptic points, situated mainly in the centres of eddies, are those stagnation points around which the motion is stable and circular. The hyperbolic points, situated mainly between and around eddies, are unstable ones with the directions along which waters converge to such a point and another directions along which they diverge. The stagnation points are moving Eulerian features and may undergo bifurcations in the course of time. In spite of non-stationarity of the velocity field some of them may exist for weeks and much more.

We have used for a comparison and verification tracks of surface drifters that are available at the site http://aoml.noaa.gov/phod/dac.

3   Results and discussion

3.1   Northward transport of subtropical water and advection velocity field

Figure 2a shows the density of tracks of tracers launched along 37° N and across all the latitudes in the central JS for the whole period of integration. The density is shown in greyscale in the logarithmic scale, $\log_{10} N_\varphi$. The magenta areas in Fig. 2a along the coastal line indicate that the AVISO grid cells there touch the land, and we did not compute trajectories there. An uneven density of points in Fig. 2a means that the northward transport of subtropical waters is meridionally inhomogeneous with "gates" with increased density of points. The gates are such spatial intervals along a given zonal line across which subtropical tracers prefer to cross.

Any tracer, as a passive particle, is able to cross the fixed latitude in the northward direction if the northward component of the velocity field is nonzero at its location. In Fig. 2b we plot distribution of the northward component of the AVISO velocity field averaged over the whole period of integration as follows:

$$\langle v_+(\lambda, \varphi) \rangle = \frac{1}{n} \sum \theta(v(\lambda, \varphi)) v(\lambda, \varphi), \tag{3}$$

where $v_+(\lambda, \varphi)$ is a northward (positive) component of the velocity at the point (λ, φ), $\theta(v)$ the Heaviside function and n the number of days in the period from 2 January 1993 to 15 June 2015. Comparing the Lagrangian representation in Fig. 2a with the Eulerian one in Fig. 2b, it is clear that areas with increased density of points in Fig. 2a correlate well with areas with increased average values of the northward component of the AVISO velocity field in Fig. 2b. Thus, the northward transport of subtropical waters in the central JS is determined mainly by the local advection velocity field, more precisely by local values of its northward component.

The greater is that northward component at a given point and the longer is the period of time when it is positive, the more tracers are able to cross the corresponding latitude.

The density difference in some meridional ranges in Fig. 2a may be very large because of the logarithmic-scale representation. There are even some places in the northern frontal area where the northward transport has not been observed during all the simulation period, from 1993 to 2015. They are marked by magenta rectangles in Fig. 2a. One "forbidden" zone is situated in the deep Japan Basin with the centre at about 41.5° N, 134.2° E, and another one is situated to the south off Vladivostok from 43 to 41° N approximately along the 132° E meridian. We stress that they are forbidden only to northward transport of tracers but can be and really are open to transport in other directions.

The "forbidden" zones exist due to long-term peculiarities of the advection velocity field there. The zone to the south off Vladivostok exists due to a quasi-permanent southward jet approximately along the meridian 132° E from 43 to 40° N (VMJ in Fig. 1). It turns to the east at about 40° N and contributes to the eastward transport. In fact, the northward velocity is practically zero in this area (see Fig. 2b) and, therefore, the northward transport is absent. The other "forbidden" zone exists due to two factors: the presence of a quasi-permanent anticyclonic eddy with the centre at about 41.3° N, 134° E in the deep Japan Basin (AC in Fig. 1) and the eastward zonal jet blocking northward transport across it. Topographically constrained anticyclonic eddies with the centre at about 41–41.5° N, 134–134.5° E have been regularly observed there (Takematsu et al., 1999; Talley et al., 2006; Prants et al., 2015).

3.2   Transport pathways of subtropical water and its intrusions across the Subpolar Front

Now let us look more carefully at the meridional distribution of subtropical tracers crossed the Subpolar Front for the whole period of simulation. We choose for reference four zonal lines along the AVISO grid at 42.125, 41.875, 40.125 and 39.875° N. They are shown in Fig. 3 by solid curves with superimposed meridional distributions of the averaged northward AVISO velocity (arrows). The number of crossings of those latitudes by the available 333 drifters is shown by dashed curves. The correspondence between the peaks in the meridional distributions of the tracers, drifters and the averaged northward AVISO velocity is rather good for all the chosen zonal lines, confirming their direct connection. However, the comparison with drifters should be taken with care because of a comparatively small number of available drifters. Drifters are not ideal passive tracers, and their motion is subjected to submesoscale features which were not caught by altimetry-derived data. Moreover, the drifters have not been launched at the zonal line 37° N like artificial tracers in simulation. Their launch sites for more than 20 years have been distributed rather randomly over the basin.

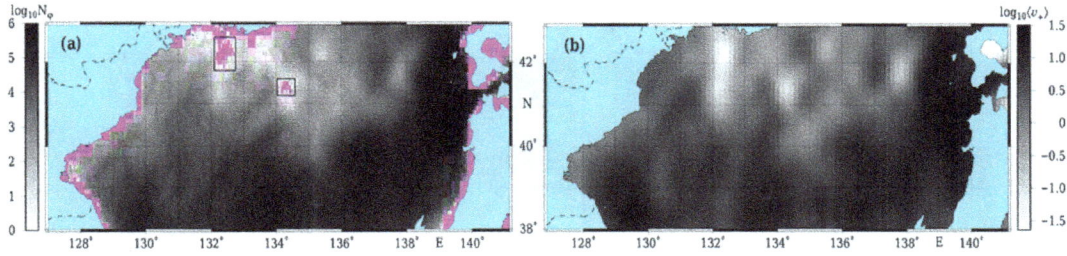

Figure 2. (a) The logarithmic-scale density of tracks of the tracers crossing all the latitudes φ in the central JS, N_φ, from 2 January 1993 to 15 June 2015. The rectangular magenta areas are forbidden zones where the northward transport has not been observed during the whole integration period. The magenta areas near the coast mean that the AVISO grid cells there touch the land, and we did not compute trajectories there. The tracers have been launched weekly along the zonal line at 37° N from 2 January 1993 to 15 June 2013. **(b)** Distribution of the averaged northward component of the AVISO velocity field $\langle v_+(\lambda, \varphi)\rangle$ in the logarithmic-scale averaged over the same period.

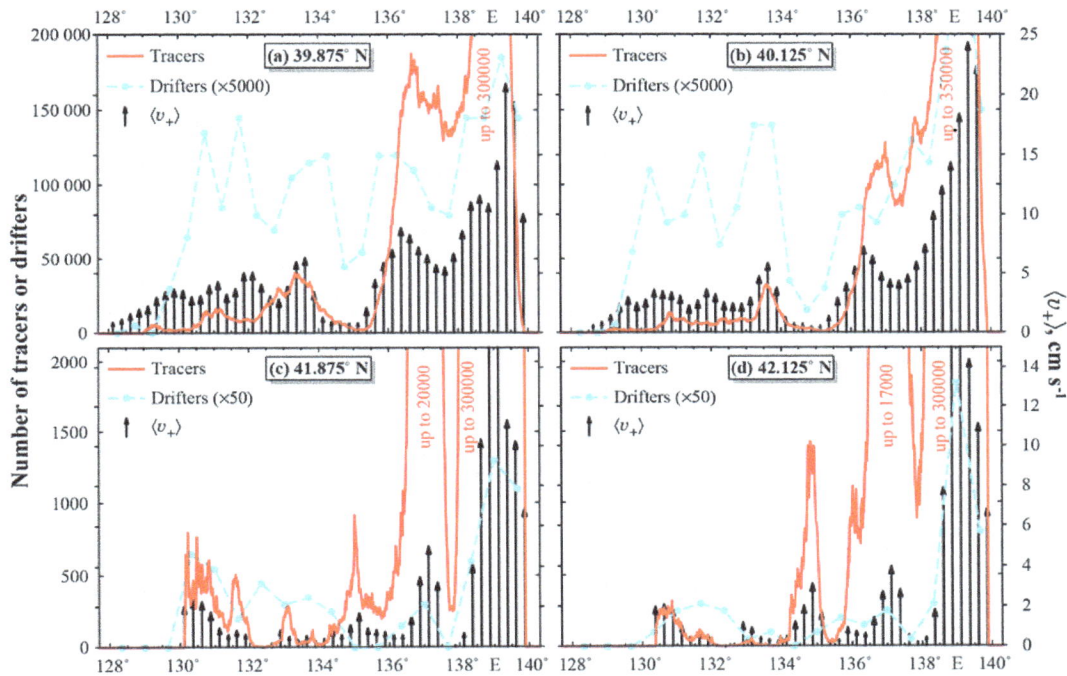

Figure 3. Meridional distributions of the number of tracers which crossed indicated zonal lines (solid curves), of the averaged northward component of the AVISO velocity in cm s^{-1} (arrows) and of the number of crossings of those zonal lines by available drifters (dashed curves). The period of observation is from 2 January 1993 to 15 June 2015.

The local maxima and minima of the distribution functions correspond to gates and conditional barriers, respectively. The very eastern, 138–140° E, and western, 129–131° E, gates are provided mainly by the near-shore branch of the Tsushima Warm Current and the East Korean Warm Current, respectively. The central gate, 133–137° E, probably exists due to topographically constrained features over the Yamato Rise there (see Fig. S1 in the Supplement). The transport through that gate will be shown to be enhanced due to a specific disposition of frontal eddies regularly observed there. The intervals between the gates may be called "conditioned barriers" because of a comparatively small number of tracers

crossing zonal lines there, and because they used to "open" for comparatively short time intervals.

Figure 4a, b shows in accordance with task 2 at which final longitudes λ_f the tracers, launched with the initial longitudes λ_0 at the line 37° N, reached the zonal lines 40 and 42° N for the whole period of integration. The meridional distribution of the number of tracers with pronounced peaks which crossed the zonal line 42° N for the same period is plotted in Fig. 4c. This zonal line was divided into eight meridional intervals numbered by the roman numerals in Fig. 4b and c with horizontal straight lines running via local minima at the distribution in Fig. 4c.

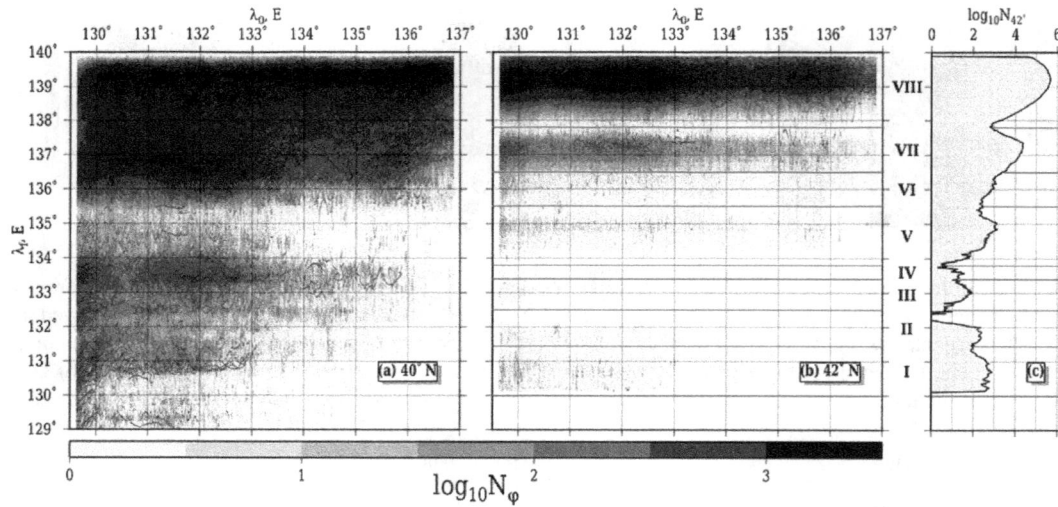

Figure 4. Density plots show in the logarithmic scale how many and at which final longitudes λ_f the tracers with initial longitudes λ_0 were able to cross the zonal lines **(a)** 40° N and **(b)** 42° N for the whole simulation period. The tracers were launched weekly at the line 37° N from 2 January 1993 to 15 June 2013. **(c)** Meridional distribution of the number of tracers which crossed the zonal line 42° N for the whole simulation period. This line is divided into eight intervals numbered by Roman numerals.

The Tsushima Warm Current contributes mainly to the eastern peak VIII in the distribution in Fig. 4c. Black across all the range of initial longitudes λ_0 in Fig. 4b indicates that fluid particles, crossing eventually the line 42° N through the gate 138–140° E, could have any value of the initial longitude λ_0 at the zonal line 37° N. They could reach that gate in different ways: either to be initially trapped by the near shore branch or to be advected by the offshore branch and then to enter the near-shore branch. Moreover, those particles could be involved initially in the East Korean Warm Current and then be transported to the east along the Subpolar Front to eventually join the Tsushima Warm Current. Thus, the subtropical tracers, crossing the gate VIII, may have rather distinct values of some Lagrangian indicators, e.g. travelling time and distance passed.

There is a narrow barrier, the white strip in Fig. 4b between gates VIII and VII, with the centre at the local minimum at 137.8° E in Fig. 4c. A comparatively small number of tracers have been able to cross the line 42° N there for the whole simulation period. Gate VII between 136 and 137.8° E (Fig. 4b, c) provides northward transport of subtropical tracers by means of a quasi-permanent vortex pair located there. The number of subtropical tracers passing through this gate is much smaller than that passing through gate VIII (remember the logarithmic scale in Fig. 4). Only a small number of tracers, launched initially at the very eastern part of the zonal line 37° N, were able to cross the line 42° N through that gate, because most of the eastern tracers passed through the gate VIII to be captured by the near-shore branch of the Tsushima Warm Current. Most of the tracers passing through gate VII came from the western and central parts of the material line at 37° N. The numbers of subtropical tracers passing

through the central and western gates are much smaller as compared with those passing by the eastern ones. We distinguish two central gates V and III, 134–135.5 and 132.5–133.5° E, respectively, and the western gates I and II (Fig. 4c) in the range 130–132.5° E. It follows from Fig. 4b that the western and central gates collect subtropical tracers mainly from the western part of the initial zonal line, from 129 to 133° E. In other words, water parcels from its eastern part (133–137° E) practically do not pass through those gates at the latitude 42° N. Thus, the western part of the initial material line at 37° N contributes to all the peaks in the tracer distribution 42° N, whereas its eastern part contributes mainly to the Tsushima peak.

To visualise the transport paths by which subtropical tracers reach the northern frontal area we compute so-called tracking maps in Fig. S4 in the Supplement showing where the subtropical tracers, which crossed eventually the zonal line 42° N, wandered for the whole integration period.

The $T-\lambda_f$ plots in Figs. S2 and S3 in the Supplement show when and at which longitudes the tracers, launched weekly at the zonal line 37° N from 2 January 1993 to 15 June 2013, reached the zonal lines 40 and 42° N, respectively. This was designated in Sect. 2 as task 3. As an example, we show in Fig. 5 a typical $T-\lambda_f$ plot for the tracers crossed eventually the zonal lines 40 and 42° N in the period from 1 March 1995 to 1 March 1996. This demonstrates the eastern gates VIII and VII (Fig. 4) through which the subtropical tracers cross the corresponding latitudes. The locations of the central and western gates fluctuate in time, and some gates may be even closed for a while to the northward transport. The patchiness in the plot means that subtropical tracers prefer to cross the zonal lines in specific places (note the peaks in Figs. 3) and

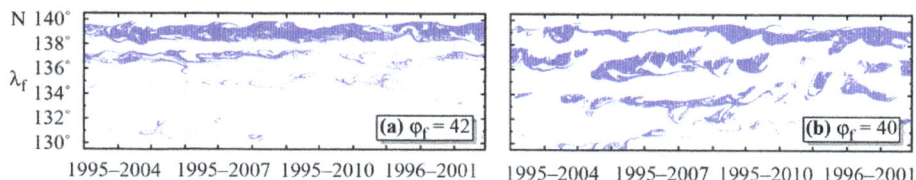

Figure 5. The $T-\lambda_f$ plots show when and at which longitudes the tracers, launched at the zonal line 37° N, eventually crossed the zonal lines **(a)** 40° N and **(b)** 42° N in the period from 1 March 1995 to 1 March 1996.

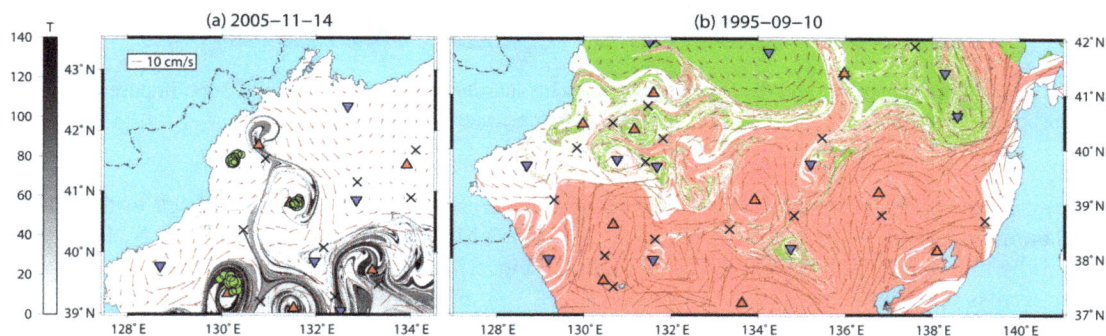

Figure 6. (a) The Lagrangian map documents intrusions of subtropical water to the southern coast of Russia through the western gate. Greyscale shows travelling time T in days that it took for subtropical tracers to reach their locations on the map from latitude 37° N to the dates shown. "White" tracers are those which did not come from latitude 37° N for the integration period, 140 days. Locations of available drifters are shown by full circles for 1 day before and after the dates indicated. **(b)** The drift map documents a streamer-like northward transport of subtropical water across the front through a central gate with the help of the cyclone with the centre at 41.5° N, 134.4° E. Red and green code the waters that entered the studied area for 2 years through its southern and northern boundaries, respectively. White indicates the tracers arriving at the coast.

during specific time intervals. Any patch with a large number of tracers somewhere, for example at the central meridional gate, means that a water mass proportional to the size of this patch passed through the central gate across a given latitude during the period of time proportional to its zonal size. Thus, the northward transport of subtropical water across the Subpolar Front occurs in a proportion-like manner. Specific oceanographic conditions may arise in a given area and at a given time which produce a large-scale intrusion of subtropical water to the north by means of mesoscale eddies present there.

To document intrusion of subtropical water there, we compute the backward-in-time Lagrangian maps (for a recent review of backward-in-time techniques see Prants, 2015). This is a realisation of task 4 in Sect. 2. The basin, shown in Fig. 6a, is seeded with a large number of tracers for each of which we compute the time required for a tracer to reach its location on the map on a fixed date from the latitude 37° N. This is what is known as a residence–time map (Lipphardt et al., 2006; Uleysky et al., 2007; Hernández-Carrasco et al., 2013). The travelling time T in days is shown in greyscale. The map in Fig. 6a illustrates the mechanism of penetration of subtropical water northward through the western gate. A vortex street with four anticyclones is formed in the fall of 2005 to the north of the Subpolar Front in the western part of

the sea. Their centres are marked in Fig. 6a by the triangles at coordinates 39.1° N, 131.5° E; 39.3° N, 130.1° E; 40.8° N, 131.4° E and 41.7° N, 130.8° E. Subtropical "grey" tracers propagate along the unstable manifolds of the three hyperbolic points between and around the eddies to the north (a simple description of the notion of stable and unstable manifolds in fluid flows can be found e.g. in Prants, 2014). The hyperbolic points are marked by crosses in Fig. 6a with the coordinates 39.2° N, 130.8° E; 40.3° N, 130.5° E and 41.6° N, 130.9° E. Thus, the vortex street provides an intrusion of subtropical water towards the southern coast of Russia. The evidence of at least two anticyclones in the AVISO velocity field is confirmed by tracks of two available drifters. Their locations are shown in Fig. 6a by full circles for 1 day before and after the date indicated on the map. Drifter no. 56739 has been trapped by the anticyclone with centre at 39.3° N, 130.1° E and drifter no. 56746 by the anticyclone with centre at 40.8° N, 131.4° E. We have found similar episodes with penetration of subtropical waters far to the north to the coast of Russia through the western gate in different years. Peripheries of mesoscale eddies in the ocean are known to be transport pathways larvae, fish and other marine organisms (see e.g. Cotte et al., 2010; Prants, 2013; Prants et al., 2014c; and references therein). In our case they might be transport for

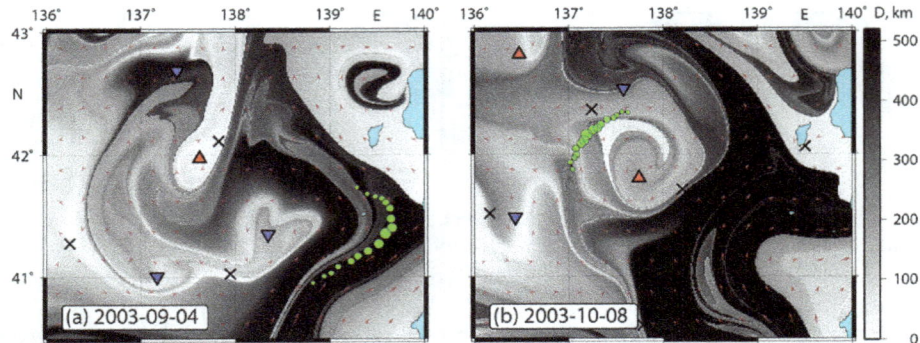

Figure 7. The drift maps in **(a)** September and **(b)** October of 2003 with snapshots of the drifter's track superimposed show how the vortex pair facilitates transport of subtropical tracers to the northwest through the eastern gate.

heat-loving organisms to reach the southern coast of Russia (Ivankova and Samuilov, 1979).

An example of the intrusion of subtropical water through the central gate across the Subpolar Front is shown in Fig. 6b with another kind of Lagrangian map, the so-called backward-in-time drift maps (Prants et al., 2011a, 2014a) computed as part of task 4. The red and green colours in the backward-in-time drift maps code the waters that entered the studied area for 2 years through its southern and northern boundaries, respectively. At the beginning of September 1995 a mesoscale cyclonic eddy to the north of the Subpolar Front with centre at about 41.5° N, 134.4° E "grabbed" some subtropical water at its southern periphery and pulled it to the north. In the course of time the streamer-like intrusion of subtropical tracers reached latitude 42° N moving to the north (Fig. 6b).

The transport of subtropical waters through the eastern gate VII (see Figs. 3 and 4) occurs mainly due to the existence of a quasi-permanent vortex pair labelled AC-C in the mean field in Fig. 1. This provides a propulsion of some subtropical tracers to the northwest whereas most of them, propagating along the eastward frontal jet, join with the Tsushima Warm Current and flow out to the Pacific through the Tsugaru Strait. The maps in the Supplement (Figs. S5 and S6) document a typical situation with a propulsion of subtropical water to the northwest in September–October 2003. The study and analysis of Lagrangian drift maps, computed for the whole observation period, have shown that frontal eddies facilitated the northward transport of subtropical water across the Subpolar Front via the central and eastern gates.

To illustrate how this quasi-permanent vortex pair works we show in Fig. 7 the drift map for tracers distributed over the area and advected for 2 months backward in time starting from the dates indicated. The values of displacements of the tracers, D, in km are shown in greyscale. The black tracers have been displaced for the same time considerably as compared to the white ones. To verify our simulation we show in Fig. 7 positions of drifter no. 35660 by full circles for 2 days before and after the date indicated with their size

increasing in time. The entire track of that drifter, launched on 2 May 2003 at the point 34.925° N, 129.3° E, is shown in Fig. S7 in the Supplement.

At the beginning of September 2003 (Fig. 7a) the vortex pair at the entrance to the gate VII consists of an anticyclone with the centre at about 42° N, 137.7° E and a cyclone at 41.25° N, 138.35° E. The cyclone pulls some subtropical water from the eastward frontal jet round its northern periphery in a streamer-like manner (see the black tongue in Fig. 7a). Then this water is wound by the anticyclone round its southern periphery and propelled northeast. This is confirmed by snapshots of the track of drifter no. 35660 for September–October 2003 (see Fig. S6 in the Supplement). Being at the beginning of September in the main stream (Fig. 7a), it has drifted round the cyclone for the first half of September, then round the anticyclone for the second half of September and at the beginning of October. Eventually drifter no. 35660 crossed the latitude 42° N (Fig. 7b) and moved to the north lugged by modified subtropical waters.

3.3 The effect of possible altimetry errors on statistical features of Lagrangian transport

It has been shown statistically that the average northward component of the AVISO velocity field dictates preferred near-surface transport pathways of subtropical waters in the central JS. The ability of satellite altimetry to accurately measure sea level anomalies has vastly improved over the last decade. However, there are still some measurement errors due to different reasons that lead to errors in the velocity field provided by AVISO.

In this section we discuss the possible effect of errors in the altimetry field on our simulation results. The AVISO velocity field has errors as compared with a "true" velocity field. The difference could be simulated by adding a noise $\Delta(u, v)$ in the velocity data. The question is how reliable are our statistical simulation results based on an imperfect AVISO velocity field? All the simulation results, based on the average AVISO velocity as in Fig. 1, are supposed to be reliable because the errors are averaged out for 22 years. As to other simulation results, they depend on possible noise Δv in the AVISO northward component v_+ which could, in principle, change the results but only if the noise were strong enough to change the direction of the meridional velocity, i.e. if $\Delta v > |v|$. If the average AVISO northward component $\langle v_+ \rangle$ is large enough as in the areas with dominated northward currents, we do not expect that it would be changed there significantly under the influence of noise. So, locations of the preferred transport pathways are not expected to be changed significantly.

If the average AVISO northward component $\langle v_+ \rangle$ is small, then two options are possible.

1. It is small due to domination of a southward current somewhere, i.e. $v_- \gg \Delta v$. It is clear that possible noise has practically no effect on northward transport in this case. For example, the forbidden zone in Fig. 2a to the south off Vladivostok, where northward transport has not been observed during the whole observation period, should be located there at any realistic level of noise because it exists due to the domination of a sufficiently strong southward jet (VMJ in Fig. 1).

2. The average AVISO northward component $\langle v_+ \rangle$ is small due to a smallness of the absolute velocity, i.e. $\sqrt{u^2 + v^2} \sim \Delta v$. In this case northward and southward transports are equalised, and they are small if the noise is small enough. Such a situation is unlikely along the Subpolar Front because of the presence of numerous mesoscale eddies along the front where the absolute velocities are not small.

The influence of possible errors in altimetry-derived velocity field on concrete mesoscale features has been studied by Harrison and Glatzmaier (2012), Hernández-Carrasco et al. (2011) and Keating et al. (2011) by analysing how an additional noise in the advection equations might change Lagrangian coherent structures revealed by the finite-time and finite-size Lyapunov techniques. Strongly attracting and repelling individual Lagrangian coherent structures in the California Current System have been shown to be robust to perturbations of the velocity field of over 20 % of the maximal regional velocity (Harrison and Glatzmaier, 2012). Individual trajectories have been shown to be sensitive to small and moderate noisy variations in the velocity field but statistical characteristics and large-scale structures like mesoscale eddies and jets are not (Cotte et al., 2010; Hernández-Carrasco et al., 2011; Keating et al., 2011).

4 Conclusions

The main results of altimetry-based simulation and analysis of the northward near-surface Lagrangian transport of subtropical water across the Japan Sea frontal zone for the period from 2 January 1993 to 15 June 2015 are the following.

1. A methodology to simulate and analyse Lagrangian large-scale transport in frontal areas is developed (tasks 1–4 in Sect. 2).

2. There are "forbidden" zones in the Japan Sea where the northward transport has not been found during all the observation period (the rectangles in Fig. 2a). The "forbidden" zone to the south of Vladivostok exists due to a quasi-permanent southward jet there (VMJ in Fig. 1). The other "forbidden" zone exists due to the presence of a quasi-permanent topographically constrained anticyclonic eddy with centre at about 41.3° N, 134° E in the deep Japan Basin and the eastward zonal jet blocking northward transport there (AC in Fig. 1).

3. Northward near-surface Lagrangian transport of subtropical water across the Subpolar Front has been statistically shown to be meridionally inhomogeneous with specific gates and barriers in the frontal zone whose locations are determined by the local advection velocity field (the pronounced peaks in Figs. 3 and 4).

4. The transport through the gates has been shown to occur by a portion-like manner, i.e. those gates "open" during specific time intervals (a patchiness in Fig. 5 and Figs. S2 and S3 in the Supplement).

5. The gates "open" due to suitable dispositions of mesoscale frontal eddies facilitating propagation of subtropical waters to the north. It is documented for the western, central and eastern gates with the help of different kinds of Lagrangian maps and validated by some tracks of available drifters (the intrusions of subtropical tracers around the eddies in Figs. 6, 7, and Figs. S5 and S6 in the Supplement). In particular, invasion of tropical and subtropical marine organisms in the northern part of the sea, to the southern coast of Russia, can be explained by the presence of vortex streets at the western gate (Fig. 6).

Competing interests. The authors declare that they have no conflict of interest.

Acknowledgements. This work was supported by the Russian Science Foundation (project no. 16–17–10025). A publication cost is covered, in part, by the Office of Naval Research grant no. N00014-16-1-2492. The altimeter products were distributed by AVISO with support from CNES.

Edited by: A. Turiel

References

Budyansky, M. V., Goryachev, V. A., Kaplunenko, D. D., Lobanov, V. B., Prants, S. V., Sergeev, A. F., Shlyk, N. V., and Uleysky, M. Y.: Role of mesoscale eddies in transport of Fukushima-derived cesium isotopes in the ocean, Deep Sea Res. Part I, 96, 15–27, doi:10.1016/j.dsr.2014.09.007, 2015.

Chang, K.-I., Teague, W., Lyu, S., Perkins, H., Lee, D.-K., Watts, D., Kim, Y.-B., Mitchell, D., Lee, C., and Kim, K.: Circulation and currents in the southwestern East/Japan Sea: Overview and review, Prog. Oceanogr., 61, 105–156, doi:10.1016/j.pocean.2004.06.005, 2004.

Choi, B.-J., Haidvogel, D. B., and Cho, Y.-K.: Nonseasonal sea level variations in the Japan/East Sea from satellite altimeter data, J. Geophys. Res.-Ocean., 109, C12028, doi:10.1029/2004jc002387, 2004.

Cotte, C., d'Ovidio, F., Chaigneau, A., Levy, M., Taupier-Letage, I., Mate, B., and Guinet, C.: Scale-dependent interactions of Mediterranean whales with marine dynamics, Limnol. Oceanogr., 56, 219–232, doi:10.4319/lo.2011.56.1.0219, 2010.

Danchenkov, M., Lobanov, V., Riser, S., Kim, K., Takematsu, M., and Yoon, J.-H.: A History of Physical Oceanographic Research in the Japan/East Sea, Oceanography, 19, 18–31, doi:10.5670/oceanog.2006.41, 2006.

Harrison, C. S. and Glatzmaier, G. A.: Lagrangian coherent structures in the California Current System – sensitivities and limitations, Geophys. Astrophys. Fluid Dynam., 106, 22–44, doi:10.1080/03091929.2010.532793, 2012.

Hernández-Carrasco, I., López, C., Hernández-García, E., and Turiel, A.: How reliable are finite-size Lyapunov exponents for the assessment of ocean dynamics?, Ocean Model., 36, 208–218, doi:10.1016/j.ocemod.2010.12.006, 2011.

Hernández-Carrasco, I., López, C., Orfila, A., and Hernández-García, E.: Lagrangian transport in a microtidal coastal area: the Bay of Palma, island of Mallorca, Spain, Nonlin. Processes Geophys., 20, 921–933, doi:10.5194/npg-20-921-2013, 2013.

Isoda, Y.: Warm eddy movements in the eastern Japan Sea, J. Oceanogr., 50, 1–15, doi:10.1007/bf02233852, 1994.

Ito, M., Morimoto, A., Watanabe, T., Katoh, O., and Takikawa, T.: Tsushima Warm Current paths in the southwestern part of the Japan Sea, Prog. Oceanogr., 121, 83–93, doi:10.1016/j.pocean.2013.10.007, 2014.

Ivankova, V. N. and Samuilov, A. E.: New fish species for the USSR waters and an invasion of heat-loving fauna in the north-western part of the Japan Sea, Voprosy Ihtiologii, 19, 449–550, 1979.

Keating, S. R., Smith, K. S., and Kramer, P. R.: Diagnosing Lateral Mixing in the Upper Ocean with Virtual Tracers: Spatial and Temporal Resolution Dependence, J. Phys. Oceanogr., 41, 1512–1534, doi:10.1175/2011JPO4580.1, 2011.

Kim, T. and Yoon, J.-H.: Seasonal variation of upper layer circulation in the northern part of the East/Japan Sea, Cont. Shelf Res., 30, 1283–1301, doi:10.1016/j.csr.2010.04.006, 2010.

Lee, D.-K. and Niiler, P.: Eddies in the southwestern East/Japan Sea, Deep Sea Res. Part I, 57, 1233–1242, doi:10.1016/j.dsr.2010.06.002, 2010.

Lee, D.-K. and Niiler, P. P.: The energetic surface circulation patterns of the Japan/East Sea, Deep Sea Research Part II, 52, 1547–1563, doi:10.1016/j.dsr2.2003.08.008, 2005.

Lipphardt, B. L., Small, D., Kirwan, A. D., Wiggins, S., Ide, K., Grosch, C. E., and Paduan, J. D.: Synoptic Lagrangian maps: Application to surface transport in Monterey Bay, J. Marine Res., 64, 221–247, doi:10.1357/002224006777606461, 2006.

Mitchell, D. A., Teague, W. J., Wimbush, M., Watts, D. R., and Sutyrin, G. G.: The Dok Cold Eddy, J. Phys. Oceanogr., 35, 273–288, doi:10.1175/jpo-2684.1, 2005.

Nikitin, A. A., Lobanov, V. B., and Danchenkov, M. A.: Possible pathways for transport of warm subtropical waters to the area of the Far Eastern Marine Reserve, Izvestiya TINRO, 131, 41–53, 2002.

Park, K.-A., Chung, J. Y., and Kim, K.: Sea surface temperature fronts in the East (Japan) Sea and temporal variations, Geophys. Res. Lett., 31, L07304, doi:10.1029/2004gl019424, 2004.

Prants, S., Ponomarev, V., Budyansky, M., Uleysky, M., and Fayman, P.: Lagrangian analysis of the vertical structure of eddies simulated in the Japan Basin of the Japan/East Sea, Ocean Model., 86, 128–140, doi:10.1016/j.ocemod.2014.12.010, 2015.

Prants, S. V.: Dynamical systems theory methods to study mixing and transport in the ocean, Phys. Scripta, 87, 038115, doi:10.1088/0031-8949/87/03/038115, 2013.

Prants, S. V.: Chaotic Lagrangian transport and mixing in the ocean, Eur. Phys. J. Spec. Top., 223, 2723–2743, doi:10.1140/epjst/e2014-02288-5, 2014.

Prants, S. V.: Backward-in-time methods to simulate large-scale transport and mixing in the ocean, Phys. Scripta, 90, 074054, doi:10.1088/0031-8949/90/7/074054, 2015.

Prants, S. V., Budyansky, M. V., Ponomarev, V. I., and Uleysky, M. Y.: Lagrangian study of transport and mixing in a mesoscale eddy street, Ocean Model., 38, 114–125, doi:10.1016/j.ocemod.2011.02.008, 2011a.

Prants, S. V., Uleysky, M. Y., and Budyansky, M. V.: Numerical simulation of propagation of radioactive pollution in the ocean from the Fukushima Dai-ichi nuclear power plant, Dokl. Earth Sci., 439, 1179–1182, doi:10.1134/S1028334X11080277, 2011b.

Prants, S. V., Ponomarev, V. I., Budyansky, M. V., Uleysky, M. Y., and Fayman, P. A.: Lagrangian analysis of mixing and transport of water masses in the marine bays, Izvestiya, Atmos. Ocean. Phys., 49, 82–96, doi:10.1134/S0001433813010088, 2013.

Prants, S. V., Budyansky, M. V., and Uleysky, M. Y.: Lagrangian study of surface transport in the Kuroshio Extension area based on simulation of propagation of Fukushima-derived radionuclides, Nonlin. Processes Geophys., 21, 279–289, doi:10.5194/npg-21-279-2014, 2014a.

Prants, S. V., Budyansky, M. V., and Uleysky, M. Y.: Lagrangian fronts in the ocean, Izvestiya, Atmos. Ocean. Phys., 50, 284–291, doi:10.1134/s0001433814030116, 2014b.

Prants, S. V., Budyansky, M. V., and Uleysky, M. Y.: Identifying Lagrangian fronts with favourable fishery conditions, Deep Sea Res. Part I, 90, 27–35, doi:10.1016/j.dsr.2014.04.012, 2014c.

Shin, C.-W.: Characteristics of a Warm Eddy Observed in the Ulleung Basin in July 2005, Ocean Polar Res., 31, 283–296, doi:10.4217/opr.2009.31.4.283, 2009.

Takematsu, M., Ostrovski, A. G., and Nagano, Z.: Observations of Eddies in the Japan Basin Interior, J. Oceanogr., 55, 237–246, doi:10.1023/a:1007846114165, 1999.

Talley, L., Min, D.-H., Lobanov, V., Luchin, V., Ponomarev, V., Salyuk, A., Shcherbina, A., Tishchenko, P., and Zhabin, I.: Japan/East Sea Water Masses and Their Re-lation to the Sea's Circulation, Oceanography, 19, 32–49, doi:10.5670/oceanog.2006.42, 2006.

Uleysky, M. Y., Budyansky, M. V., and Prants, S. V.: Effect of dynamical traps on chaotic transport in a meandering jet flow, Chaos, 17, 043105, doi:10.1063/1.2783258, 2007.

Yoon, J.-H. and Kim, Y.-J.: Review on the seasonal variation of the surface circulation in the Japan/East Sea, J. Marine Syst., 78, 226–236, doi:10.1016/j.jmarsys.2009.03.003, 2009.

7

Non-Gaussian data assimilation of satellite-based leaf area index observations with an individual-based dynamic global vegetation model

Hazuki Arakida[1], Takemasa Miyoshi[1,2,3], Takeshi Ise[4], Shin-ichiro Shima[1,5], and Shunji Kotsuki[1]

[1]RIKEN Advanced Institute for Computational Science, Kobe, 650-0047, Japan
[2]Department of Atmospheric and Oceanic Science, University of Maryland, College Park, MD 20742, USA
[3]Application Laboratory, Japan Agency for Marine-Earth Science and Technology, Yokohama, 236-0001, Japan
[4]Field Science Education and Research Center, Kyoto University, Kyoto, 606-8502, Japan
[5]Graduate School of Simulation Studies, University of Hyogo, Kobe, 650-0047, Japan

Correspondence to: Hazuki Arakida (hazuki.arakida@riken.jp) and Takemasa Miyoshi (takemasa.miyoshi@riken.jp)

Abstract. We developed a data assimilation system based on a particle filter approach with the spatially explicit individual-based dynamic global vegetation model (SEIB-DGVM). We first performed an idealized observing system simulation experiment to evaluate the impact of assimilating the leaf area index (LAI) data every 4 days, simulating the satellite-based LAI. Although we assimilated only LAI as a whole, the tree and grass LAIs were estimated separately with high accuracy. Uncertain model parameters and other state variables were also estimated accurately. Therefore, we extended the experiment to the real world using the real Moderate Resolution Imaging Spectroradiometer (MODIS) LAI data and obtained promising results.

1 Introduction

The terrestrial biosphere is an important part of the Earth system model (ESM) to simulate the carbon and water cycles. However, terrestrial biosphere models tend to have large uncertainties, for example, in phenology (Richardson et al., 2012; Murray-Tortarolo et al., 2013) and in spatial distributions of plant species (Cheaib et al., 2012). Recently, data assimilation (DA) methods which incorporate observation data into models have been applied to terrestrial biosphere models to reduce the uncertainties in the state variables and model parameters (Luo et al., 2011; Peng et al., 2011). Previous studies have successfully applied the ensemble Kalman filter (e.g., Evensen, 2003; Williams et al., 2005; Quaife et al., 2008; Stöckli et al., 2011) or adjoint method (e.g., Kaminski et al., 2013; Kato et al., 2013) to the "static" vegetation models, but studies with the "dynamic" global vegetation models (DGVMs) are still limited (Luo et al., 2011; Peng et al., 2011), although Hartig et al. (2012) pointed out the importance.

The static vegetation models are time independent and do not include the vegetation succession process (Peng, 2000). Alternatively, DGVMs include the vegetation succession process and can simulate carbon and water cycle changes linking to the vegetation shift under the changing climate. Specifically, "individual-based" DGVMs simulate local interactions among individual plants such as competitions for light and water, so that the model can simulate the vegetation succession more explicitly (Smith et al., 2001; Sato et al., 2007). Garreta et al. (2010) pioneered to apply DA to an individual-based DGVM for paleoclimate, but no study has been published thus far to assimilate fine timescale data from satellites and ground stations using an individual-based DGVM. If the initial vegetation structure and the model parameters of an individual-based DGVM are estimated more accurately by assimilating the fine timescale data, the uncertainties of the simulated future vegetation would be greatly reduced.

This study explores the ability to assimilate frequent satellite-based leaf area index (LAI) data with an individual-based DGVM known as the SEIB-DGVM, which stands for the spatially explicit individual-based DGVM (Sato et al., 2007). We developed a non-Gaussian ensemble DA system with the SEIB-DGVM based on a particle filter approach. Although the particle filter is an existing, well-known approach, this is the first attempt to apply it to an individual-based DGVM with frequent LAI data. Therefore, we focus on the methodological development in this study and perform a series of numerical experiments at a single location with only a couple of plant functional types (PFTs) as the first step. It would be numerically straightforward to extend it to the global scale in future studies, since the local-scale experiments can be performed in parallel for different locations. In the present study, we first perform idealized simulation experiments to investigate how well we can estimate the model parameters associated with phenology by assimilating the LAI data every 4 days, simulating the satellite-based LAI product from the Moderate Resolution Imaging Spectroradiometer (MODIS) aboard the Terra and Aqua spacecrafts. We also investigate to what extent assimilating the LAI data could improve the estimates of the state variables such as GPP (gross primary production), RE (ecosystem respiration), NEE (net ecosystem exchange), and biomass, the most fundamental variables for carbon cycle and vegetation states. Sensitivities to the filter settings such as the random perturbation sizes and particle sizes are also investigated. Following the idealized experiments, we perform an experiment using the real MODIS LAI observation data to see how well the proposed approach performs in the real world.

2 Method

2.1 SEIB-DGVM

The SEIB-DGVM simulates establishment, growth, and decay of the individuals of prescribed PFTs within a spatially explicit virtual forest (Sato et al., 2007), forced by climate conditions such as air temperature, soil temperature, cloudiness, precipitation, humidity, and winds. We used version 2.71 (Sato and Ise, 2012) but with minimal modifications for DA. The model simulates daily states, but the original model outputs were only once per year. Outputs are needed for DA once every 4 days; thus, we modified the model code to output the model states every 4 days. In addition, the original model code assumed running for many years continuously, and the initial seed for the random number generator was fixed. As a result, in this study, we stop the model every 4 days, and the same seed is repeated every time when we start the model. Therefore, we modified the model code to randomly generate the seed for the random number generator every time when we initiate the model. Other modifications are summarized in Appendix A.

The size of the model state space is determined by the prognostic variables for tree, grass, forest as a whole, and soil. Each individual tree has 13 prognostic variables such as biomass of root, leaf and trunk, and we assume that up to 300 trees can exist in the forest area. Therefore, the number of tree variables is less than or equal to 3900 (i.e., 300×13). As for grass, the forest area is divided into 30 by 30 grid cells, and each grid cell has four variables such as biomass of root and leaf. Hence, the number of grass variables is fixed at 3600 (i.e., $30 \times 30 \times 4$). In addition, forest as a whole has eight prognostic variables such as snow and soil carbon mass, and finally, soil moisture (one variable) is defined for 30 soil layers. Therefore, the number of state variables is between 3638 (no tree, i.e., $0 + 3600 + 8 + 30$) and 7538 (300 trees, i.e., $3900 + 3600 + 8 + 30$).

Among the various model outputs ranging from individual tree height to soil water content (Sato et al., 2007, with updated information available from the package of version 2.71), we focus on LAI because it is the key to the vegetation model, and because previous studies show a promise in assimilating satellite-based LAI data with a static vegetation model (Stöckli et al., 2011) and a "non-individual-based" DGVM (Demarty et al., 2007). We extend the previous studies to assimilate the LAI data with the individual-based DGVM.

2.2 Particle filter-based DA

Individual-based DGVMs include highly non-linear processes such as occasional establishment and death of individual plants. These processes produce and eliminate state variables, and the phase space changes time to time. DA methods that have been used in geophysical applications usually assume that the state variables are defined uniquely for the given dynamical system and that the phase space dimension stays the same. The widely used ensemble Kalman filter, for example, finds the best linear combination of the ensemble with optimal fit to the observations, but it is not trivial to define a linear combination or even the ensemble mean for the variables missing in some ensemble members. Therefore, it would not be trivial to apply the widely used DA methods to individual-based DGVMs.

Alternatively, particle filters run independent parallel simulations or particles and represent the probability density function (PDF) explicitly by assigning probability to each particle. Therefore, particle filters can handle non-Gaussianity and non-linearity explicitly, and can be applied to the individual-based DGVMs in a straightforward manner (e.g., Garreta et al., 2010) even though the phase space dimension is different for each particle.

Here, we adopt a particle filter approach known as sequential importance resampling (SIR; Fig. 1) (Gordon et al., 1993). Although the method is not efficient for large dimensional systems (e.g., Bickel et al., 2008; Snyder et al., 2008, 2015; Snyder, 2012), we tested this well-known method as

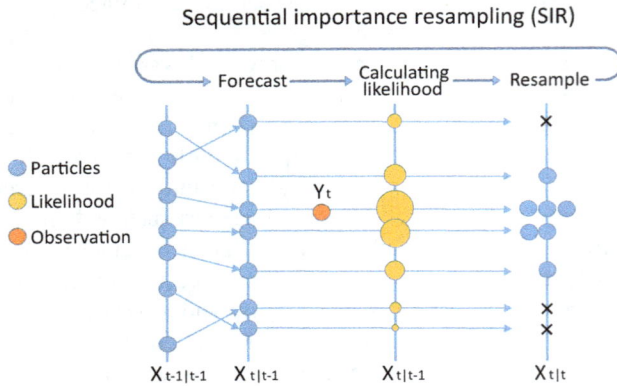

Figure 1. Schematic showing the SIR particle filter method. The size of the circles corresponds to the assigned probability.

the first attempt to construct the DA system with SEIB-DGVM. First, n parallel simulations are performed, and each simulation is considered as a particle representing the true state of the system with equal probability. Next, likelihood $l_t^{(i)}$ is calculated for each particle using the Gaussian likelihood function:

$$l_t^{(i)} = p\left(y_t | x_{t|t-1}^{(i)}\right) = \frac{1}{\sqrt{2\pi \cdot \sigma^2}} \exp\left\{-\frac{(y_t - x_{t|t-1}^{(i)})^2}{2 \cdot \sigma^2}\right\} \quad (1)$$

for $i = 1, \ldots, n$.

Here, $x_{t|t-1}^{(i)}$ denotes the simulated LAI of the ith particle at time t from the previous time step $t-1$, y_t the observed LAI at time t, and σ the observation error standard deviation. Since the prior probability is uniform, Bayes' rule gives that the posterior probability of the ith particle is proportional to $l_t^{(i)}$, i.e., the particles closer to the observation have more probability. Next, we resample the particles, so that each particle has equal probability. The particles with more probability (larger $l_t^{(i)}$) are duplicated, and the particles with less probability (smaller $l_t^{(i)}$) are removed. If n is sufficiently large, we can evaluate the posterior PDF accurately. Each resampled particle represents the true state of the system with equal probability and acts as the initial particle for the next time step. This Bayesian framework is repeated.

2.3 OSSE and the real-world experiment

We first perform a series of idealized observing system simulation experiments (OSSEs). The OSSE (e.g., Atlas, 1997) is a widely used approach in meteorological DA to test the general performance of a DA system and to evaluate the impact of specific observing systems. OSSE has the nature run, which is usually generated by running a simulation for a certain period. Observation data are simulated from the nature run by applying the observation operator, i.e., converting the

model variables to the observed variables. Here, we add artificial random noise to simulate the observation error. DA experiments are initiated from the state independent of the nature run, and the simulated observations are assimilated. The resulting analyses and subsequent forecasts are compared with the nature run to evaluate the performance of DA. Once an OSSE is done, it is straightforward to extend the OSSE to the real world by simply replacing the simulated observations with the real-world observations.

3 OSSE

3.1 Experimental design

To generate the nature run, the SEIB-DGVM was initialized with the bare ground (i.e., no plant at the beginning) and was run for 107 years using the climate forcing data from 2001 to 2010 available at the SEIB-DGVM web page (http://seib-dgvm.com/). Here, the 10-year forcing data are repeated for the 107-year simulation, and the last 7 years from years 101 to 107 use the actual climate forcing of 2001 to 2007; thus, we refer to years 101 to 107 as 2001 to 2007. The daily climate data were generated by the procedure of Sato and Ise (2012) with updated information available at the SEIB-DGVM web page, based on the monthly Climate Research Unit observation-based data (CRU-TS3.22 0.5° monthly climate time series) (Harris et al., 2014) and the daily data from the National Centers for Environmental Prediction (NCEP)/National Center for Atmospheric Research (NCAR) reanalysis (Kalnay et al., 1996). We chose the study area at one of the AsiaFlux sites, the Siberia Yakutsk larch forest site at Spasskaya Pad, the middle basin of the Lena River ($62°15'18''$ N, $129°14'29''$ E). The observed climate data at this site were not directly used in this study, but these data may have been included in the NCEP/NCAR reanalysis. Field-observed carbon flux data are available as the ground truth to verify the DA results at this site. Forced by the climate data, the SEIB-DGVM simulates the vegetation shifts from the bare ground to a grassland, and then to a forest. The two PFTs, the boreal deciduous needleleaf trees and C3 grass, are the dominant PFTs in this study area. Therefore, we do not consider the other PFTs in this study following Sato et al. (2010). We call these two PFTs simply "tree" and "grass".

The nature run (Fig. 2a) was performed with the "true" parameter values Pmax = 15 µmolCO$_2$ m^{-2} s^{-1} and Dor = 230 DOY (day of year) for tree and Pmax = 9 µmolCO$_2$ m^{-2} s^{-1} and Dor = 270 DOY for grass, where Pmax and Dor stand for the maximum photosynthesis rate and the start date of the dormancy, respectively (Fig. 2b). Hereafter, we omit the units for Pmax (µmolCO$_2$ m^{-2} s^{-1}) and Dor (DOY) for simplicity. The LAI observations for the last 4 years from 2004 to 2007 were created by adding independent Gaussian random noise to the LAI values from the nature run (Fig. 2a)

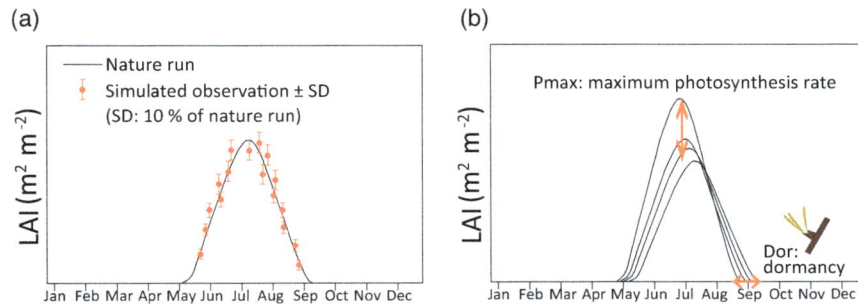

Figure 2. Schematic illustrations of the nature run, observations, and model parameter sensitivities. **(a)** Time series of LAI ($m^2\ m^{-2}$) for the nature run (black), simulated observations (red dots), and their error standard deviations (SD, red error bars). **(b)** Time series of LAI with different Pmax and Dor values. The perturbed parameters (Pmax and Dor for tree and grass) cause differences between the particles.

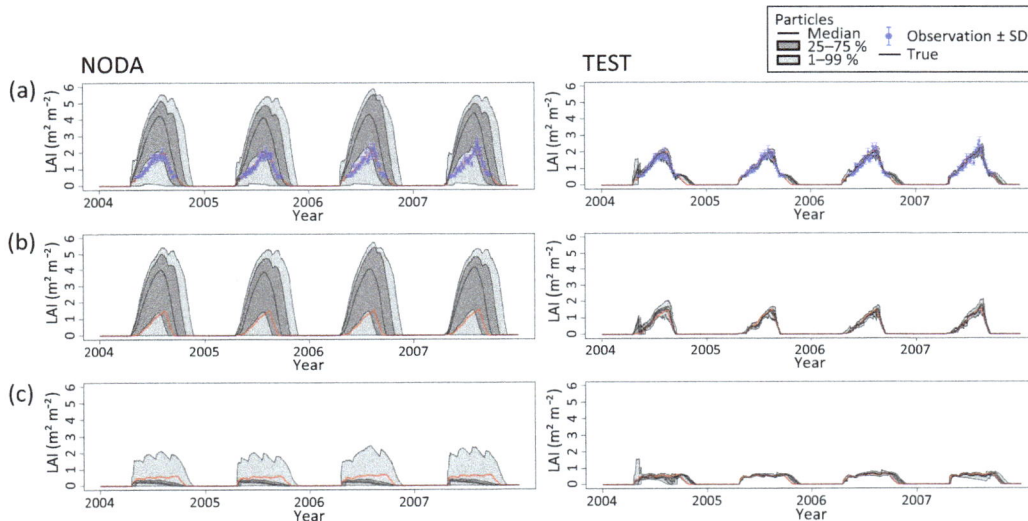

Figure 3. Time series of LAI for **(a)** tree plus grass, **(b)** tree, and **(c)** grass for an experiment without DA (NODA, left) and an experiment with DA (TEST, right). Dark and light gray areas indicate the quartiles and 1–99 % quantiles of the particles as shown in the legend. Thick black curves indicate the medians. Blue dots with error bars indicate the observations and their error standard deviations, and red lines indicate the nature run.

every 4 days, simulating the MODIS LAI product. Here, the observation error standard deviation was given by 10 % of the nature run LAI value. The observed LAI values of less than 0.5 were not used for DA because the MODIS data for the real-world experiment did not include LAI values of less than 0.5. There are too few data with real MODIS LAI values of less than 0.5, and we assign the missing value in preprocessing. Since the LAI is observed only for values of 0.5 or larger, the LAI observation exists only in the summer season.

Next, 8000 particles (parallel simulations) were generated with uniformly perturbed parameters: Pmax = [0, 60] for tree, Pmax = [0, 15] for grass, and Dor = [200, 300] for both. Here, [a, b] denotes random draws from the uniform distribution between a and b. These initial perturbation sizes are based on the previous studies (Kolari et., 2006; Zeng et al., 2011; Zhao et al., 2015; Takagi et al., 2015). We ran 8000

parallel simulations for 103 years for spin-up from the bare ground using the same climate forcing data as the nature run. In the course of the vegetation succession, these randomly perturbed parameter sets result in a variety of LAI simulations (Fig. 2b).

The 8000 particles at the end of the 103-year spin-up runs are used as the initial conditions for DA. The simulated LAI observations are assimilated every 4 days. The nature run and particle filter use the same climate forcing data, so that the difference comes from the model parameter values. The particles continue to be the free runs until the first LAI observation is assimilated in the summer season. The state variables and model parameters are estimated together at DA, and the model systematic errors associated with the model parameters are corrected by DA with parameter estimation. No explicit bias correction is applied. To avoid the exact

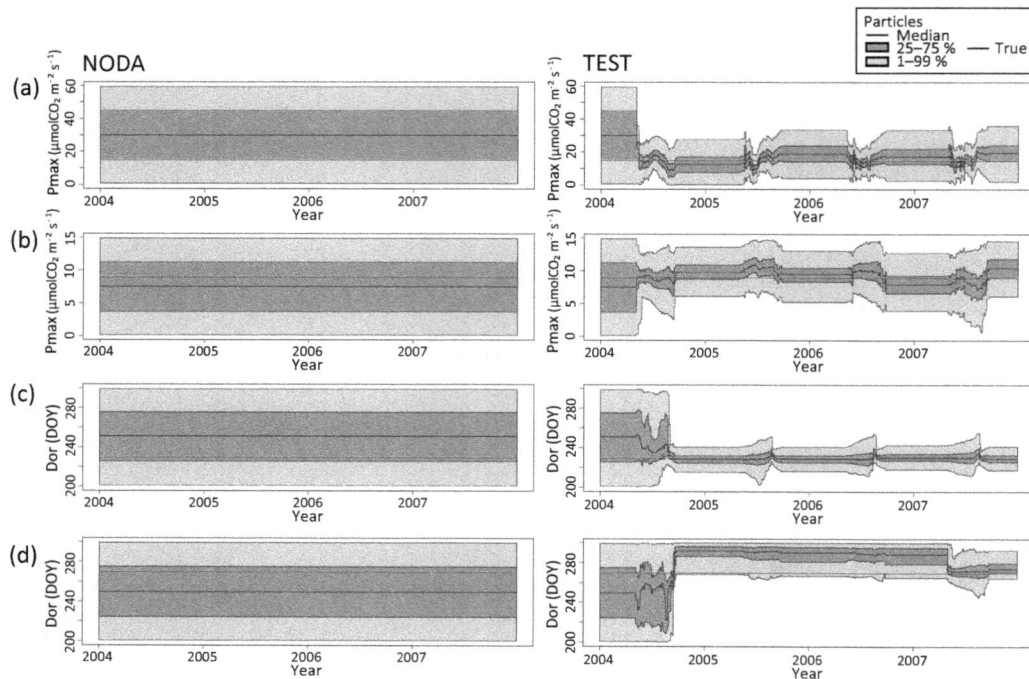

Figure 4. Similar to Fig. 3 but for the model parameters: (**a**) Pmax for tree, (**b**) Pmax for grass, (**c**) Dor for tree, and (**d**) Dor for grass. There is no observation for these parameters.

duplications after resampling, the model parameters Pmax and Dor are randomly perturbed for the duplicated particles. The random perturbations avoid particle degeneracy, which usually causes filter divergence. After some tuning, we found proper perturbation sizes that work for stable filtering without causing particle degeneracy, especially for biomass which is found to be the most sensitive to the perturbation sizes. Here, random draws [−4, 4] are added to Pmax for tree and to Dor for both tree and grass, and [−1, 1] are added to Pmax for grass because the initial Pmax perturbation size for grass is a quarter of that of tree. The sensitivity to the resampling perturbation sizes will be discussed in the next section. In the case that these perturbed parameters exceed the corresponding initial parameter range, the excess value was bounced back from the limits. To assess the impact of DA, we also perform an experiment without DA ("NODA" hereafter) and compare it to the experiment with DA ("TEST" hereafter).

3.2   Results

Figure 3 shows the time series of LAI for NODA (left) and TEST (right). The observations (Fig. 3a, blue dots with error bars) cannot distinguish the tree and grass, but the model simulates LAIs for tree and grass separately (Fig. 3b, c). Although the particles without DA are widely spread (left, gray areas), DA makes the particles much narrower (right) and consistent with the nature run (red curves). With DA, the me-

dian of the particles for tree is almost identical to the nature run for the entire 4 years (Fig. 3b, right). As for grass, the median of the particles is also very close to the nature run with DA, but in the first 3 years the dormancy period is delayed (Fig. 3c, right).

The model parameters are estimated accurately (Fig. 4). There is no direct observation of these parameters, so that the estimations are purely due to DA of the LAI observations. Although the particles of the NODA experiment are uniformly distributed (Fig. 4, left), DA makes the particles close to the true parameters (Fig. 4, right). Since we assimilated the LAI only when it was 0.5 or larger, DA has an impact only in the summer season when the leaves grow. It takes 1–4 years until the true values fall within the quartiles of the particles. The Pmax estimates for both tree and grass show occasional jumps but tend to stay around the true values (Fig. 4a, b). Dor for tree seems the most accurate and stable after the dormancy period of the first year (Fig. 4c). Dor for grass takes the longest; the estimation is not accurate until the dormancy of the fourth year (Fig. 4d). This may be related to the previous results showing the erroneous estimates of the grass LAI near the dormancy period in the first 3 years (Fig. 3c). The systematic errors in NODA come from the uncertain parameter settings. TEST can estimate the parameters through DA and can reduce the systematic errors. This is different from the bias-correction strategy of the first guess.

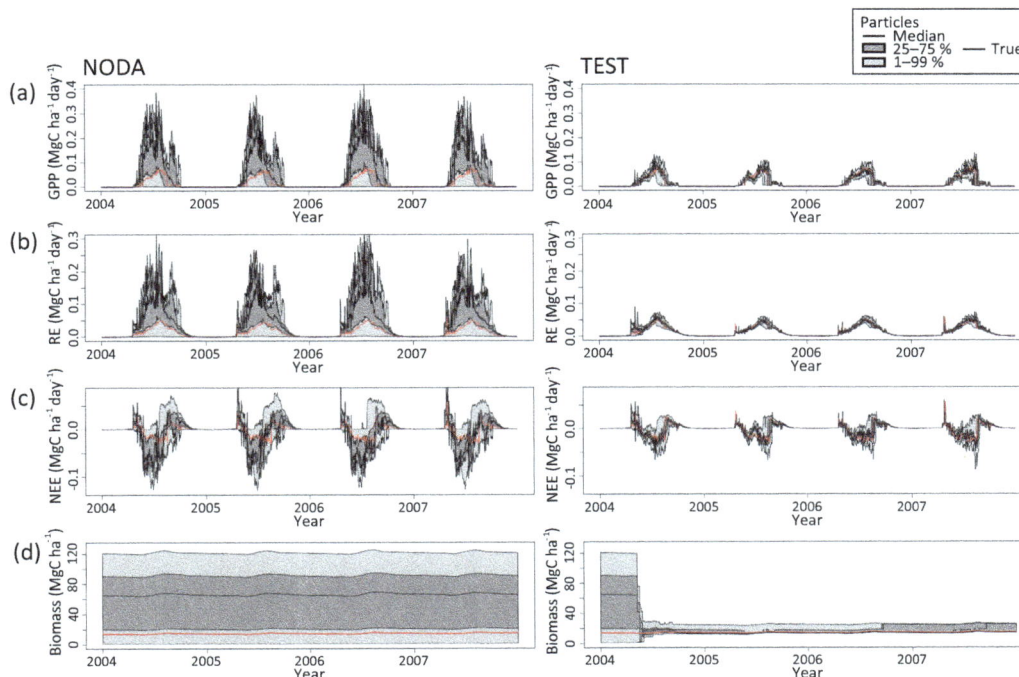

Figure 5. Similar to Figs. 3 and 4 but for unobserved model variables: **(a)** GPP, **(b)** RE, **(c)** NEE, and **(d)** biomass.

Table 1. Parameter settings for TEST, OSSE2 and OSSE3.

OSSEs	Pmax for tree	Pmax for grass	Dor for tree	Dor for grass
TEST	15	9	230	270
OSSE2	20	12	220	260
OSSE3	25	7	210	280

Other model variables such as GPP, RE, NEE, and biomass show large improvements (Fig. 5). Although the particles of the NODA experiment are widely spread, DA with only LAI observations greatly reduces the uncertainties for the four variables, and the estimations are generally reasonable.

4 Sensitivity experiments for OSSE

4.1 Sensitivity to the nature run

To investigate the sensitivity to the choice of the nature run, we performed two additional OSSEs, which we call "OSSE2" and "OSSE3", by generating different nature runs with different parameter sets (Table 1). The random numbers for the observation errors are also different. The other settings follow the TEST experiment.

The results show that both OSSE2 and OSSE3 perform well in general. Namely, the LAI and parameters are es-

timated generally well (Fig. 6). We find the main difference between OSSE2 and OSSE3 in the parameters for grass (Fig. 6c, e). OSSE3 shows significantly larger uncertainties for the parameters for grass. In OSSE2, the Pmax value for grass is larger and produces more grass LAI. Since grass starts to grow earlier and stays longer than tree, it is critical to have LAI observations near the emerging and falling periods for estimating the grass parameters. Due to the larger Pmax value for grass in OSSE2, LAI can be observed with the observing threshold of LAI of 0.5 near the emerging and falling periods. By contrast, in OSSE3, the Pmax value for grass is smaller, and the small grass LAI of less than 0.5 cannot be observed. We can see this in the LAI time series (Fig. 6a, right) near the tails in the spring and fall seasons every year. The uncertainties of LAI are not reduced year by year, corresponding to the large uncertainties of the grass parameters. In the summer, LAI becomes larger mostly due to trees, so that the tree parameters can be estimated well.

4.2 Sensitivity to the initial perturbation size

Here, we investigate the sensitivity to the initial perturbation sizes with particle sizes ranging from 1000 to 16 000. Table 2 shows the three initial perturbation settings: small, moderate, and large. For the TEST experiment, the moderate initial perturbation sizes were used. We perform additional sensitivity experiments with the small and large initial perturbation sizes. Except for the initial perturbation sizes and the particle size, the experiments follow the TEST experiment.

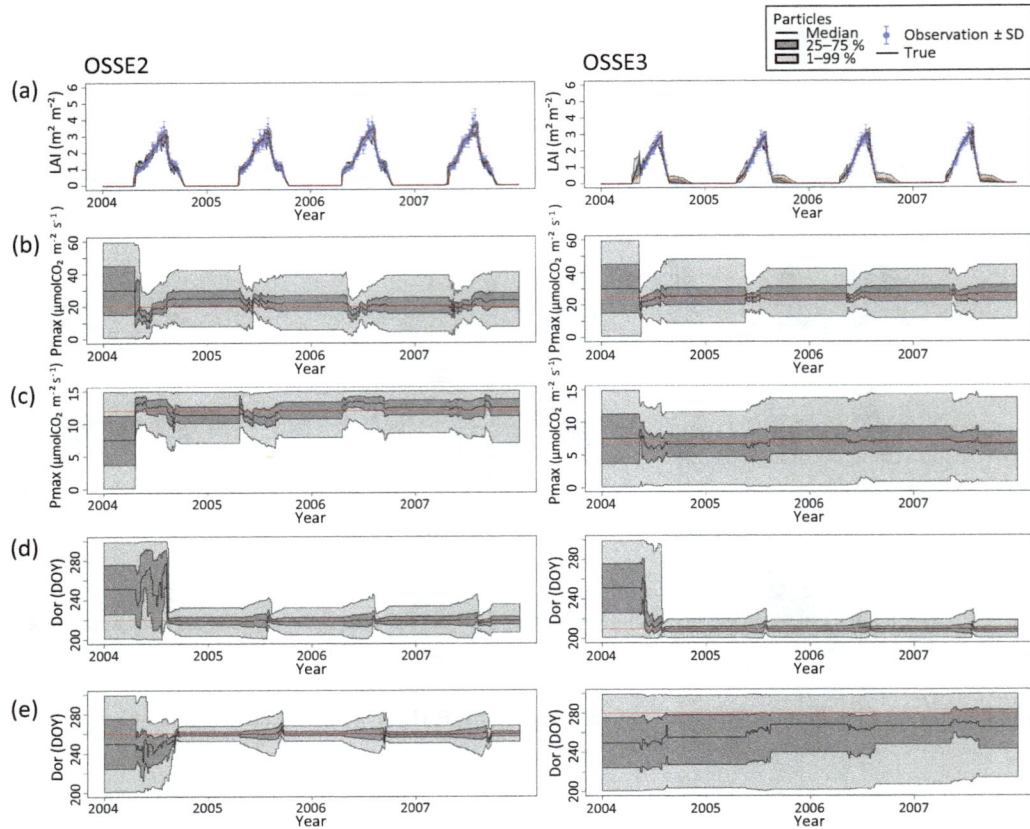

Figure 6. Similar to Figs. 3 and 4 but for OSSE2 (left) and OSSE3 (right). **(a)** Time series of LAI for tree plus grass, **(b)** Pmax for tree, **(c)** Pmax for grass, **(d)** Dor for tree, and **(e)** Dor for grass.

Table 2. Initial perturbation settings.

Initial perturbation sizes	Pmax for tree	Pmax for grass	Dor for tree	Dor for grass
Small	[0, 20]	[0, 10]	[200, 250]	[250, 300]
Moderate	[0, 60]	[0, 15]	[200, 300]	[200, 300]
Large	[0, 120]	[0, 30]	[150, 350]	[150, 350]

Table 3 shows the mean absolute errors (MAEs) and the widths of the 1–99 % quantiles, respectively, averaged over a year in 2007. We consider that the filter diverges when the MAE is larger than the half width of the 1–99 % quantiles, as shown by the italic font in the tables. The results show that the filter diverges for biomass in 10 out of 15 experiments. The five experiments that do not diverge are (4000; small), (8000; small), (16 000; small), (8000; moderate) = TEST, and (16 000; moderate), where (;) denotes (particle size; initial perturbation sizes). The (1000; large) experiment causes filter divergence for most variables and parameters. The (2000; large) experiment shows filter diver-

gence for Dor for grass in addition to biomass. Sampling a wider interval with a smaller particle size generally reduces the particle density, or the effective number of the particles, so that the results seem to be reasonable.

4.3 Sensitivity to the resampling perturbation size

Here, we investigate the sensitivity to the resampling perturbation sizes with particle sizes ranging from 500 to 16 000, in a similar way as the previous subsection. Resampling perturbations add random perturbations to Pmax and Dor when resampling and avoid particle degeneracy. Table 4 shows the three resampling perturbation settings: small, moderate, and large. For the TEST experiment, the moderate resampling perturbation sizes were used.

Table 5 shows similar tables as Table 3 but for the sensitivity to the resampling perturbation sizes. We use the similar notation of (;) denoting (particle size; resampling perturbation setting). The results show that the filter diverges for biomass in 13 out of 18 experiments. The five experiments that do not diverge are (4000; large), (8000; moderate) = TEST, (8000; large), (16 000; moderate), and (16 000; large). The (500; small) experiment is the most unstable, with

Table 3. Results for the sensitivity experiments on the initial perturbation size: (a) mean absolute error (MAE) and (b) the widths of the 1–99 % quantiles, averaged over a year in 2007. Italic font shows the filter divergence. Bold letters show the TEST experiment (8000 particles with moderate initial perturbations).

(a)

Particle sizes	Initial perturbation sizes	Pmax for tree	Pmax for grass	Dor for tree	Dor for grass	LAI (m² m⁻²)	Biomass (MgC ha⁻¹)	GPP (MgC ha⁻¹ day⁻¹)	RE (MgC ha⁻¹ day⁻¹)	NEE (MgC ha⁻¹ day⁻¹)
1000	Small	0.76	0.75	0.43	3.68	0.03	2.46	0.003	0.001	0.002
	Moderate	3.49	1.59	0.64	0.85	0.02	4.75	0.003	0.001	0.003
	Large	*9.18*	*0.59*	*33.41*	*35.70*	*0.10*	*0.21*	*0.007*	*0.002*	*0.006*
2000	Small	0.66	0.52	0.51	3.83	0.03	1.36	0.003	0.001	0.002
	Moderate	3.49	2.05	0.71	3.11	0.04	3.29	0.003	0.001	0.002
	Large	3.30	1.21	1.01	59.52	0.10	9.22	0.003	0.002	0.003
4000	Small	0.97	0.26	0.79	2.38	0.03	1.16	0.002	0.001	0.002
	Moderate	3.55	1.20	0.63	3.02	0.03	2.80	0.003	0.002	0.003
	Large	3.46	0.69	1.23	2.46	0.03	0.39	0.003	0.002	0.003
8000	Small	0.69	0.30	0.78	2.30	0.03	0.24	0.002	0.001	0.002
	Moderate	**2.88**	**1.10**	**0.73**	7.83	**0.03**	**3.72**	**0.003**	**0.001**	**0.003**
	Large	*3.36*	*0.99*	*1.18*	*3.20*	*0.04*	*1.60*	*0.003*	*0.001*	*0.003*
16000	Small	2.92	1.01	0.72	6.91	0.03	0.80	0.003	0.001	0.003
	Moderate	3.22	1.13	0.65	2.16	0.03	3.08	0.003	0.001	0.003
	Large	3.02	1.03	0.39	6.26	0.03	1.72	0.003	0.001	0.002

(b)

Particle sizes	Initial perturbation sizes	Pmax for tree	Pmax for grass	Dor for tree	Dor for grass	LAI (m² m⁻²)	Biomass (MgC ha⁻¹)	GPP (MgC ha⁻¹ day⁻¹)	RE (MgC ha⁻¹ day⁻¹)	NEE (MgC ha⁻¹ day⁻¹)
1000	Small	14.93	5.10	23.44	28.26	0.17	4.63	0.013	0.005	0.012
	Moderate	24.18	8.02	23.60	26.80	0.18	0.42	0.017	0.006	0.014
	Large	*17.92*	*7.47*	*19.61*	*26.97*	*0.11*	*0.29*	*0.010*	*0.005*	*0.009*
2000	Small	16.68	5.36	27.77	30.34	0.18	0.93	0.014	0.005	0.012
	Moderate	24.08	9.05	26.24	29.50	0.20	0.63	0.018	0.007	0.015
	Large	27.30	9.23	25.97	54.89	0.24	0.62	0.017	0.006	0.014
4000	Small	15.72	4.60	24.18	28.76	0.14	4.89	0.012	0.005	0.011
	Moderate	27.11	8.62	27.73	28.68	0.20	0.70	0.018	0.007	0.015
	Large	27.07	8.12	27.59	29.29	0.19	0.60	0.016	0.006	0.014
8000	Small	16.02	4.50	25.38	30.14	0.15	7.27	0.012	0.005	0.011
	Moderate	**28.32**	**9.29**	**26.23**	**33.99**	**0.20**	**11.40**	**0.017**	**0.008**	**0.015**
	Large	27.47	9.37	26.60	31.75	0.21	0.71	0.017	0.007	0.014
16000	Small	27.66	9.28	27.18	48.79	0.22	8.44	0.017	0.007	0.015
	Moderate	28.47	8.85	27.86	31.91	0.21	6.53	0.017	0.008	0.015
	Large	28.76	8.93	25.77	47.88	0.21	2.08	0.017	0.006	0.014

Table 4. Resampling perturbation settings.

Resampling perturbation sizes	Pmax for tree	Pmax for grass	Dor for tree	Dor for grass
Small	$[-2, 2]$	$[-0.5, 0.5]$	$[-2, 2]$	$[-2, 2]$
Moderate	$[-4, 4]$	$[-1, 1]$	$[-4, 4]$	$[-4, 4]$
Large	$[-8, 8]$	$[-2, 2]$	$[-8, 8]$	$[-8, 8]$

more variables and parameters showing filter divergence. Resampling perturbations act as variance inflation in the ensemble filters (e.g., Anderson and Anderson, 1999). It is known that variance inflation generally stabilizes the filter, and the results obtained here seem to be consistent. With 4000 particles or more, the parameters and state variables except for biomass were estimated accurately, although the filter collapsed for biomass with smaller perturbations even with large particle sizes.

5 Real-world experiment

5.1 Experimental settings

Here, the OSSE is extended to the real world by replacing the simulated observations with the real observations. The sensitivity results in the previous section showed that the settings used for the TEST experiment provided stable filter performance; therefore, we follow the TEST experiment here with the moderate initial and resampling perturbation sizes and with 8000 particles.

Since the OSSE used the actual climate forcing in 2004 to 2007, we used the quality-controlled MODIS LAI product of MCD15A3 for those years with flagged as "good quality", "Terra or Aqua", "detectors apparently fine for up to 50 %", "significant clouds not present", and "main method used with or without saturation". We took the median of the LAI observations in the 10 km radius from the study site (62°15′18″ N, 129°14′29″ E). There are a number of missing data in the quality-controlled MODIS data. Therefore, if the number of the data in the 10 km radius is less than 300, we set these data as the missing data for DA. Since the MODIS data resolution is 1 km, the 10 km radius area contains about 314 data. The observation error standard deviations are assigned to each LAI datum in the original MODIS product (Knyazikhin et al., 1999). We rely on the estimate of the observation error standard deviations and take the median of the error standard deviations in the same way as getting the LAI data. The observation error standard deviation is used in the particle filter when computing the likelihood function (Eq. 1).

The model-simulated NEE was validated with the field observation data at this AsiaFlux site (Ohta et al., 2001, 2008, 2014). The data were quality controlled by the steady-state test as indicated by the quality flag 0. Although the model

simulates daily-average NEE, the field observation data represent instantaneous NEE every 30 min. The observation data are missing frequently, and it is not trivial to derive daily averages. Therefore, the raw data are compared with the DA results directly. This allows only a rough verification about whether or not the simulated NEE is in a reasonable range, but this is the only possible verification with an independent source.

5.2 Results

Figures 7, 8, and 9 show similar time series to Figs. 3, 4, and 5, respectively, but with the real MODIS LAI observations. Although the particles of the NODA experiments are widely spread, DA makes the particles much narrower (right) for all variables and parameters. With DA, the median of LAI is very close to the observations, within the range of the observation error standard deviations (Fig. 7a). The grass and tree LAIs are estimated separately (Fig. 7b, c), but there is no direct observation or other verification truth to compare with. This is similar to the model parameters (Fig. 8) and other model variables (Fig. 9) except for NEE, for which direct field observation data are available. As in the OSSE results, the range of uncertainties for NEE is reduced significantly by DA (Fig. 9c). Since the field observations are made instantaneously every 30 min, the observation values (red) appear to have a wider range. However, the SEIB-DGVM simulates only daily-average NEE, and it is not straightforward to compare the outputs from SEIB-DGVM with the field observations. We still find that the median of NEE becomes closer to the observations, particularly near the dormancy period. The simulated NEE generally stays within the reasonable range compared with the field observations. In general, the particle filter shows promising results with the real MODIS LAI data.

6 Conclusion

We assimilated the satellite-based MODIS LAI data using a non-Gaussian ensemble DA system with the SEIB-DGVM based on the SIR particle filter approach. To the best of the authors' knowledge, this is the first study to assimilate the fine timescale satellite data with an individual-based DGVM. We found that DA performed generally well both for the OSSE and real-world experiments. Although we assimilated only LAI as a whole, the tree and grass LAIs were estimated separately. This suggests that the satellite-based DA reduced the uncertainties in the initial vegetation structure of the individual-based DGVM toward the simulation of future vegetation change. Another notable result includes that the model parameters of the individual-based DGVM were estimated successfully and that the uncertainties in the unobserved model variables relevant to carbon cycle and vegetation states were also reduced significantly. Similarly to the previous studies with a static vegetation model (Stöckli et

Table 5. Similar to Table 3 but for the sensitivity experiments on the resampling perturbation sizes.

(a)

Particle sizes	Resampling perturbation sizes	Pmax for tree	Pmax for grass	Dor for tree	Dor for grass	LAI ($m^2\ m^{-2}$)	Biomass ($MgC\ ha^{-1}$)	GPP ($MgC\ ha^{-1}\ day^{-1}$)	RE ($MgC\ ha^{-1}\ day^{-1}$)	NEE ($MgC\ ha^{-1}\ day^{-1}$)
500	Small	2.59	1.68	1.74	15.74	0.08	1.47	0.003	0.001	0.002
	Moderate	2.28	0.66	1.34	5.57	0.03	1.43	0.003	0.001	0.003
	Large	10.13	1.25	2.55	5.29	0.04	0.35	0.004	0.002	0.003
1000	Small	0.80	0.58	1.09	1.41	0.02	4.62	0.002	0.002	0.003
	Moderate	3.49	1.59	0.64	0.85	0.02	4.75	0.003	0.001	0.003
	Large	5.00	2.17	2.78	8.56	0.04	4.78	0.004	0.002	0.004
2000	Small	1.84	0.42	0.47	7.94	0.02	0.59	0.002	0.001	0.002
	Moderate	3.49	2.05	0.71	3.11	0.04	3.29	0.003	0.001	0.002
	Large	6.84	1.26	2.60	4.26	0.05	0.79	0.004	0.002	0.003
4000	Small	1.41	0.67	0.56	0.36	0.02	2.94	0.002	0.002	0.002
	Moderate	3.55	1.20	0.63	3.02	0.03	2.80	0.003	0.002	0.003
	Large	6.74	1.17	1.66	6.94	0.05	2.15	0.004	0.002	0.003
8000	Small	1.23	0.37	0.61	0.71	0.02	3.20	0.002	0.001	0.002
	Moderate	**2.88**	**1.10**	**0.73**	**7.83**	**0.03**	**3.72**	**0.003**	**0.001**	**0.003**
	Large	7.78	1.56	1.26	7.46	0.05	3.44	0.004	0.002	0.003
16000	Small	1.46	0.45	0.77	1.70	0.02	3.15	0.002	0.001	0.002
	Moderate	3.22	1.13	0.65	2.16	0.03	3.08	0.003	0.001	0.003
	Large	5.67	1.71	1.50	6.07	0.05	1.36	0.004	0.002	0.003

(b)

Particle sizes	Resampling perturbation sizes	Pmax for tree	Pmax for grass	Dor for tree	Dor for grass	LAI ($m^2\ m^{-2}$)	Biomass ($MgC\ ha^{-1}$)	GPP ($MgC\ ha^{-1}\ day^{-1}$)	RE ($MgC\ ha^{-1}\ day^{-1}$)	NEE ($MgC\ ha^{-1}\ day^{-1}$)
500	Small	12.41	3.98	11.29	18.21	0.13	0.37	0.012	0.004	0.010
	Moderate	25.82	7.47	23.68	31.83	0.18	0.44	0.015	0.006	0.013
	Large	47.51	11.38	52.32	40.18	0.24	0.70	0.022	0.008	0.018
1000	Small	14.31	4.17	13.85	16.57	0.14	0.46	0.010	0.004	0.009
	Moderate	24.18	8.02	23.60	26.80	0.18	0.42	0.017	0.006	0.014
	Large	44.68	11.57	53.50	45.94	0.24	0.60	0.022	0.009	0.018
2000	Small	14.10	4.40	12.69	31.61	0.16	0.46	0.010	0.004	0.009
	Moderate	24.08	9.05	26.24	29.50	0.20	0.63	0.018	0.007	0.015
	Large	46.88	12.23	49.80	46.06	0.25	0.82	0.023	0.009	0.018
4000	Small	14.27	5.03	13.62	15.53	0.14	0.51	0.011	0.004	0.009
	Moderate	27.11	8.62	27.73	28.68	0.20	0.70	0.018	0.007	0.015
	Large	43.99	12.93	50.11	45.44	0.25	6.28	0.023	0.010	0.019
8000	Small	15.30	5.07	13.78	16.72	0.15	0.67	0.011	0.004	0.009
	Moderate	**28.32**	**9.29**	**26.23**	**33.99**	**0.20**	**11.4**	**0.017**	**0.008**	**0.015**
	Large	45.92	12.70	53.92	44.86	0.25	8.39	0.024	0.010	0.020
16000	Small	15.36	5.33	13.70	23.18	0.15	2.13	0.011	0.004	0.010
	Moderate	28.47	8.85	27.86	31.91	0.21	6.53	0.017	0.008	0.015
	Large	44.05	12.27	52.21	46.10	0.25	7.76	0.024	0.010	0.019

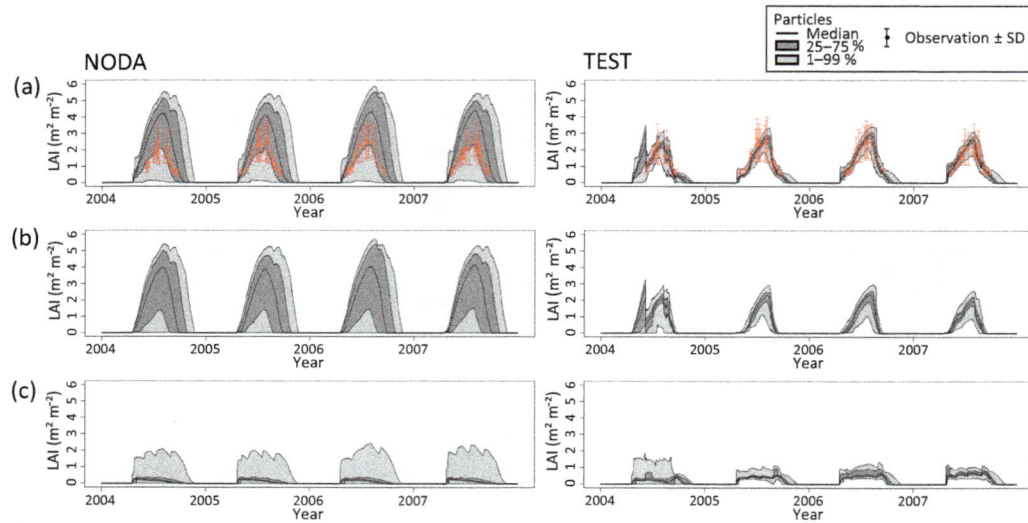

Figure 7. Similar to Fig. 3, showing LAI for **(a)** tree plus grass, **(b)** tree, and **(c)** grass but for the real-world experiment. Red dots with error bars indicate the observations and their error standard deviations.

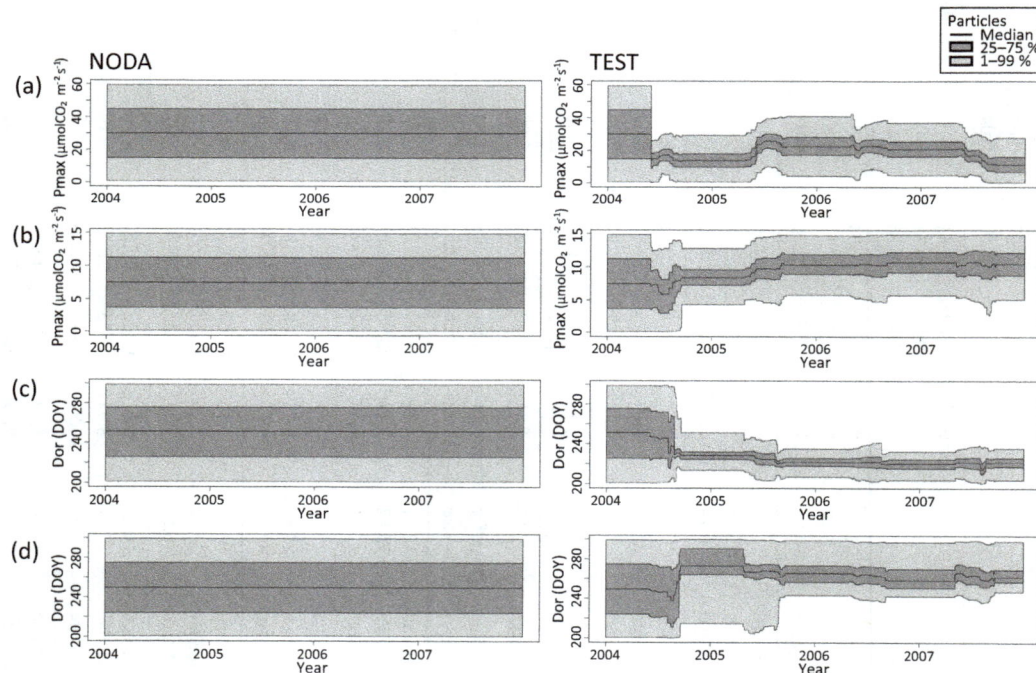

Figure 8. Similar to Fig. 4, showing the model parameters: **(a)** Pmax for tree, **(b)** Pmax for grass, **(c)** Dor for tree, and **(d)** Dor for grass but for the real-world experiment.

al., 2011) and a non-individual-based DGVM (Demarty et al., 2007), the results in the present study also suggest that LAI is the key to DA for phenology and carbon dynamics.

Generally, particle filters do not work well in high-dimensional problems (e.g., Bickel et al., 2008; Snyder et al., 2008, 2015; Snyder, 2012). The SEIB-DGVM has several thousand state variables, but we applied random pertur-

bations to only four model parameters in the particle filter. The four model parameters, i.e., Pmax and Dor for tree and grass, control the leaf season and photosynthesis rate of the forest as a whole. Therefore, the effective degrees of freedom of the estimation problem would be substantially lower than the number of variables of the SEIB-DGVM. This may be why the particle filter worked well in this study.

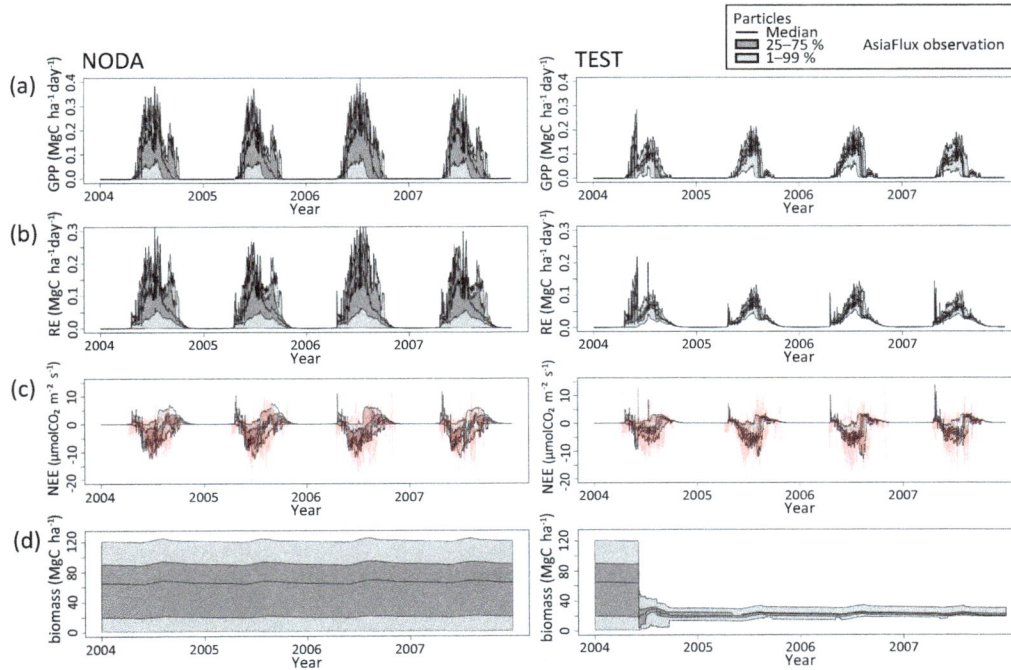

Figure 9. Similar to Fig. 5, showing the unobserved model variables: **(a)** GPP, **(b)** RE, **(c)** NEE, and **(d)** biomass but for the real-world experiment. Red lines indicate the direct field observations made instantaneously every 30 min at the AsiaFlux site, while the model simulates only the daily averages.

Additional sensitivity experiments revealed general robustness but some sensitivities to the nature run, initial and resampling perturbation sizes, and particle size, particularly for biomass, which tends to show particle degeneracy. When resampling, the random perturbations were applied only to Pmax and Dor, and not to model state variables. This contributes to reduce the variety of vegetation structures such as tree densities and tree heights due to the frequent DA every 4 days (not shown). This tends to cause particle degeneracy for biomass even with large particle sizes when the resampling perturbation size is small (Table 5). When the resampling perturbation size is relatively large, degeneracy of the vegetation structure is mitigated to some extent. Therefore, in this study, we tuned the resampling perturbation sizes to avoid the filter collapse for biomass, and found that the "moderate" perturbation size with 8000 particles is a reasonable choice. However, the moderate perturbation size may be large for variables other than biomass, and this may be why the estimated parameters show occasional jumps. Adding resampling perturbations to other variables in addition to Pmax and Dor would be better. Also, since resampling perturbations affect the particle spread strongly, the DA technique does not necessarily provide accurate estimates of the errors. In future studies, we will explore more effective resampling methods to avoid the filter collapse for biomass and to represent error estimates more accurately.

As a potential limitation, it is important to note that we have made strong assumptions in OSSE. For example, the only source of model imperfections was the model parameter uncertainties of the four parameters. It was also assumed that the observation error statistics were perfectly known. These conditions would have never been met in the real-world experiment.

As the first step, this study focused on the methodological development of the data assimilation system with SEIB-DGVM and estimated only four parameters of two PFTs using LAI observations at a single location. As a next step, more parameters and distributions of more diverse PFTs should be considered at different locations. Local-scale experiments can be performed in parallel for different locations since the satellite-based LAI observations are available globally. The simulation with the initial states and parameter sets obtained from the SEIB-DGVM-based DA system would be expected to improve the estimates of the carbon cycle changes over the globe.

Appendix A

Table A1. List of modifications to SEIB-DGVM ver.2.71.

Modification to main.f90	

SUBROUTINE main_loop
   Initialize variables: parameters (Pmax for tree and grass, Dor for tree and grass) are read here.
   Wild fire subroutines: fire function was excluded.

Modification to metabolic.f90

SUBROUTINE photosynthesis_condition:
  ce_water (no dimension, the minimum value): limitation on photosynthesis via soil water:
  $x = \min(1.0, \max(0.001, \text{stat\_water}(p))) \rightarrow x = \min(1.0, \max(0.1, \text{stat\_water}(p)))$
SUBROUTINE leaf_season:
  Days_leaf_shed (days): day length required for full leaf drop $\rightarrow$ from 14 to 30
  Days_release_larch (days): days required for full release of stock energy for larch $\rightarrow$ from 7 to 60
  Checker (foliage $\rightarrow$ dormancy)
    case (1): if $(x < 7.0)$ flag$(p) =$ .true. $\rightarrow$ if (DOY $> =$ Dor_f) flag$(p) =$ .true. (DOY: day of the year, Dor_f: Dor for tree)
    case (5:6): if $(y > 0.01)$ flag$(p) =$ .false. $\rightarrow$ if (DOY $> =$ Dor_g) flag$(p) =$ .true. (DOY: day of the year, Dor_g: Dor for grass)
    If (dfl_leaf_onset$(p) \rightarrow$ Days_foliation_min) flag$(p) =$ .false. $\rightarrow$ comment out
  Checker (dormancy $\rightarrow$ foliage)
    case (1): if $(x > = 65.0)$ flag$(p) =$ .true. $\rightarrow$ if (DOY $> = 110$) flag$(p) =$ .true (DOY: day of the year)
    case (5:6): if $(y < = 0.01)$ flag$(p) =$ .false. $\rightarrow$ if (DOY $> = 110$) flag$(p) =$ .true (DOY: day of the year)
  Gradual release of stock energy: (for bug fix)
    IF (dfl_leaf_onset$(p) > =$ day_length_release) cycle $\rightarrow$ IF (dfl_leaf_onset$(p) > =$ (day_length_release-1)) cycle
SUBROUTINE maintenance_resp:
  Herbaceous PFT source 1: (for bug fix[a])
    mass_combust = mass_combust + mass_required $\rightarrow$ mass_combust = mass_combust + mass_required $\cdot x$
    npp$(p) =$ npp$(p) -$ mass_required $\rightarrow$ npp$(p) =$ npp$(p) -$ mass_required $\cdot x$
SUBROUTINE growth_wood:
  Delay_from_foliation (days): delay of stem growth and reproduction process after foliation $\rightarrow$ from 21 to 0[a]

Modification to parameter.txt

TO_f (times yr^{-1}): turn over time for foliage (grass)	from 0.50 to 3.19[b]
TO_r (times yr^{-1}): turn over time for root (tree). We set the same value as the other boreal tree PFTs.	from 0.16 to 0.42
ALM1 (m^2 m^{-1}): allometry index of LA vs. dbh of sapwood (tree)	from 6000 to 0[b]
ALM3 (g dm m^{-3}): allometry index of trunk mass (tree)	from 0 to 700 000[b]
FR_ratio (g dm g dm^{-1}): ratio of leaf mass vs. root mass (tree)	from 0.17 to 0.35[b]
FR_ratio (g dm g dm^{-1}): ratio of leaf mass vs. root mass (grass)	from 0.33 to 0.10[a]
SLA (one sided m^2 g dm^{-1}): specific leaf area (tree)	from 0.014 to 0.010[b]
SLA (one sided m^2 g dm^{-1}): specific leaf area (grass)	from 0.015 to 0.020[b]
Topt0 (°C): optimum temperature (tree)	from 20.0 to 21.0[b]
Tmin (°C): minimum temperature (tree). We set the same value as the other boreal tree PFTs.	from 5.0 to -4.0
Tmax (°C): maximum temperature (tree)	from 35.0 to 38.0[b]
GS_b2 (no dimension): parameters of stomatal conductance (grass)	from 3.0 to 5.0[a]
M1 (no dimension): asymptotic maximum mortality rate (tree)	from 0.003 to 0.001[b]
TC_min (°C): minimum coldest month temperature for persisting (tree and grass)	from -1000.0 to -45.0[b]
GDD_min (5 °C base): minimum degree-day sum for establishment (tree)	from 350 to 250[b]
Est_scenario: scenario for establishment for tree. Only specified woody PFT was set to establish.	[a]
Est_pft_OnOff: establish switch for tree. Only boreal deciduous needleleaf tree was set to establish.	[a]

without footnote: modifications in this study, [a] H. Sato (personal communications, 2014), [b] Sato et al. (2016).

Competing interests. The authors declare that they have no conflict of interest.

Acknowledgements. The authors thank Hisashi Sato, the main developer of SEIB-DGVM, for useful discussions. The authors also thank reviewers Matthias Morzfeld and Malaquias Pena Mendez for their careful reviews and constructive comments that helped improve the manuscript significantly. The source code of SEIB-DGVM and the climate forcing data are available at http://seib-dgvm.com/. MODIS LAI product of MCD15A3 was retrieved from the online data pool, courtesy of the NASA Land Processes Distributed Active Archive Center (LP DAAC), USGS/Earth Resources Observation and Science (EROS) Center, Sioux Falls, South Dakota (https://lpdaac.usgs.gov/data_access/data_pool). Carbon flux data were retrieved from the AsiaFlux database (https://db.cger.nies.go.jp/asiafluxdb/).

Edited by: Amit Apte

References

Anderson, J. L. and Anderson, S. L.: A Monte Carlo implementation of the nonlinear filtering problem to produce ensemble assimilations and forecasts, Mon. Weather Rev., 127, 2741–2758, 1999.

Atlas, R.: Atmospheric observations and experiments to assess their usefulness in data assimilation, J. Meteorol. Soc. Jpn., 75, 111–130, 1997.

Bickel, P., Li, B., and Bengtsson, T.: Sharp failure rates for the bootstrap particle filter in high dimensions, in: Pushing the Limits of Contemporary Statistics: Contributions in Honor of Jayanta K. Ghosh, IMS Collections, 3, edited by: Clarke, B. and Ghosal, S., Institute of Mathematical Statistics, Beachwood, Ohio, USA, 318–329, 2008.

Cheaib, A., Badeau, V., Boe, J., Chuine, I., Delire, C., Dufrêne, E., François, C., Gritti, E. S., Legay, M., Pagé, C., Thuiller, W., Viovy, N., and Leadley, P.: Climate change impacts on tree ranges: model intercomparison facilitates understanding and quantification of uncertainty, Ecol. Lett., 15, 533–544, 2012.

Demarty, J., Chevallier, F., Friend, A. D., Viovy, N., Piao, S., and Ciais, P.: Assimilation of global MODIS leaf area index retrievals within a terrestrial biosphere model, Geophys. Res. Lett., 34, L15402, https://doi.org/10.1029/2007GL030014, 2007.

Evensen, G.: The ensemble Kalman filter: Theoretical formulation and practical implementation, Ocean. Dynam., 53, 343–367, 2003.

Garreta, V., Miller, P. A., Guiot, J., Hély, C., Brewer, S., Sykes, M. T., and Litt, T.: A method for climate and vegetation reconstruction through the inversion of a dynamic vegetation model, Clim. Dynam., 35, 371–389, 2010.

Gordon, N. J., Salmond, D. J., and Smith, A. F. M.: Novel approach to nonlinear/non-Gaussian Bayesian state estimation, IEE Proc. F, 140, 107–113, 1993.

Harris, I., Jones, P. D., Osborn, T. J., and Lister, D. H.: Updated high-resolution grids of monthly climatic observations – the CRU TS3.10 Dataset, Int. J. Climatol., 34, 623–642, 2014.

Hartig, F., Dyke, J., Hickler, T., Higgins, S. I., O'Hara, R. B., Scheiter, S., and Huth, A.: Connecting dynamic vegetation models to data – an inverse perspective, J. Biogeogr., 39, 2240–2252, 2012.

Kalnay, E., Kanamitsu, M., Kistler, R., Collins, W., Deaven, D., Gandin, L., Iredell, M., Saha, S., White, G., Woollen, J., Zhu, Y., Chelliah, M., Ebisuzaki, W., Higgins, W., Janowiak, J., Mo, K. C., Ropelewski, C., Wang, J., Leetmaa, A., Reynolds, R., Jenne, R., and Joseph, D.: The NCEP/NCAR 40-year reanalysis project, B. Am, Meteorol. Soc., 77, 437–471, 1996.

Kaminski, T., Knorr, W., Schürmann, G., Scholze, M., Rayner, P. J., Zaehle, S., Blessing, S., Dorigo, W., Gayler, V., Giering, R., Gobron, N., Grant, J. P., Heimann, M., Hooker-Stroud, A., Houweling, S., Kato, T., Kattge, J., Kelley, D., Kemp, S., Koffi, E. N., Köstler, C., Mathieu, P.-P., Pinty, B., Reick, C. H., Rödenbeck, C., Schnur, R., Scipal, K., Sebald, C., Stacke, T., Terwisscha van Scheltinga, A., Vossbeck, M., Widmann, H., and Ziehn, T.: The BETHY/JSBACH Carbon Cycle Data Assimilation System: experiences and challenges, J. Geophys. Res.-Biogeo., 118, 1414–1426, 2013.

Kato, T., Knorr, W., Scholze, M., Veenendaal, E., Kaminski, T., Kattge, J., and Gobron, N.: Simultaneous assimilation of satellite and eddy covariance data for improving terrestrial water and carbon simulations at a semi-arid woodland site in Botswana, Biogeosciences, 10, 789–802, https://doi.org/10.5194/bg-10-789-2013, 2013.

Knyazikhin, Y., Glassy, J., Privette, J. L., Tian, Y., Lotsch, A., Zhang, Y., Wang, Y., Morisette, J. T., Votava, P., Myneni, R. B., Nemani, R. R., and Running, S. W.: MODIS Leaf Area Index (LAI) and Fraction of Photosynthetically Active Radiation Absorbed by Vegetation (FPAR) product (MOD15) Algorithm, Theoretical Basis Document, NASA Goddard Space Flight Center, Greenbelt, MD 20771, USA, 1999.

Kolari, P., Pumpanen, J., Kulmala, L., Ilvesniemi, H., Nikinmaa, E., Grönholm, T., and Hari, P.: Forest floor vegetation plays an important role in photosynthetic production of boreal forests, Forest Ecol. Manag., 211, 241–248, 2006.

Luo, Y., Ogle, K., Tucker, C., Fei, S., Gao, C., LaDeau, S., Clark, J. S., and Schimel, D. S.: Ecological forecasting and data assimilation in a data-rich era, Ecol. Appl., 21, 1429–1442, 2011.

Murray-Tortarolo, G., Anav, A., Friedlingstein, P., Sitch, S., Piao, S., Zhu, Z., Poulter, B., Zaehle, S., Ahlström, A., Lomas, M., Levis, S., Viovy, N., and Zeng, N.: Evaluation of land surface models in reproducing satellite-derived LAI over the high-latitude Northern Hemisphere. Part I: Uncoupled DGVMs, Remote Sens., 5, 4819–4838, 2013.

Ohta, T., Hiyama, T., Tanaka, H., Kuwada, T., Maximov, T. C., Ohata, T., and Fukushima, Y.: Seasonal variation in the energy and water exchanges above and below a larch forest in eastern Siberia, Hydrol. Process., 15, 1459–1476, 2001.

Ohta, T., Maximov, T. C., Dolman, A. J., Nakai, T., van der Molen, M. K., Kononov, A. V., Maximov, A. P., Hiyama, T., Iijima, Y., Moors, E. J., Tanaka, H., Toba, T., and Yabuki, H.: Interannual variation of water balance and summer evapotranspiration in an eastern Siberian larch forest over a 7-year period (1998–2006), Agr. Forest Meteorol., 148, 1941–1953, 2008.

Ohta, T., Kotani, A., Iijima, Y., Maximov, T. C., Ito, S., Hanamura, M., Kononov, A. V., and Maximov, A. P.: Effects of waterlogging on water and carbon dioxide fluxes and environmental variables in a Siberian larch forest, 1998–2011, Agr. Forest Meteorol., 188, 64–75, 2014.

Peng, C.: From static biogeographical model to dynamic global vegetation model: a global perspective on modelling vegetation dynamics, Ecol. Model., 135, 33–54, 2000.

Peng, C., Guiot, J., Wu, H., Jiang, H., and Luo, Y.: Integrating models with data in ecology and palaeoecology: advances towards a model–data fusion approach, Ecol. Lett., 14, 522–536, 2011.

Quaife, T., Lewis, P., De Kauwe, M., Williams, M., Law, B. E., Disney, M., and Bowyer, P.: Assimilating canopy reflectance data into an ecosystem model with an Ensemble Kalman Filter, Remote Sens. Environ., 112, 1347–1364, 2008.

Richardson, A. D., Anderson, R. S., Arain, M. A., Barr, A. G., Bohrer, G., Chen, G., Chen, J. M., Ciais, P., Davis, K. J., Desai, A. R., Dietze, M. C., Dragoni, D., Garrity, S. R., Gough, C. M., Grant, R., Hollinger, D. Y., Margolis, H. A., McCaughey, H., Migliavacca, M., Monson, R. K., Munger, J. W., Poulter, B., Raczka, B. M., Ricciuto, D. M., Sahoo, A. K., Schaefer, K., Tian, H., Vargas, R., Verbeeck, H., Xiao, J., and Xue, Y.: Terrestrial biosphere models need better representation of vegetation phenology: results from the North American Carbon Program Site Synthesis, Glob. Change Biol., 18, 566–584, 2012.

Sato, H. and Ise, T.: Effect of plant dynamic processes on African vegetation responses to climate change: Analysis using the spatially explicit individual-based dynamic global vegetation model (SEIB-DGVM), J. Geophys. Res., 117, G03017, https://doi.org/10.1029/2012JG002056, 2012.

Sato, H., Itoh, A., and Kohyama, T.: SEIB–DGVM: A new Dynamic Global Vegetation Model using a spatially explicit individual-based approach, Ecol. Model., 200, 279–307, 2007.

Sato, H., Kobayashi, H., and Delbart, N.: Simulation study of the vegetation structure and function in eastern Siberian larch forests using the individual-based vegetation model SEIB-DGVM, Forest Ecol. Manag., 259, 301–311, 2010.

Sato, H., Kobayashi, H., Iwahana, G., and Ohta, T.: Endurance of larch forest ecosystems in eastern Siberia under warming trends, Ecol. Evol., 6, 5690–5704, 2016.

Smith, B., Prentice, I. C., and Sykes, M. T.: Representation of vegetation dynamics in the modelling of terrestrial ecosystems: comparing two contrasting approaches within European climate space, Global Ecol. Biogeogr., 10, 621–637, 2001.

Snyder, C.: Particle filters, the "optimal" proposal and high-dimensional systems, ECMWF Seminar on Data Assimilation for Atmosphere and Ocean, ECMWF, Shinfield, UK, 6-9 September 2011, 161–170, 2012.

Snyder, C., Bengtsson, T., Bickel, P., and Anderson, J.: Obstacles to high-dimensional particle filtering, Mon. Weather Rev., 136, 4629–4640, 2008.

Snyder, C., Bengtsson, T., and Morzfeld, M.: Performance bounds for particle filters using the optimal proposal, Mon. Weather Rev., 143, 4750–4761, 2015.

Stöckli, R., Rutishauser, T., Baker, I., Liniger, M. A., and Denning, A. S.: A global reanalysis of vegetation phenology, J. Geophys. Res., 116, G03020, https://doi.org/10.1029/2010JG001545, 2011.

Takagi, K., Hirata, R., Ide, R., Ueyama, M., Ichii, K., Saigusa, N., Hirano, T., Asanuma, J., Li, S-G., Machimura, T., Nakai, Y., Ohta, T., and Takahashi, Y.: Spatial and seasonal variations of CO_2 flux and photosynthetic and respiratory parameters of larch forests in East Asia, Soil Sci. Plant Nutr., 61, 61–75, 2015.

Williams, M., Schwarz, P. A., Law, B. E., Irvine, J., and Kurpius, M. R.: An improved analysis of forest carbon dynamics using data assimilation, Glob. Change Biol., 11, 89–105, 2005.

Zeng, H., Jia, G., and Epstein, H.: Recent changes in phenology over the northern high latitudes detected from multi-satellite data, Environ. Res. Lett., 6, 1–11, 2011.

Zhao, J., Zhang, H., Zhang, Z., Guo, X., Li, X., and Chen, C.: Spatial and temporal changes in vegetation phenology at middle and high latitudes of the Northern Hemisphere over the past three decades, Remote Sens., 7, 10973–10995, 2015.

8

Ocean swell within the kinetic equation for water waves

Sergei I. Badulin[1,2] **and Vladimir E. Zakharov**[1,2,3,4,5]

[1]P. P. Shirshov Institute of Oceanology of the Russian Academy of Sciences, Moscow, Russia
[2]Laboratory of Nonlinear Wave Processes, Novosibirsk State University, Novosibirsk, Russia
[3]Department of Mathematics, University of Arizona, Tucson, USA
[4]P. N. Lebedev Physical Institute of the Russian Academy of Sciences, Moscow, Russia
[5]Waves and Solitons LLC, Phoenix, Arizona, USA

Correspondence to: Sergei I. Badulin (badulin.si@ocean.ru)

Abstract. Results of extensive simulations of swell evolution within the duration-limited setup for the kinetic Hasselmann equation for long durations of up to 2×10^6 s are presented. Basic solutions of the theory of weak turbulence, the so-called Kolmogorov–Zakharov solutions, are shown to be relevant to the results of the simulations. Features of self-similarity of wave spectra are detailed and their impact on methods of ocean swell monitoring is discussed. Essential drop in wave energy (wave height) due to wave–wave interactions is found at the initial stages of swell evolution (on the order of 1000 km for typical parameters of the ocean swell). At longer times, wave–wave interactions are responsible for a universal angular distribution of wave spectra in a wide range of initial conditions. Weak power-law attenuation of swell within the Hasselmann equation is not consistent with results of ocean swell tracking from satellite altimetry and SAR (synthetic aperture radar) data. At the same time, the relatively fast weakening of wave–wave interactions makes the swell evolution sensitive to other effects. In particular, as shown, coupling with locally generated wind waves can force the swell to grow in relatively light winds.

1 Physical models of ocean swell

Ocean swell is an important constituent of the field of surface gravity waves in the sea and, more generally, of the sea environment as a whole. Swell is usually defined as a fraction of a wave field that does not depend (or depends slightly) on local wind. Being generated in confined stormy areas, these waves can propagate long distances of many thousands of miles, thus influencing vast ocean stretches. For example, swell from the Roaring Forties in the Southern Ocean can traverse the Pacific and reach distant shores of California and Kamchatka. Predicting swell as a part of sea wave forecast remains a burning problem for maritime safety and marine engineering.

Pioneering works by Barber and Ursell (1948), Munk et al. (1963), and Snodgrass et al. (1966) discovered a rich physics of the phenomenon and gave the first examples of accurate measurements of magnitudes, periods and directional spreading of swell. All the articles contain thorough discussions of the physical background of swell generation, attenuation and interaction with other types of ocean motions. A fascinating story of a grand experiment on ocean swell has been presented to a wide audience in the documentary "Waves across the Pacific" (can be found at https://www.youtube.com/watch?v=MX5cKoOm6Pk).[1]

Nonlinear wave–wave interactions have been sketched by Snodgrass et al. (1966) as a novelty introduced by the milestone papers by Phillips (1960) and Hasselmann (1962). A possible important role of these interactions at high swells for relatively short times of evolution has been outlined and evaluated. The first estimates of the observed rates of swell attenuation were carried out by Snodgrass et al. (1966) based on observation at near-shore stations. Their *e*-folding scale of about 4000 km (the distance at which an exponentially decaying wave height decreases by a factor of *e*) is consistent with some of today's results of the satellite tracking of swell (Ardhuin et al., 2009, 2010; Jiang et al., 2016) and with the

---

[1]The authors are thankful to Gerbrant van Vledder, Delft University of Technology, for this reference.

treatment of these results within the model of swell attenuation due to coupling with the turbulent atmospheric layer (e.g., Tsimring, 1986; Kantha, 2006). The alternative semi-empirical model of Babanin (2006) predicts a quite different algebraic law and stronger swell attenuation at shorter distances from the swell source (Young et al., 2013). Note that the effect of the decay of a monochromatic wave due to turbulent wave flow is found to be quadratic in wave amplitude, i.e., to be of lower-order nonlinearity than in the non-dissipative theory of weakly nonlinear water waves.

It should be stressed that a number of theoretical and numerical models, including those mentioned above, treat swell as a quasi-monochromatic wave and, thus, ignore nonlinear interactions of the swell harmonics themselves and the swell coupling with locally generated wind waves. The latter effect can be essential, as observations and simulations clearly show (e.g., Kahma and Pettersson, 1994; Pettersson, 2004; Young, 2006; Badulin et al., 2008b, and references therein). Usually the swell continues to be considered a superposition of harmonics that do not interact with each other and, thus, can be described by the well-known methods of the linear theory of waves (e.g., Ewans, 1998; Ewans et al., 2004). Many features of the observed swell can be related to such models. For example, the observed effect of linear growth of the swell frequency at a site can be explained as an effect of dispersion of a linear wave packet over a long time and successfully used for relating these observations to stormy areas that generate the swell (e.g., Barber and Ursell, 1948; Ewans et al., 2004).

Synthetic aperture radar (SAR) allows for a spatial resolution of up to tens of meters (e.g., Ardhuin et al., 2010; Young et al., 2013). Satellite altimeters measure wave height averaged over a snapshot of a few square kilometers. These snapshots are adequate for currently known methods of statistical description of waves in research and application models. These can be used for swell tracking in combination with other tools (e.g., wave models as in Jiang et al., 2016). Retracking of swells allows us, first, to relate the swell events to their probable sources – stormy areas – and, secondly, the swell transformation enables us to estimate the effects of other motions of the atmosphere and ocean – seasonal wind activity (e.g., Chen et al., 2002), wave–current interaction (e.g., Beal et al., 1997) and bathymetry effects (Young et al., 2013). Such a work requires adequate physical models of swell propagation and transformation. This paper aims to narrow the gap.

Meanwhile, the linear treatment remains quite restrictive and cannot explain important features of swell. The observed swell spectra exhibit frequency downshift which is not predicted by deterministic linear or weakly nonlinear models of narrow-banded wave guide evolution (e.g., data of Snodgrass et al., 1966, and comments on these data by Henderson and Segur, 2013). Moreover, these spectra show invariance of their shapes that is unlikely to appear in a linear dispersive wave system. These noted features are common for wave spectra described by the kinetic equation for water waves, the so-called Hasselmann (1962) equation.

In this paper we present results of extensive simulations of ocean swell within the Hasselmann equation for deep water waves. The simplest duration-limited setup has been chosen to obtain numerical solutions for the duration of up to 2×10^6 s (about 23 days) for typical parameters of ocean swell (wavelengths 150–400 m, wave periods 10–16 s, initial significant heights 3–15 m).

We analyze the simulation results within the framework of the theory of weak turbulence (Zakharov et al., 1992). The slowly evolving swell solutions appear to be quite close to the stationary Kolmogorov–Zakharov spectra. We give a short theoretical introduction and present estimates of the basic constants of the theory in the next section. In Sect. 3 we relate results of simulations to properties of the self-similar solutions of the kinetic equation. Zaslavskii (2000) was the first to present the self-similar solutions for swell assuming the angular narrowness of the swell spectra and stated explicit analytical results. In fact, more general consideration, in the spirit of Badulin et al. (2002, 2005a), leads to important findings and raises questions independently of the assumption of angular narrowness.

We demonstrate the fact that is usually ignored: the power-law swell attenuation within the conservative kinetic equation. We show that it does not contradict the observations mentioned above. We also reveal a remarkable feature of collapsing the swell spectra onto an angular distribution that depends weakly on initial angular spreading. Such universality can be of great value for modeling swell and developing methods for its monitoring (Delpey et al., 2010).

We conclude this paper with a discussion of how to apply this model. Evidently, the setup of duration-limited evolution is quite restrictive and does not reflect essential features of ocean swell when wave dispersion and spatial divergence play a key role. At the same time, wave–wave interactions remain of importance independently of the setup. The weakening of swell evolution is not directly related to abatement of wave–wave interactions which are able to effectively restore perturbations of these quasi-stationary states (Zakharov and Badulin, 2011). On the contrary, this favors coupling of the quasi-stationary swell with the ocean environment. In particular, the locally generated wind-driven waves can switch the swell attenuation to swell amplification. This effect can be considered for interpretation of recent observations of swell from space ("negative" dissipation in the words of Jiang et al., 2016). Many problems of adequate physical description of swell in the ocean are still open. This paper is an attempt to reveal essential features of swell evolution within the simplest model of the kinetic Hasselmann equation.

2 Solutions for ocean swell

2.1 The Kolmogorov–Zakharov solutions

In this section we reproduce previously reported theoretical results on the evolution of swell as a random field of weakly interacting wave harmonics. We apply the statistical theory of wind-driven seas (Zakharov, 1999) to the sea swell, whose description with this approach is usually considered questionable. A random wave field is described by the kinetic equation derived by Hasselmann (1962) for weakly nonlinear deep water waves in the absence of dissipation and external forcing:

$$\frac{\partial N_k}{\partial t} + \nabla_k \omega_k \nabla_r N_k = S_{nl}. \tag{1}$$

Equation (1) is written for the spectral density of wave action $N(k, x, t) = E(k, x, t)/\omega(k)$ ($E(k, x, t)$ is the wave energy spectrum and the wave frequency obeys the linear dispersion relation $\omega = \sqrt{g|k|}$). Subscripts for ∇ correspond to the two-dimensional gradient operator in the corresponding space of coordinates x and wavevectors k (i.e., $\nabla_r = (\partial/\partial x, \partial/\partial y)$).

The right-hand term S_{nl} describes the effect of wave–wave resonant interactions and can be written in an explicit form (see the Appendices in Badulin et al., 2005a, for a collection of the formulas). The cumbersome term S_{nl} causes many problems for wave modeling whenever Eq. (1) is extensively used. Nevertheless, for the deep water case, one has a key property of homogeneity

$$S_{nl}[\kappa k, \upsilon N_k] = \kappa^{19/2} \upsilon^3 S_{nl}[k, N_k] \tag{2}$$

that helps in acquiring important analytical results. Stretching in κ times in the wave scale or in υ times in the wave action, where κ, υ are positive, leads to simple re-scaling of the collision term, S_{nl}. This important property gives a clue for constructing power-law stationary solutions of the kinetic equation, i.e., solutions for the equation

$$S_{nl} = 0. \tag{3}$$

Two isotropic stationary solutions of Eq. (3) correspond to constant fluxes of wave energy and action in wave scales. The direct cascade solution (Zakharov and Filonenko, 1966) in terms of the frequency spectrum of energy

$$E^{(1)}(\omega, \theta) = 2C_p \frac{P^{1/3} g^{4/3}}{\omega^4} \tag{4}$$

introduces the basic Kolmogorov constant C_p and describes the energy transfer to infinitely short waves with constant flux P. The wave action transfer in the opposite direction of long waves is described by the inverse cascade solution (Zakharov and Zaslavsky, 1982) with wave action flux Q and another Kolmogorov constant C_q:

$$E^{(2)}(\omega, \theta) = 2C_q \frac{Q^{1/3} g^{4/3}}{\omega^{11/3}}. \tag{5}$$

Note that key features of the isotropic Kolmogorov–Zakharov solution Eqs. (4) and (5) are reproduced quite well by means of direct numerical simulations based on the integro-differential Zakharov equation (Annenkov and Shrira, 2006) or on the primitive Euler equations (Onorato et al., 2002).

An approximate weakly anisotropic Kolmogorov–Zakharov solution has been obtained by Kats and Kontorovich (1974) as an extension of Eq. (4):

$$E^{(3)}(\omega, \theta) = 2\frac{P^{1/3} g^{4/3}}{\omega^4} \left(C_p + C_m \frac{gM}{\omega P} \cos\theta + \dots \right). \tag{6}$$

It associates the wave spectrum anisotropy with the constant spectral flux of wave momentum M and the so-called second Kolmogorov constant C_m. As is seen from Eq. (6), the solution anisotropy vanishes as $\omega \to \infty$: wave spectra become isotropic for short waves. The whole set of the Kolmogorov–Zakharov (KZ) solutions (Eqs. 4–6) can be treated naturally within the dimensional approach: these are just particular cases of solutions of the form

$$E^{(KZ)}(\omega) = \frac{P^{1/3} g^{4/3}}{\omega^4} G(\omega Q/P, gM/(\omega P), \theta), \tag{7}$$

where G is a function of dimensionless arguments scaled by spectral fluxes of wave energy P, action Q and momentum M.

Originally, solutions (Eqs. 4–6) were derived in particularly sophisticated and cumbersome ways. Later on, simpler and more physically transparent approaches were presented (Zakharov and Pushkarev, 1999; Balk, 2000; Pushkarev et al., 2003, 2004; Badulin et al., 2005a; Zakharov, 2010). These more general approaches allow us to find higher-order terms of the anisotropic Kolmogorov–Zakharov solutions (Eq. 6). In particular, they predict the next term to be proportional to $\cos 2\theta/\omega^2$, which is the second angular harmonics of the stationary solution (Eq. 6).

Swell solutions evolve slowly with time and, thus, give a good opportunity for discussing features of the KZ solutions (or, alternatively, the KZ solutions can be used as a reference case for the swell studies). One of the key points of this discussion is the question of uniqueness or universality of the swell solutions that can be treated in the context of general KZ solutions (Eq. 7). The principal terms of the general Kolmogorov–Zakharov solutions (Eqs. 4–6) have the clear physical meanings of total fluxes of wave action, Eq. (5), energy, Eq. (4), and momentum, Eq. (6), and do not refer to specific initial conditions. This is not the case for the higher-order terms. The link between these additional terms with inherent properties of the collision integral S_{nl} and/or with specific initial conditions is a subject of further studies.

2.2 Self-similar solutions of the kinetic equation

The homogeneity property Eq. (2) is extremely useful for studies of non-stationary (inhomogeneous) solutions of the kinetic equation. Approximate self-similar solutions for reference cases of duration- and fetch-limited development of the wave field can be obtained under the assumption of dominance of the wave–wave interaction term S_{nl} (Pushkarev et al., 2003; Zakharov, 2005; Badulin et al., 2005a; Zakharov and Badulin, 2011). These solutions exhibit the so-called incomplete or second-type self-similarity (e.g., Barrenblatt, 1979). In terms of frequency–angle dependencies of wave action spectra, one has correspondingly for the duration- and fetch-limited cases (Badulin et al., 2005a, 2007; Zakharov et al., 2015)

$$N(\omega, \theta, \tau) = a_\tau \tau^{p_\tau} \Phi_{p_\tau}(\xi, \theta), \tag{8}$$

$$N(\omega, \theta, \chi) = a_\chi \chi^{p_\chi} \Phi_{p_\chi}(\zeta, \theta), \tag{9}$$

with dimensionless time τ and fetch χ:

$$\tau = t/t_0; \qquad \chi = x/x_0. \tag{10}$$

Dimensionless arguments of shape functions $\Phi_{p_\tau}(\xi)$ and $\Phi_{p_\chi}(\zeta)$ in Eqs. (8) and (9) contain free scaling parameters b_τ and b_χ and exponents of frequency downshifting q_τ and q_χ:

$$\xi = b_\tau \omega^2 \tau^{-2q_\tau}; \qquad \zeta = b_\chi \omega^2 \chi^{-2q_\chi}. \tag{11}$$

The homogeneity property (Eq. 2) dictates "magic relations" (in the words of Pushkarev and Zakharov, 2015, 2016) between exponents p_τ, q_τ and p_χ, q_χ

$$p_\tau = \frac{9q_\tau - 1}{2}; \qquad p_\chi = \frac{10q_\chi - 1}{2}. \tag{12}$$

Additional "magic relations" coming from the homogeneity property (Eq. 2) fix a link between the amplitude scales a_τ, a_χ and the bandwidth scales b_τ, b_χ of the self-similar solutions (Eqs. 8–11):

$$a_\tau = b_\tau^{19/4}; \qquad a_\chi = b_\chi^{5/2}. \tag{13}$$

Thus, "magic relations" (Eqs. 12 and 13) reduce the number of free parameters of the self-similar solutions (Eqs. 8 and 9) from four (two exponents and two coefficients) to only two: a dimensionless exponent p_τ (p_χ) and an amplitude of the solution a_τ (a_χ).

The shape functions $\Phi_{p_\tau}(\xi, \theta)$ and $\Phi_{p_\chi}(\zeta, \theta)$ in Eqs. (8) and (9) are specified by solutions of a nonlinear boundary problem for an integro-differential equation in self-similar variables ξ or ζ (conditions of decay at zero and infinity) and angle θ (periodicity) (see Sect. 5.2 in Badulin et al., 2005a, for details). These solutions reveal relatively narrow angular distributions with a single pronounced maximum and remarkably weak dependence on the exponent of wave growth p_ξ, (p_χ), as simulations show (e.g., Badulin et al., 2008a). This feature of quasi-universality (in the words of Badulin et al., 2005a) of the solutions of a nonlinear problem can be treated within a diffusion approximation for the kinetic equation (Zakharov and Pushkarev, 1999, and Zakharov, 2010) as

a "survival" of very few eigenfunctions – angular harmonics of the corresponding linear boundary problem. As will be shown below, the weakly anisotropic KZ solution (Eq. 6) represents a principal angular harmonic of such a decomposition.

Two-lobe patterns can be observed beyond the spectral peak as local maxima at oblique directions or as "shoulders" in wave frequency spectra. Their appearance within the kinetic equation approach is generally associated with wave generation by wind (e.g., Bottema and van Vledder, 2008, 2009) and/or the effect of wave–wave interactions (Banner and Young, 1994; Pushkarev et al., 2003). Numerical simulations within the potential Euler equations also show formation of the two-lobe patterns for rather short times (a few hundred spectral peak periods) of the evolution of an initially unimodal spectral distribution (Toffoli et al., 2010).

An essential approximation which is widely used both for experimentally observed and simulated wave spectra is generally treated as an important property of spectral shape invariance (terminology of Hasselmann et al., 1976) or the spectra quasi-universality (in the words of Badulin et al., 2005a). In fact, such "invariance" does not suppose a point-by-point matching of properly normalized spectral shapes. The proximity of integrals of the shape functions Φ_{p_τ}, Φ_{p_χ} in a range of wave growth rates p_τ, p_χ appears to be sufficient, in particular, for formulating efficient semi-empirical parameterizations of wind-wave growth in terms of integral values (e.g., Hasselmann et al., 1976). Consistent analysis within the weak turbulence approach that used this important approximation has recently lead to a remarkable theoretically based relationship (Zakharov et al., 2015):

$$\mu^4 \nu = \alpha_0^3. \tag{14}$$

Here wave steepness μ is estimated from total wave energy E and spectral peak frequency ω_p:

$$\mu = \frac{E^{1/2} \omega_p^2}{g}. \tag{15}$$

The "number of waves" ν in a spatially homogeneous wind sea (i.e., for the duration-limited case) is defined as follows:

$$\nu = \omega_p t. \tag{16}$$

For spatial (fetch-limited) wave growth, the coefficient of proportionality C_f in the equivalent expression $\nu = C_f |k_p| x$ (k_p being the wavevector of the spectral peak) is close to the ratio between the phase and group velocities $C_{ph}/C_g = 2$. A universal constant $\alpha_0 \approx 0.7$ is a counterpart of the constants C_p, C_q of the stationary Kolmogorov–Zakharov solutions (Eqs. 4 and 5) and has a similar physical meaning of a ratio between wave energy and the energy spectral flux (in power $1/3$). A remarkable feature of the universal wave growth law (Eq. 14) is its independence of wind speed. This wind-free paradigm based on intrinsic scaling of wave development is shown to be a useful tool of analysis of wind-wave growth (Zakharov et al., 2015). Below we demonstrate its effectiveness for interpreting swell simulations.

2.3 Self-similarity of swell solutions

The self-similar solution for swell is just a member of a family of solutions (Eqs. 8 and 9) with special values of temporal or spatial rates:

$$p_\tau = 1/11; \quad q_\tau = 1/11, \tag{17}$$

$$p_\chi = 1/12; \quad q_\chi = 1/12. \tag{18}$$

Exponents (Eqs. 17 and 18) provide conservation of the total wave action for its evolution in time (duration-limited setup) or in space (fetch-limited):

$$N = \int\limits_{0}^{+\infty} \int\limits_{-\pi}^{+\pi} N(\omega, \theta) \mathrm{d}\omega \mathrm{d}\theta = \text{const}. \tag{19}$$

By contrast, total energy

$$E = \int \omega N(\boldsymbol{k}) \mathrm{d}\boldsymbol{k} \tag{20}$$

and wave momentum

$$\mathbf{K} = \int \boldsymbol{k} N(\boldsymbol{k}) \mathrm{d}\boldsymbol{k} \tag{21}$$

are only formal constants of motion of the Hasselmann equation and decay with time t or fetch x

$$E \sim t^{-1/11}; \quad K_x \sim t^{-2/11}, \tag{22}$$

$$E \sim x^{-1/12}; \quad K_x \sim x^{-2/12}. \tag{23}$$

The swell decay (Eqs. 22 and 23) reflects a basic feature of the kinetic equation for water waves: energy Eq. (20) and momentum Eq. (21) are not conserved (see Zakharov et al., 1992; Pushkarev et al., 2003, and references herein). The wave action is the only true integral of the kinetic Eq. (1).

The swell solution manifests another general feature of evolving spectra: the downshifting of the spectral peak frequency (or other characteristic frequency), i.e.,

$$\omega_\mathrm{p} \sim t^{-1/11}; \quad \omega_\mathrm{p} \sim x^{-1/12}. \tag{24}$$

The universal law of wave evolution (Eq. 14) is, evidently, valid for the self-similar swell solution as well with a minor difference in the value of the constant α_0. As soon as this constant is expressed in terms of the integrals of the shape functions Φ_{p_τ}, Φ_{p_χ} and the swell spectrum shape differs essentially from ones of the growing wind seas, this constant appears to be less than α_0 of the growing wind seas.

The theoretical background presented above is used below for analysis of results of simulations.

3 Swell simulations

3.1 Simulation setup

Simulations of ocean swell require special care. First of all, calculations for quite long periods of time (up to 2×10^6 s in our case) should be accurate enough in order to capture relatively slow evolution of solutions and, thus, be able to relate results to the theoretical background presented above. Duration-limited evolution of the swell has been simulated with the Pushkarev et al. (2003) version of the code based on the WRT algorithm (Webb, 1978; Tracy and Resio, 1982). Features of the code and numerical setups have been described in previous papers (Badulin et al., 2002, 2004, 2005a, b, 2007; Zakharov et al., 2007; Badulin et al., 2008a, 2013; Pushkarev and Zakharov, 2015, 2016). The frequency resolution for log-spaced grid has been set to $(\omega_{n+1} - \omega_n)/\omega_n = 1.03128266$. It corresponds to 128 grid points in the frequency range $0.02 - 1$ Hz (approximately 1.5 to 3850 m wavelength).

Standard angular resolution $\Delta\theta = 10°$ has been taken as adequate for the goals of our study. A control series of runs with angular resolution $\Delta\theta = 5°$ showed very close but still quantitatively different shaping of wave spectra (see discussion below), while differences in integral parameters (wave height, period, total momentum) did not exceed 1 % after 2×10^6 s of evolution.

Initial conditions were similar in all series of simulations: spectral density of action in wavenumber space was almost constant in a box of the wavenumber modulo and angles. A slight modulation (5 % of the box height) and a low pedestal outside the box (6 orders less than the maximal value) have been set in order to stimulate wave–wave interactions since the collision integral S_{nl} vanishes for $N(\boldsymbol{k}) = \text{const}$:

$$N(\boldsymbol{k}) = \begin{cases} N_0(1 + 0.05\cos^2(\theta/2)), & |\theta| < \Theta/2, \omega_l < \omega < \omega_h \\ 10^{-6}N_0, & \text{otherwise}. \end{cases} \tag{25}$$

In Eq. (25) the references to angle θ ($\cos\theta = k_x/|\boldsymbol{k}|$) and wave frequency ω are used for conciseness of the expression for spatial wave action spectrum $N(\boldsymbol{k})$. The default values ω_l and ω_h corresponding to wave periods 10 and 2.5 s have been used for the most cases providing sufficient space for spectral evolution to low frequencies (spectra downshifting) and for stability of calculations at high frequencies for the default cutoff frequency $f_c = 1$ Hz.

Dissipation was absent in the runs. Free boundary conditions were applied at the high-frequency end of the domain of calculations: generally, short-term oscillations of the spectrum tail do not lead to instability; i.e., the resulting solutions can be regarded as ones corresponding to conditions of decay at infinitely small scales ($N(\boldsymbol{k}) \to 0$ when $|\boldsymbol{k}| \to \infty$).

Calculations with a hyper-viscosity (e.g., Pushkarev et al., 2003) or a diagnostic tail at the high-frequency range of the spectrum (Gagnaire-Renou et al., 2010) do not affect results quantitatively compared to our simulations without any dis-

Table 1. Initial parameters of the simulation series.

ID	Θ	N (m^2 s)	H_s (m)
sw030	30°	0.720	4.63
sw050	60°	0.719	4.6
sw170	180°	0.714	4.74
sw230	240°	0.721	4.67
sw330	330°	0.722	4.79

sipation. Thus, these "non-conservative" options can mimic successfully the effect of energy leakage at $|\boldsymbol{k}| \to \infty$ in our formally non-dissipative problem. Very strong dissipation at less than 10 grid points at the very end of frequency domain suppresses the spectral level and, simultaneously, reduces the overall energy dissipation at these points. Thus, the effect on the evolution of the energy-containing part of the solution appears to be quite weak and depends slightly on the particular form and magnitude of the hyper-viscosity. In some cases, the hyper-viscosity option that suppresses high-frequency noise can accelerate calculations. In a sense, it is equivalent to reducing an effective number of grid points. Test runs with the reduced frequency domain (cutoff up to $f_c = 0.6$ Hz, 112 grid points) did not show an essential quantitative difference with the default option ($f_c = 1$ Hz, 128 grid points).

In contrast to wind-driven waves where wind speed is an essential physical parameter that gives a useful physical scale, the swell evolution is determined by initial conditions only, i.e., by N_0 (dimension of wave action spectral density $[N(\boldsymbol{k})] = [\text{Length}^4 \cdot \text{Time}]$), a characteristic frequency (sideband $[\omega_l, \omega_h]$) and angular spreading Θ within the setup (Eq. 25). We tried different combinations of these parameters. Three frequency bands [0.026–0.09], [0.058–0.25], and [0.1–0.4] Hz have been chosen to generate swell with wavelengths approximately 200, 300, and 400 m at the final stages of evolution. The angular spreading Θ was set at 30, 50, 170, 230 and 330°. Initial significant wave heights H_s were taken as approximately 4.8, 8, 10, 12, and 18 m. As will be detailed below, an abrupt fall in wave energy occurred at the very first hours of evolution (up to 50 % for the first 1 h). Thus, the above high values of H_s can be accepted as realistic values for sea swell. In total, more than 30 combinations of wave height, frequency range and angular spreading have been simulated for the duration of at least 10^6 s. In some cases, for high amplitudes and narrow angular spreadings, simulations have failed because of strong numerical instability.

Below we focus on the series of Table 1 where initial wave heights were fixed (within 2 %) at approximately 4.8 m and angular spreading varied from very narrow, $\Theta = 30°$, to almost isotropic, $\Theta = 330°$ (Eq. 25). The frequency range of the initial perturbations was $0.1 - 0.4$ Hz. The simulations have been carried out for duration 2×10^6 s with angular res-

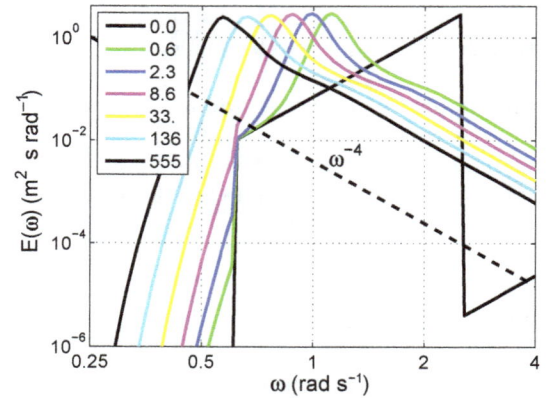

Figure 1. Frequency spectra of energy at different times (legend, in hours) for case sw330 ($\Theta = 330°$).

olution $\Delta\theta = 10°$ and checked for series sw030 and sw330 with $\Delta\theta = 5°$.

3.2 Self-similar features of swell

Evolution of swell spectra with time is shown in Fig. 1 for case sw330 of Table 1. The example shows a strong tendency to self-similar shaping of wave spectra. This remarkable feature has been demonstrated and discussed for swell in previous works (Badulin et al., 2005a; Benoit and Gagnaire-Renou, 2007; Gagnaire-Renou et al., 2010) for special parameters that provided relatively fast evolution of rather short and unrealistically high waves. In our simulations, we start with the mean wave period of about 3 s that corresponds to the end of calculations of Badulin et al. (2005a, see Fig. 8 therein). The initial spectrum evolves very quickly and keeps a characteristic shape for less than 1 h, when wave steepness falls dramatically below $\mu = 0.15$ ($T_p \approx 6$s), while wave height loses only about 20 % of its initial value (see Fig. 1, green curve for $t = 0.6$ h). For 555 h, the spectral peak period reaches 11.4 s (the corresponding wavelength $\lambda \approx 200$ m) and wave steepness becomes $\mu = 0.022$. The final significant wave height $H_s \approx 2.8$ m is essentially less than its initial value 4.8 m. All these values can be considered typical ones for ocean swell.

The dependence of key wave parameters on time is shown in Fig. 2 for different runs of Table 1. Power-law dependencies of self-similar solutions (Eqs. 17, 18 and 22–24) are shown by dashed lines. In Fig. 2a, b total wave energy E and the spectral peak frequency ω_p show good correspondence to power laws of the self-similar solutions (Eq. 8). By contrast, the power-law decay of the x-component of wave momentum K_x depends essentially on angular spreading of initial wave spectra. While for narrow spreading (runs sw030 and sw050) there is no visible deviation from the $K_x \sim t^{-2/11}$ law, wide-angle cases clearly show these deviations. The "almost isotropic" solution for sw330 tends quite slowly to the theoretical dependency of wave momentum K_x (Eq. 23). The

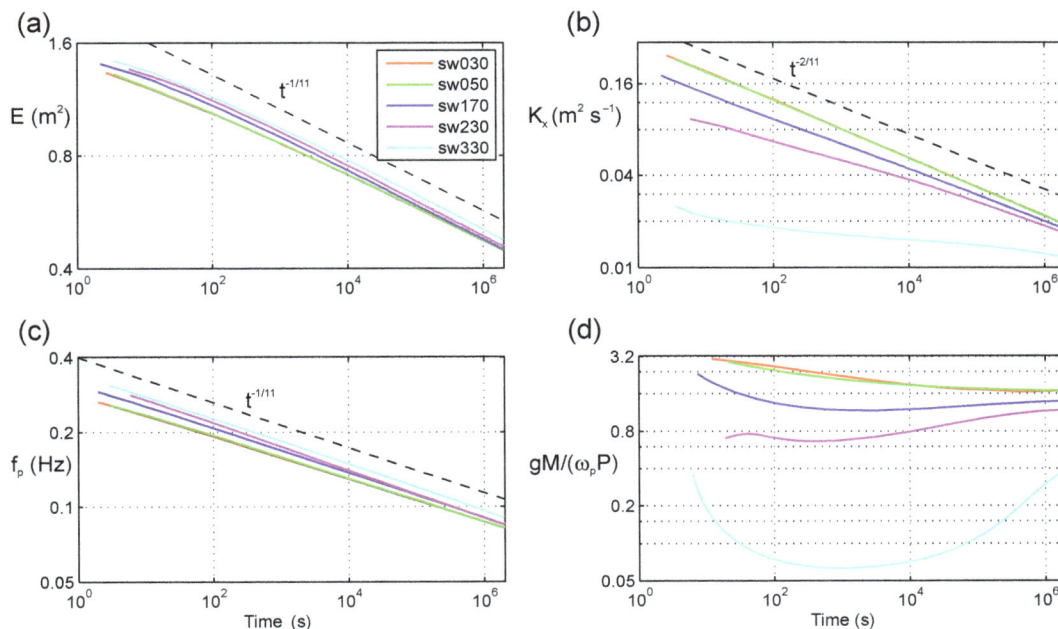

Figure 2. Evolution of wave parameters for runs of Table 1 (in the legend): **(a)** total energy E; **(b)** total wave momentum K_x; **(c)** frequency $f_p = \omega_p/(2\pi)$ of the energy spectra peak; **(d)** estimate of the parameter of anisotropy A (Eq. 26). Dashed lines show asymptotic power laws (Eqs. 22 and 24)

duration of more than 3 weeks appears "too short": one can see a transitional behavior when wave spectra evolve from the "almost isotropic" state to an inherent distribution with a pronounced anisotropy.

A simple quantitative estimate of the "degree of anisotropy" is given in Fig. 2d. Evolution of the dimensionless parameter of anisotropy in terms of the approximate Kolmogorov–Zakharov solution (Eq. 6) by Kats and Kontorovich (1974) is shown for all the cases of Table 1. We introduce the parameter of anisotropy A as follows:

$$A = \frac{gM}{\omega_p P}, \tag{26}$$

where the total energy flux P (energy flux at $\omega \to \infty$) is estimated from the evolution of total energy:

$$P = -\frac{dE}{dt}. \tag{27}$$

Similarly, the total wave momentum (Eq. 21) provides an estimate of its flux as follows:

$$M = -\frac{dK_x}{dt}. \tag{28}$$

Spectral peak frequency ω_p has been used for the definition of "degree of anisotropy", A (Eq. 26). Different scenarios are seen in Fig. 2d depending on angular spreading of wave spectra. Nevertheless, a general tendency to a universal behavior at very large times (more than 2×10^6 s) looks quite plausible.

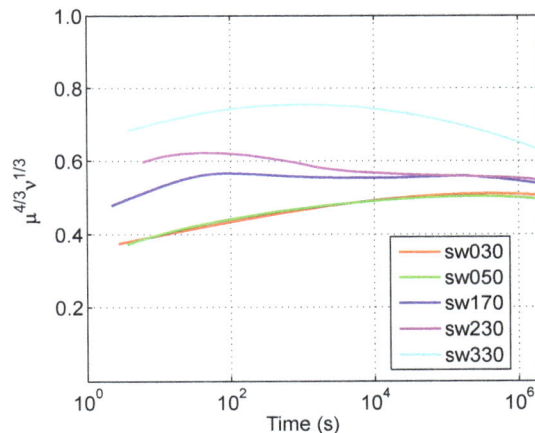

Figure 3. Evolution of the left-hand side of the invariant (Eq. 14) $(\mu^4 \nu)^{1/3}$ for runs of Table 1 (in the legend).

Similar dispersion of runs depending on anisotropy of initial distributions is seen in Fig. 3 when tracing the invariant of the self-similar solutions (Eq. 14). Again, like in Fig. 2b, 2×10^6 s are not sufficient to demonstrate the validity of a relationship (Eq. 14) in full. A limit α_0 (Eq. 14) is very likely reached at larger times. This limit is a bit less (by approximately 15 %) than one for growing wind seas: $\alpha_0 \approx 0.7$. Again, the "almost isotropic" solution shows its stronger departure from the rest of the series. The differences are better seen in angular distributions rather than in normalized

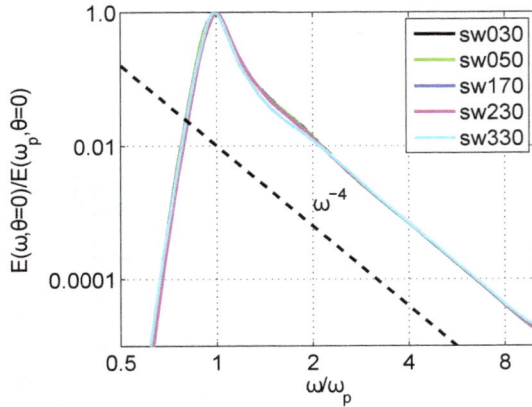

Figure 4. Normalized frequency spectra for direction $\theta = 0°$ after 11.5 days (approximately 10^6 s) of swell evolution for runs of Table 1 (see the legend).

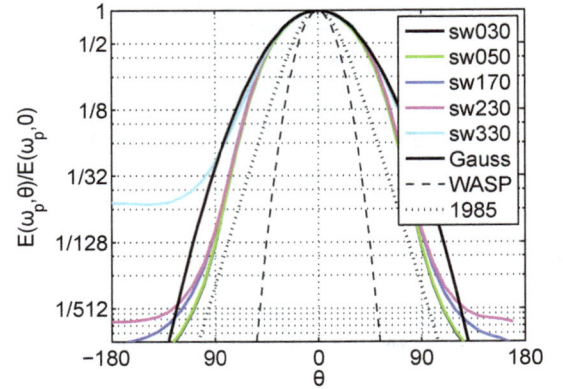

Figure 5. Normalized dependence of swell energy spectra on angle at peak frequency ω_p after 11.5 days (approximately 10^6 s) of swell evolution for the runs of Table 1 (see the legend). Dashed line – Gaussian distribution (Eq. 29) with dispersion $\sigma_\Theta = 35°$; dotted line – growing sea (Eq. 30) and Eq. 9.1–9.2 of Donelan et al. (1985); dashed line – wrapped-normal fit of Ewans et al. (2004, Table 11.2, Fig. 11.8).

spectral shapes (Fig. 4) when we are trying to check self-similarity features of the solutions in the spirit of Badulin et al. (2005a); Benoit and Gagnaire-Renou (2007).

3.3   Directional spreading of swell spectra

Despite a significant difference in the runs in integral characteristics of the swell anisotropy (e.g., Fig. 2b, d), the resulting spectral distributions still show pronounced features of universality, as is seen in frequency spectra (Fig. 4). As will be shown below, this universality of swell spectra is seen in angular distributions as well. This is of importance in the context of remarks of Sect. 2.2: while the shape functions Φ_{p_τ}, Φ_{p_χ} of self-similar solutions (Eqs. 8 and 9) are not unique, there is likely a mechanism of their selection that supports the universality of the swell spectral distributions. Within a linear theory, it could be treated as survival of the only eigenfunction or, more prudently, of very few eigenmodes of the problem. As mentioned in Sect. 2.2. this "linear" treatment can be used with some reservations for our problem, which is heavily nonlinear in terms of wave spectra but allows for a quasi-linear analysis in terms of spectral fluxes (see Zakharov and Pushkarev, 1999; Pushkarev et al., 2003).

The only physical mechanism of the mode selection in the swell problem is nonlinear relaxation to an inherent state due to four-wave resonant interactions. This relaxation generally occurs at essentially shorter timescales than ones of wind pumping and wave dissipation (Zakharov and Badulin, 2011). There is no contradiction with the present vision of the sea wave balance in the above statement. The effect of nonlinear interactions on wave spectra is 2-fold: firstly, it supports an inherent shaping of the spectra by very fast feedback to its perturbation and, secondly, it is responsible for relatively slow nonlinear cascading within this inherent shaping.

Normalized sections of spectra at the peak frequency ω_p are shown in Fig. 5 for runs of Table 1 at $t = 10^6$ s (approx. 11.5 days). "The almost isotropic" run sw330 shows a relatively high pedestal of about 2 % of the maximal value, while other series have a background of 1 order less. At the same time, the core of all distributions is quite close to a Gaussian shape

$$y_{\text{gauss}} = \exp\left(-\frac{\theta^2}{2\sigma_\Theta^2}\right) \tag{29}$$

with half-width $\sigma_\Theta = 35°$ (dashed curve in Fig. 5). Experimentally based spreading functions are represented in Fig. 5 by two reference curves. For growing wind seas the dependence by Donelan et al. (1985, Eq. 9.2)

$$y_{1985} = \text{sech}^2(\beta\theta); \quad \beta = 2.28 \tag{30}$$

gives almost twice as narrow a distribution (dotted line in Fig. 5). The wrapped-normal fit of angular distribution for one of the case of the West Africa Swell Project (see Table 11.2 and Fig. 11.8 in Ewans et al., 2004) with standard deviation $\sigma_\Theta \approx 14.3°$ gives a sharper distribution shown by a dashed curve.

Evolution of directional spreading in time is shown in absolute values in Fig. 6 for three runs: the most anisotropic case sw030 (Fig. 6a, b), weakly anisotropic initial state sw230 (Fig. 6c, d) and "the almost isotropic" run sw330 (Fig. 6e, f). In the left column the angular spreading at peak frequency shows remarkably close patterns for the first two cases: peak values at large times differ by a few percent only. The weakly anisotropic case sw230 (initial angular spreading 230° with essential counter-propagating fraction) reaches

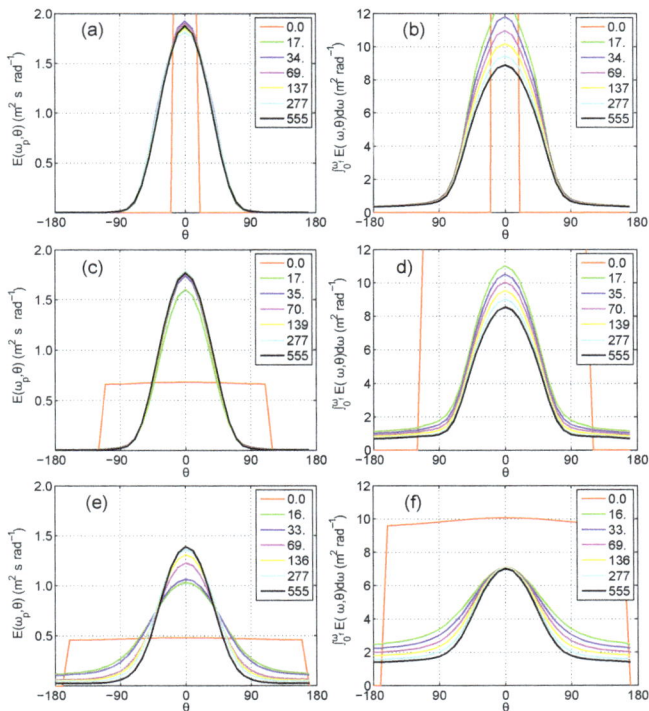

Figure 6. Angular spreading of the swell spectra at different times (in hours; see the legend). Left column – wave spectra at peak frequency; right – integral of wave spectra in frequency as a function of direction. **(a, b)** run sw030 of Table 1 – strong initial anisotropy; **(c, d)** run sw230 – weak anisotropy; **(e, f)** "almost anisotropic" run sw330.

its almost saturated state for a couple of days only (cf. the curves at $t = 17\,\text{h}$ and $t = 35\,\text{h}$). Similar proximity of these two cases can be observed for integrals of spectra in frequency as shown in the right column of Fig. 6, i.e., for values

$$\mathcal{E}(\theta) = \int\limits_0^{\omega_c} E(\omega, \theta)\mathrm{d}\omega. \tag{31}$$

Self-similar solutions (Eq. 8) predict a power-law decay of magnitude of $\mathcal{E}$ with time which is what we see in Fig. 6b, d for the first two cases. The behavior of "the almost isotropic" case sw330 is qualitatively different. The relatively strong adjustment to a narrow directional spreading occurs in the course of the entire duration $2 \times 10^6\,\text{s}$. The duration appears to be too short to reach a self-similar regime resembling cases sw030 and sw230.

The effect of sharpening of angular distributions of the run sw330 in Fig. 6e, f requires additional comments. First, it manifests a transitional nature of the case sw330 when a solution is rather far from its self-similar asymptotics. Secondly, this case illustrates the above statement of the paragraph on two scales of wave spectra evolution. The angular adjustment occurs at relatively short temporal scales as compared with

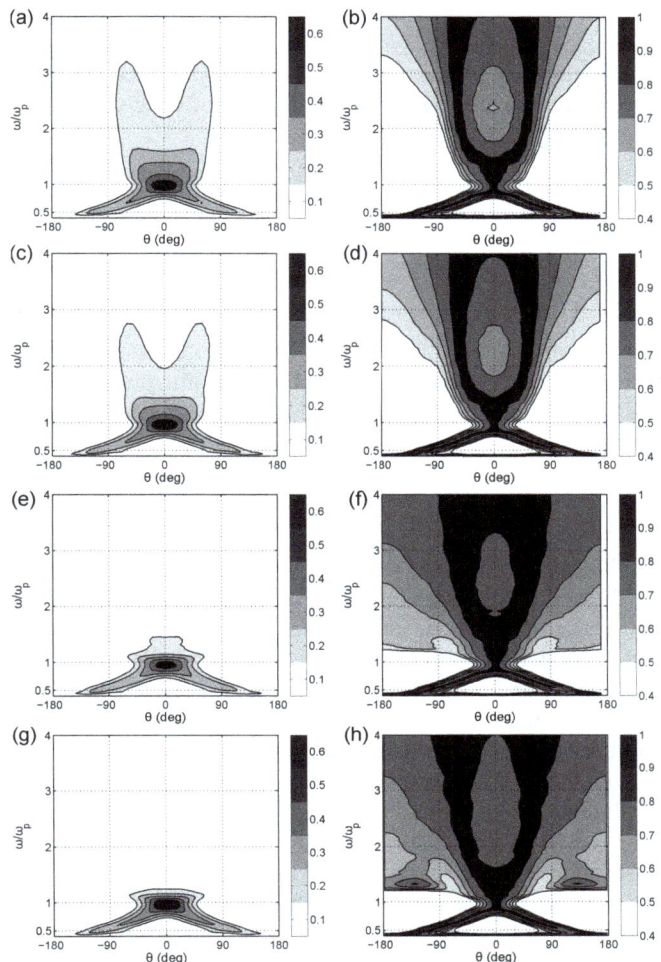

Figure 7. Isolines of spreading functions for different runs (see Table 1): **(a, b)** sw030; **(c, d)** sw170; **(e, f)** sw330; **(g, h)** run sw330 with finer resolution at angle $\Delta\theta = 5°$. Left column – definition (Eq. 32); right – directional distribution (Eq. 33).

slow evolution of integral parameters (cf. Fig. 2). This adjustment is provoked by excursion of initially "almost isotropic" distribution from an anticipated "inherent state" that, thus, stimulates wave–wave interactions as a mechanism of relaxation. The example demonstrates ability of wave–wave interactions to effectively rebuild directional distributions. Note that in some cases, say, in the problem of relaxation of wave field to sudden changes of wind direction the wave–wave interactions are considered ineffective as compared to relaxation "due mainly to imbalance $S_{in} < S_{diss}$" (e.g., Young et al., 1987, S_{in} – wind input, S_{diss} – wave dissipation).

3.4 Bi-modality of swell spectra

Bi-modality of directional spreading of ocean swell is widely discussed for experimental data as a possible result of swell evolution (e.g., Ewans, 1998, 2001; Ewans et al., 2004). Our simulations encounter this effect as a persistent feature of swell spectra. Figure 7 represents directional spreading of swell spectra in two ways. The left column shows directional distribution function $H(\omega,\theta)$ in the spirit of widely used definition (e.g., Ewans, 1998)

$$E(\omega,\theta) = \overline{E}(\omega)H(\omega,\theta), \quad H(\omega,\theta) \geq 0,$$

$$\int\limits_{-\pi}^{\pi} H(\omega,\theta)\mathrm{d}\omega = 1. \tag{32}$$

An alternative representation in the right column of Fig. 7 uses spectral densities normalized by their maxima at a fixed frequency to trace "ridges" of the surface $\widetilde{E}(\omega,\theta)$ defined as follows (cf. Eq. 1 in Young et al., 1995):

$$\widetilde{E}(\omega,\theta) = E(\omega,\theta)/\max_{-\pi<\theta\leq\pi}(E(\omega,\theta)). \tag{33}$$

Both representations reveal bi-modality of swell spectra fairly well for all cases of Table 1. "Narrow" initial spectrum sw030 and "wide" one sw170 evolve to very close X-shaped side-lobe patterns (Fig. 7a, c). Pronounced side lobes are seen both above and below the spectral peak frequency. The directional distribution function $H(\omega,\theta)$ (Eq. 32) does not show a similar pattern for "the almost isotropic" case sw330 (Fig. 7e, g), but the X-shapes are seen fairly well in the "ridge" representation (Eq. 33) for all the cases. Directional spreading for run sw330 is shown for simulations with standard angular resolution $\Delta\theta = 10°$ (Fig. 7e, f) and with fine one $\Delta\theta = 5°$ (Fig. 7g, h). Higher resolution makes "ridges" sharper and allows for resolving of more details of the directional distribution. In particular, side lobes appear for counter-propagating waves at $\theta \approx \pm 3\pi/4$ and $\omega/\omega_{\mathrm{p}} \approx 5/4$. At the same time, the standard angular resolution in our simulations $\Delta\theta = 10°$ seems to be adequate for the bi-modality phenomenon.

The patterns similar to ones of Fig. 7 have been obtained in simulations of the Hasselmann equation for wind-driven waves with the exact term of nonlinear transfer S_{nl} by Banner and Young (1994); Young et al. (1995) at formally finer resolution $\Delta\theta = 6.67°$. It should be noted that directions beyond the cone $\theta = \pm 120°$ have not been taken into account to speed up calculations in the cited papers. It can explain the discrepancy with our results at the high-frequency end of Fig. 7f, h (cf. Plate 1 in Young et al., 1995). This point can be clarified in further studies.

An important issue of agreement of our results and findings by Banner and Young (1994) and Young et al. (1995) is the presence of low-frequency (below the spectral peak) side lobes. Experimental results by Ewans (cf. Figs. 8 and 16; 1998) show good correspondence of the directional spreading functions to numerical results at high frequencies, but do not fix any side lobes below the spectral peak.

Generally, the phenomenon of side-lobe occurrence is associated with a joint effect of wave–wave interactions and wave generation by wind (e.g., Banner and Young, 1994;

Pushkarev et al., 2003; Bottema and van Vledder, 2008). The theoretical background of Sect. 2.1 and our simulations of swell can propose an interpretation and alternative ways of advanced analysis of the effect in terms of stationary solutions of Kolmogorov–Zakharov (Eq. 7). These solutions being presented as power series of dimensionless ratios of spectral fluxes and as an extension of the approximate solution Eq. (6) by Kats and Kontorovich (1974) predict higher-order angular harmonics and can be found within the formal procedure of Pushkarev et al. (2003, 2004). This approach is not fully correct in the vicinity of the spectral peak, but still looks plausible and useful for interpretation of the effect of wave–wave interactions. Analysis of the next paragraph shows perspectives of the KZ solution paradigm.

3.5 Swell spectra vs. KZ solutions

The very slow evolution of swell in our simulations provides a chance to check the relevance of the classic Kolmogorov–Zakharov solutions (Eqs. 4–7) to the problem under study. The key feature of the swell solution from the theoretical viewpoint is its "hybrid" (in the words of Badulin et al., 2005a) nature: the inverse cascade (negative fluxes) determines the evolution of the spectral peak and its downshifting, while the direct cascade (positive fluxes) occurs at frequencies slightly (approximately 20 %) above the peak. This hybrid nature is illustrated by Fig. 8 for energy and wave momentum fluxes. In order to avoid ambiguity in the treatment of the simulation results within the weak turbulence theory, we will not discuss this hybrid nature of swell solutions, and will focus on the direct cascade regime. Thus, the general solution (Eq. 7) in the form

$$E(\omega,\theta) == \frac{P^{1/3}g^{4/3}}{\omega^4}G(0, gM/(\omega P),\theta)$$

and its approximate explicit version Eq. (6) by Kats and Kontorovich (1971, 1974) will be used below for describing the direct cascading of energy and momentum at high frequency (as compared to ω_{p}).

Two runs of Table 1, sw030 and "almost isotropic" sw330, are presented in Fig. 8 in order to show qualitative similarity of extreme cases of initial directional spreading. Positive fluxes P and M decay with time, in good agreement with power-law dependencies (Eq. 22), and have rather low variations in a relatively wide frequency range $3\omega_{\mathrm{p}} < \omega < 6\omega_{\mathrm{p}}$ in Fig. 8. For energy flux P (Fig. 8a, b) one can see good quantitative correspondence (note that the times for some curves are slightly different). Absolute values of momentum flux M as well as magnitudes of wave momentum itself (see Fig. 2) differ by more than 1 order.

The domain of quasi-constant fluxes $\omega > 3\omega_{\mathrm{p}}$ can be used for verification of the relevance of the stationary KZ solutions (Eqs. 4–6) to the quasi-stationary swell solutions. All the cases of Table 1 show very close patterns of spectral

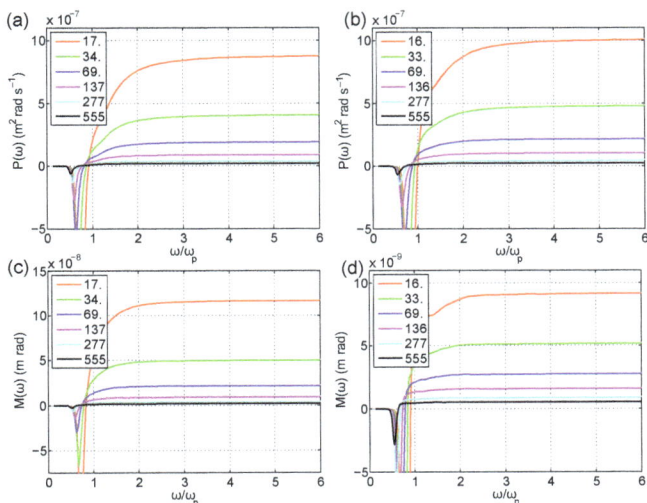

Figure 8. Top row – spectral fluxes of energy for series sw030 (**a**) and sw330 (**b**); bottom row – spectral fluxes of momentum for series sw030 (**c**) and sw330 (**d**) at different times (legend, in hours).

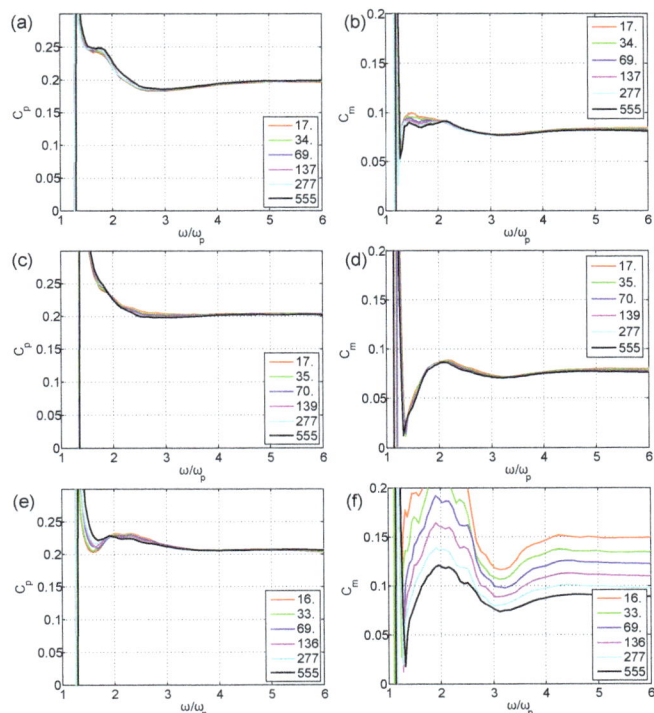

Figure 9. Left – estimates of the first Kolmogorov constant C_p; right – estimates of the second Kolmogorov constant C_m for the approximate anisotropic KZ solution (Eq. 6). (**a, b**) run sw030; (**c, d**) sw230; (**e, f**) sw330. Time in hours is given in the legend.

fluxes (e.g., Fig. 8) and, what is more important, very close estimates of Kolmogorov's constants.

The first and second Kolmogorov constants can easily be estimated for the approximate solution (Eq. 6) from combinations of along- and counter-propagating spectral densities as follows:

$$C_p = \frac{\omega^4 \left(E(\omega, 0) + E(\omega, \pi)\right)}{4g^{4/3} P^{1/3}}, \tag{34}$$

$$C_m = \frac{\omega^5 P^{2/3} \left(E(\omega, 0) - E(\omega, \pi)\right)}{4g^{7/3} M}. \tag{35}$$

These estimates provide very close values of the Kolmogorov constants for all the series of Table 1 with the only exception of "the almost isotropic" run sw330 for the second Kolmogorov constant C_m. Fig. 9 gives the first Kolmogorov constant $C_p \approx 0.21 \pm 0.01$ (slightly lower values for initially narrow distributions) and $C_m \approx 0.08 \pm 0.02$ for all the runs except sw330 (cf. Figs. 9b, d for "narrow" sw030 and "wide" sw230).

The analytic estimate gives the very close result $C_p = 0.219$ (Zakharov, 2010, Eq. 4.33). Numerical simulations by Lavrenov et al. (2002), Pushkarev et al. (2003), and Badulin et al. (2005a) missed a factor of 2 in definitions of the Kolmogorov constants (cf. our definitions in Eqs. 4–6 and Eqs. 4.29 and 4.30 in Zakharov, 2010). Taking this into account, one has the reported values $0.151 < C_p < 0.162$; $0.105 < C_m < 0.121$ in Lavrenov et al. (2002, Table 1), $0.16 < C_p < 0.23$; $0.09 < C_m < 0.14$ in Pushkarev et al. (2003, eqs. 5.3, 5.6, 5.8) and $0.19 < C_p < 0.20$ in Badulin et al. (2005a). The first experimental attempt to evaluate the first Kolmogorov constant by Deike et al. (2014) presented the value $C = 1.8 \pm 0.2 \approx 2\pi C_p$, i.e., a 2π times bigger counterpart of C_p.

While the estimates of the Kolmogorov's constants for the swell look consistent, the numerical solutions differ essentially from the approximate weakly anisotropic KZ solution (Eq. 6). The directional spreading cannot be described by the only angular harmonics as in Eq. (6); higher-order corrections are clearly seen in Fig. 7 as side lobes. Nevertheless, the robustness of the estimates of the second Kolmogorov constant C_m provides a good reference for estimates of the spectra anisotropy.

The estimates of C_m for sw330 (Fig. 9f) demonstrate a specific nonstationarity of the swell solution in terms of wave momentum flux, while the first Kolmogorov constant C_p (Fig. 9f) shows the relevance of the stationary KZ solutions to the swell problem.

4   Discussion. Swell and the ocean environment

Results of our simulations showed their fairly good correspondence to findings of the theory of wave (weak) turbulence. The relevance of these results to experimental facts seems to be a logical conclusion of this work. The issue of relevance is 2-fold. First, our results can help in explaining effects whose interpretation in terms of alternative approaches (mostly, within linear theory) is questionable. Sec-

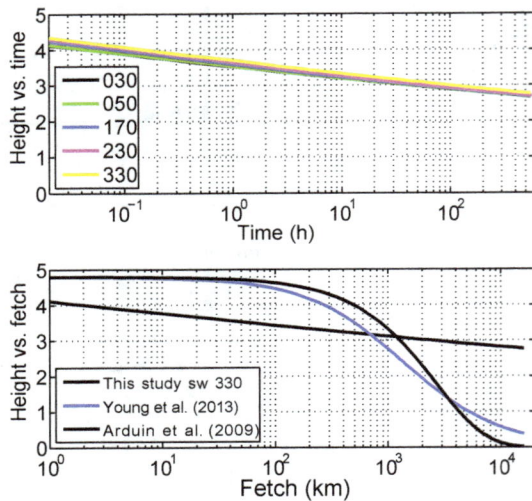

Figure 10. Top – dependence of significant wave height H_s on time for the cases of Table 1. Bottom – attenuation of swell for models Ardhuin et al. (2009); Young et al. (2013) and one of this paper (see the legend). Results of duration-limited simulations are recasted into dependencies on fetch by simple transformation (Eq. 36).

ondly, one can formulate, or, at least, sketch cases where our approach becomes invalid or requires an extension. Both aspects are considered in the final section.

Attenuation in course of long-term swell evolution is an appealing problem of the swell monitoring. We show that contribution of wave–wave interactions to this process can be important mostly at initial stages of swell evolution. The observed rates of swell attenuation in an open ocean cannot be treated within our approach for a number of reasons. First of all, the duration-limited setup of our simulations do not account for important mechanisms of frequency dispersion and spatial divergence due to sphericity of the Earth. These mechanisms can both contribute into swell attenuation together with wave–wave interactions and essentially contaminate results of observations. The intrinsic swell attenuation is, generally, small as compared to the effect of reduction (or amplification at large fetches) (see Fig. 2b in Ardhuin et al., 2009) which is accounted for within the linear model of geometrical optics whose validity is generally assumed for ocean swell.

Ocean swell for long times (fetches) likely becomes an important constituent of the ocean environment which can be heavily affected by relatively short wind-driven waves. We discuss the effect of swell amplification at rather low wind speeds and give tentative estimates based on the approach of this paper.

4.1 Swell attenuation within the kinetic equation

Dependence of wave height on time is shown in upper panel of Fig. 10 (see also Fig. 2) for the runs of Table 1. All the

runs show quantitatively close evolution. Strong drop of up to 30 % of initial value occurs within a relatively short time of about 1 day. An essential part of the wave energy leakage corresponds to this transitional stage at the very beginning of swell evolution when swell tends very rapidly to self-similar asymptotics. Afterwards, the decay becomes much slower following the power-law dependence of the self-similar solutions (Eq. 22).

For comparison with other models, and available observations, the duration-limited simulations have been recasted into dependencies of fetch through the simplest time-to-fetch transformation (e.g., Hwang and Wang, 2004; Hwang, 2006):

$$x(s) = \int_0^s C_g(\omega_p(t))dt. \tag{36}$$

The equivalent fetch is estimated as a distance covered by a wave guide travelling with the group velocity of the spectral peak component. The corresponding dependencies are shown in bottom panel in Fig. 10. Two quasi-linear models by Ardhuin et al. (2009) and Babanin (2006) predict relatively slow attenuation at fetches in a "near zone" less than 1000 km (approximately 1 day) and then gradual decay up to very few of the percentage points of initial value at final distances about 18000 km where our model shows qualitatively different weak attenuation.

It should be noted that our model describes attenuation of the ocean swell "on its own" due to wave–wave interactions without any external effects. Thus, the effect of an abrupt drop in wave amplitude at short times (fetch) should be taken into consideration above all others when discussing the possible application of our results to swell observations and physical interpretation of the experimental results.

4.2 Swell and wind-sea coupling. Arrest of weakly turbulent cascading

Extremely weak attenuation of swell due to wave–wave interactions provokes a question on robustness of this effect. A variety of physical mechanisms in the ocean environment can change the swell evolution qualitatively. The above discussion of swell attenuation presents a remarkable example of such transformation when dissipation becomes dominant. Tracking of swell events from space gives an alternative scenario of transformation when swell appears to be growing. Satellite tracks can comprise up to 30 % of cases of growing swell; "most of them are not statistically significant" (Jiang et al., 2016). Nevertheless, a possible effect of wind-sea background on long ocean swell opens an important discussion in view of theoretical (Badulin et al., 2008b) and experimental (Benilov et al., 1974; Badulin and Grigorieva, 2012) results that demonstrate swell amplification by a wind-wave background.

As noted and shown above, evolution of swell can occur at different timescales for different physical quantities.

Integrals of motion (energy, action, momentum) evolve at relatively large scales: frequency downshift and energy follows power-law dependencies 1/11 ($\omega_p \sim t^{-1/11}$ and $E \sim t^{-1/11}$). The slow evolution is supported by interactions within a wave spectra that is close to an "inherent" quasi-stationary state.

Oppositely, spectral shaping is evolving due to excursions from an "inherent state" at much shorter scales that can be estimated following Zakharov and Badulin (2011, see Eqs. 21 and 22 therein). The nonlinear relaxation rate as defined by eqs. 14–16 of the cited paper can be written as

$$\Gamma(\omega) = B\omega \left(\frac{\omega}{\omega_p} \right)^3 \mu^4 H(\omega, \theta). \tag{37}$$

Here B is a big dimensionless coefficient (e.g., $B = 22.5\pi \approx 70.7$ for an isotropic spectrum; see Zakharov and Badulin, 2011) and $H(\omega, \theta)$ is the directional distribution function (Eq. 32). The big coefficient B in (37) provides relatively fast relaxation of local excursions (in wave scales) from the slowly evolving "inherent" swell, especially, in the high-frequency domain (factor $(\omega/\omega_p)^3$ in Eq. 37). Evidence of this relaxation can be seen in the evolution of the angular distribution of run sw330, where visible transformation of angular distribution is observed for the entire duration of more than 3 weeks (Fig. 6): the non-self-similar background of the swell spectra is feeding the core of the spectral distribution.

A similar effect can be realized in the mixed sea when background of relatively short wind-driven waves feeds the swell. Total energy flux of the swell is decaying as rapidly as $dE/dt \sim t^{-12/11}$ and at sufficiently large time the associated direct cascading can be arrested by inverse cascading of wind-driven waves which fast relaxation to an "inherent" swell ensures the swell feeding. This mechanism has been analyzed numerically (Badulin et al., 2008b) and showed its remarkable efficiency.

Simple estimates of the possibility of the effect can be made in terms of balancing of two fluxes: direct cascade of swell and inverse cascade of wind-driven fraction. The swell energy leakage can be estimated from the weakly turbulent law (Badulin et al., 2007, eq.1.9) as follows:

$$\left(\frac{dE}{dt} \right)_{direct} = \frac{E^3 \omega_p^9}{\alpha_{swell}^3 g^4} = \frac{\mu_{swell}^6 C_{swell}^3}{\alpha_{swell}^3 g}. \tag{38}$$

Here swell parameters are marked by proper subscripts: $C_{swell} = g/\omega_p$ – phase velocity of the spectral peal component; μ_{swell} – swell steepness by definition (Eq. 15); and α_{swell} – self-similarity parameter (α_{ss} in Badulin et al., 2007). Similar conversion of sea state parameters to spectral flux can be done for the wind-sea fraction (see Sect. 5.1 in Badulin et al., 2007, or Table 1 in Gagnaire-Renou et al., 2011):

$$\left(\frac{dE}{dt} \right)_{inverse} \approx C_w \left(\frac{\rho_w}{\rho_w} \right)^3 \frac{U_{10}^3}{\alpha_{wind}^3 g}, \tag{39}$$

where coefficient $C_w = O(1)$ is introduced as soon as the conversion is based on dimensonal analysis and generalization of experimental results (Toba, 1972). A counterpart of α_{swell}, the self-similarity parameter α_{wind}, is approximately 2 times less in magnitude (Badulin et al., 2007). Thus, the condition of the balance of fluxes assotiated with different fractions of the mixed sea (Eqs. 38, 39) says that

$$2C_w \frac{\rho_a}{\rho_w} \frac{U_{10}}{C_{swell}} \approx \mu_{swell}^2. \tag{40}$$

For relatively short swell with period $T_p = 10\,\text{s}$ ($\lambda \approx 150\,\text{m}$) and wind speed $U_{10} = 7\,\text{m s}^{-1}$ one gets a critical swell steepness $\mu_{swell} \approx 0.03$. In other words, the mean-over-ocean wind $7\,\text{m s}^{-1}$ can balance (arrest) direct cascading of rather steep swell and, hence, provoke a growth of the swell due to absorbing short wind-driven waves. Evidently, this simple balance model gives very tentative estimate of the effect. Nevertheless, visual observations (Badulin and Grigorieva, 2012) and satellite data (Jiang et al., 2016), in our opinion, provide telling arguments for this phenomenon. Thus, "negative dissipation" of swell (in the words of Jiang et al., 2016) could find its explanation within the simple model.

The simple estimate (Eq. 40) shows the limited value of our "pure swell" model for the ocean environment. Potentially, the effect of even light wind on long-term propagation of swell can change the result qualitatively. Our pilot numerical studies (see also Badulin et al., 2008b) show the importance of the swell and wind-sea coupling. This effect will be detailed in our further studies.

5 Conclusions

We presented results of sea swell simulations within the framework of the kinetic equation for water waves (the Hasselmann equation) and treated these properties within the paradigm of the theory of weak turbulence. A series of numerical experiments (duration-limited setup, WRT algorithm) has been carried out in order to outline features of wave spectra in a range of scales usually associated with ocean swell, i.e., wavelengths larger than $100\,\text{m}$ and duration of propagation up to $2 \times 10^6\,\text{s}$ (more than 23 days). It should be stressed that the exact collision integral S_{nl} (nonlinear transfer term) has been used in all the calculations. Alternative, mostly operational approaches, like the DIA (discrete interaction approximation approach), can corrupt the results quantitatively and even qualitatively.

Key results of the study are the following.

1. A strong tendency for self-similar asymptotics is demonstrated. These asymptotics are shown to be insensitive to initial conditions in terms of evolution of integral quantities (wave energy, momentum). Moreover, universal angular distributions of wave spectra at large times have been obtained for both narrow (initial angular spreading 30°) and almost isotropic initial spectra.

Bi-modality of the spectral distributions in our simulations is found to be in agreement with previous numerical and experimental results (Banner and Young, 1994; Ewans, 2001; Ewans et al., 2004). The universality of the spectral shaping can be treated as an effect of mode selection when very few eigenmodes of the boundary problem determine the system evolution. The inherent features of wave–wave interactions are responsible for this universality making the effect of initial conditions insignificant. Generally, the self-similar swell co-exists with a background, which is far from a self-similar state.

2. The classic Kolmogorov–Zakharov (KZ) isotropic and weakly anisotropic solutions for direct and inverse cascades are shown to be relevant to slowly evolving sea swell solutions. Estimates of the corresponding KZ constants are found to agree well with previous analytical, numerical and experimental results. Thus, features of KZ solutions can be used as a reference for advanced approaches in the swell studies.

3. We show that an inherent peculiarity of the Hasselmann equation, energy and momentum leakage, can also be considered a mechanism of the sea swell attenuation. Today's models of sea swell are unlikely to account for this effect. Possible problems of the models are sketched in Sect. 3.1 when different options of simulation of the "conservative dissipation" are discussed. All these options require sufficiently large high-frequency range where the short-term oscillations in absence of dissipation or hyper-viscosity can mimic the energy leakage at $|k| \to \infty$. It should be noted that the energy decay rates of sea swell in the numerical experiments, generally, do not contradict the results of recent swell observations and modeling. These studies based on satellite data and wave model hindcasting are focused mostly on "far field" behavior of swell, generally, 1000 or more kilometers away from a stormy area. Our simulations show that a dramatic transformation of the swell occurs at shorter distances, in the "near field". The essential swell energy losses in the near field, mostly due to nonlinear transfer, is an intriguing challenge for sea wave forecasting since the very first discussions of the phenomenon within the concept of wave–wave interactions (e.g., Sect. 8a, b in Snodgrass et al., 1966). Thus, Fig. 10 outlines different domains of our model relevance rather than the model relevance for the general problem of ocean swell attenuation.

4. Long-term evolution of swell is associated with rather slow frequency downshift ($\omega_p \sim t^{-1/11}$) and energy attenuation ($E \sim t^{-1/11}$). Meanwhile, the decay of other wave field quantities is essentially faster: wave steepness is decaying as $\mu \sim t^{-5/22}$, and total spectral flux is even faster: $dE/dt \sim t^{-12/11}$. This point is of key importance in our analysis as far as we consider non-

linear cascades of wave energy as governing the physical mechanism of swell evolution. As we showed in the discussion, the weak direct cascade of swell can be arrested by relatively light winds and then swell can start to grow. In our opinion, this conclusion correlates with manifestations of swell amplification in satellite data (Jiang et al., 2016) and in visual observations (Badulin and Grigorieva, 2012). Thus, "negative dissipation" of swell (in the words of Jiang et al., 2016) could find its explanation within the simple estimate (Eq. 40) of Sect. 4.2.

5. The last conclusion uncovers deficiency of the duration-limited setup for the phenomenon of swell. An alternative setup of fetch-limited evolution ($\partial/\partial t \equiv 0$, $\nabla_r \neq 0$) introduces dispersion of wave harmonics as a competing mechanism that can change the swell evolution dramatically. Recent advances in wave modeling (Pushkarev and Zakharov, 2016) make the problem of spatial–temporal swell evolution feasible and specify the perspectives of our first step study. The theoretical background for the classic fetch-limited setup when solutions depend on the only spatial coordinate (i.e., $\partial/\partial x \neq 0$, $\partial/\partial y \equiv 0$) is sketched in Sect. 2 of this paper. The one-dimensional model adds an essential physical effect of wave dispersion. A passage to polar coordinates allows us to consider an effect of spatial divergence in formally one-dimensional problem where solutions depend on the radial coordinate but are still anisotropic in wavevector space. Self-similar solutions for this problem in the spirit of Sect. 2 can be easily found and related to numerical results. All the prospective simulations require developing effective numerical approaches. In particular, high angular resolution (not worse than 5°) could be recommended for these studies. V. Geogjaev and V. Zakharov has developed such code recently (a talk at the meeting Waves in Shallow Water Environment, 2016, Venice). We plan to use it in the swell studies.

Competing interests. The authors declare that they have no conflict of interest.

Acknowledgements. The authors are thankful for the support of Russian Science Foundation grant no. 14-22-00174. The authors are indebted to Victor Shrira and Vladimir Geogjaev for discussions and valuable comments. The authors are also grateful to Andrei Pushkarev for his assistance in simulations. The authors appreciate critical consideration of the paper by reviewers Gerbrant van Vledder and Sergei Annenkov. Their constructive feedback led to substantial revision of Sects. 3 and 4.

Edited by: V. Shrira

References

Annenkov, S. Y. and Shrira, V. I.: Direct Numerical Simulation of Downshift and Inverse Cascade forWater Wave Turbulence, Phys. Rev. Lett., 96, 204501, https://doi.org/10.1103/PhysRevLett.96.204501, 2006.

Ardhuin, F., Chapron, B., and Collard, F.: Observation of swell dissipation across oceans, Geophys. Res. Lett., 36, L06607, https://doi.org/10.1029/2008GL037030, 2009.

Ardhuin, F., Rogers, E., Babanin, A. V., Filipot, J.-F., Magne, R., Roland, A., van der Westhuysen, A., Queffeulou, P., Aouf, J.-M. L. L., and Collard, F.: Semiempirical Dissipation Source Functions for Ocean Waves. Part I: Definition, Calibration, and Validation, J. Phys. Oceanogr., 40, 1917–1941, https://doi.org/10.1175/2010JPO4324.1, 2010.

Babanin, A. V.: On a wave-induced turbulence and a wave mixed upper ocean layer, Geophys. Res. Lett., 33, L20605, https://doi.org/10.1029/2006GL027308, 2006.

Badulin, S. I. and Grigorieva, V. G.: On discriminating swell and wind-driven seas in Voluntary Observing Ship data, J. Geophys. Res., 117, https://doi.org/10.1029/2012JC007937, 2012.

Badulin, S. I., Pushkarev, A. N., Resio, D., and Zakharov, V. E.: Direct and inverse cascade of energy, momentum and wave action in wind-driven sea, in: 7th International workshop on wave hindcasting and forecasting, 92–103, Banff, October 2002, available at: https://www.researchgate.net/publication/253354120_Direct_and_inverse_cascades_of_energy_momentum_and_wave_action_in_spectra_of_wind-driven_waves (last access: 20 May 2017), 2002.

Badulin, S. I., Pushkarev, A. N., Resio, D., and Zakharov, V. E.: Self-similarity of wind-wave spectra. Numerical and theoretical studies, in: Rogue Waves 2004, edited by: Olagnon, M. and Prevosto, M., Editions IFREMER, Brest, 39, 205–214, 2004.

Badulin, S. I., Pushkarev, A. N., Resio, D., and Zakharov, V. E.: Self-similarity of wind-driven seas, Nonlin. Processes Geophys., 12, 891–945, https://doi.org/10.5194/npg-12-891-2005, 2005a.

Badulin, S. I., Pushkarev, A. N., Resio, D., and Zakharov, V. E.: Universality of wind-wave spectra and exponents of wind-wave growth, in: Geophysical Research Abstracts, 7, 01515 pp., EGU, 2005b.

Badulin, S. I., Babanin, A. V., Resio, D., and Zakharov, V.: Weakly turbulent laws of wind-wave growth, J. Fluid Mech., 591, 339–378, 2007.

Badulin, S. I., Babanin, A. V., Resio, D., and Zakharov, V.: Numerical verification of weakly turbulent law of wind wave growth, in: IUTAM Symposium on Hamiltonian Dynamics, Vortex Structures, Turbulence, Proceedings of the IUTAM Symposium held in Moscow, 25–30 August, 2006, edited by: Borisov, A. V., Kozlov, V. V., Mamaev, I. S., and Sokolovskiy, M. A., vol. 6 of IUTAM Bookseries, Springer, 175–190, ISBN: 978-1-4020-6743-3, 2008a.

Badulin, S. I., Korotkevich, A. O., Resio, D., and Zakharov, V. E.: Wave-wave interactions in wind-driven mixed seas, in: Proceed-

ings of the Rogue waves 2008 Workshop, IFREMER, Brest, France, 77–85, 2008b.

Badulin, S. I., Zakharov, V. E., and Pushkarev, A. N.: Simulation of wind wave growth with reference source functions, Geophysical Research Abstracts, 15, EGU2013-5284-1, 2013.

Balk, A. M.: On the Kolmogorov-Zakharov spectra of weak turbulence, Phys. D: Nonlin. Phenom., 139, 137–157, 2000.

Banner, M. L. and Young, I. R.: Modeling Spectral Dissipation in the Evolution of Wind Waves. Part I: Assessment of Existing Model Performance, J. Phys. Oceanogr., 24, 1550–1571, https://doi.org/10.1175/1520-0485(1994)024<1550:MSDITE>2.0.CO;2, 1994.

Barber, N. F. and Ursell, F.: The Generation and Propagation of Ocean Waves and Swell. I, Wave Periods and Velocities, Philos. T. R. Soc. Lond., 240, 527–560, 1948.

Barrenblatt, G. I.: Scaling, self-similarity, and intermediate asymptotics: Dimensional analysis and intermediate asymptotics, Plenum Press, New York/London, 1979.

Beal, R. C., Kudryavtsev, V. N., Thompson, D. R., Grodsky, S. A., Tilley, D. G., Dulov, V. A., and Graber, V. A.: The influence of the marine atmospheric boundary layer on ERS-1 synthetic aperture radar imagery of the Gulf Stream, J. Geophys. Res., 102, 5799–5814, 1997.

Benilov, A. Y., Kouznetsov, O. A., and Panin, G. N.: On the analysis of wind wave-induced disturbances in the atmospheric turbulent surface layer, Bound.-Lay. Meteorol., 6, 269–285, 1974.

Benoit, M. and Gagnaire-Renou, E.: Interactions vague-vague non-linéaires et spectre d'équilibre pour les vagues de gravité en grande profondeur d'eau, in: Proc. 18th Congrès Français de Mécanique, Grenoble (France), 2007.

Bottema, M. and van Vledder, G. P.: Effective fetch and non-linear four-wave interactions during wave growth in slanting fetch conditions, Coastal Eng., 55, 261–275, 2008.

Bottema, M. and van Vledder, G. P.: A ten-year data set for fetch- and depth-limited wave growth, Coastal Eng., 56, 703–725, 2009.

Chen, K. S., Chapron, B., and Ezraty, R.: A Global View of Swell and Wind Sea Climate in the Ocean by Satellite Altimeter and Scatterometer, J. Atmos. Ocean. Technol., 19, 1849–1859, 2002.

Deike, L., Miquel, B., Gutiérrez, P., Jamin, T., Semin, B., Berhanu, M., Falcon, E., and Bonnefoy, F.: Role of the basin boundary conditions in gravity wave turbulence, J. Fluid Mech., 781, 196–225, https://doi.org/10.1017/jfm.2015.494, 2014.

Delpey, M. T., Ardhuin, F., Collard, F., and Chapron, B.: Space-time structure of long ocean swell fields, J. Geophys. Res., 115, https://doi.org/10.1029/2009JC005885, 2010.

Donelan, M. A., Hamilton, J., and Hui, W. H.: Directional spectra of wind-generated waves, Philos. T. R. Soc. Lond., 315, 509–562, 1985.

Ewans, K., Forristall, G. Z., Prevosto, M. O. M., and Iseghem, S. V.: WASP West Africa Swell Project, Final report, Ifremer – Centre de Brest, Shell International Exploration and Production, B.V., 2004.

Ewans, K. C.: Observations of the Directional Spectrum of Fetch-Limited Waves, J. Phys. Oceanogr., 28, 495–512, 1998.

Ewans, K. C.: Directional spreading in ocean swell, in: The Fourth International Symposium on Ocean Wave Measurement and Analysis, ASCE, San Francisco, 2001.

Gagnaire-Renou, E., Benoit, M., and Forget, P.: Ocean wave spectrum properties as derived from quasi-exact computations of nonlinear wave-wave interactions, J. Geophys. Res., 115, 2156–2202, https://doi.org/10.1029/2009JC005665, 2010.

Gagnaire-Renou, E., Benoit, M., and Badulin, S. I.: On weakly turbulent scaling of wind sea in simulations of fetch-limited growth, J. Fluid Mech., 669, 178–213, 2011.

Hasselmann, K.: On the nonlinear energy transfer in a gravity wave spectrum. Part 1. General theory, J. Fluid Mech., 12, 481–500, 1962.

Hasselmann, K., Ross, D. B., Müller, P., and Sell, W.: A parametric wave prediction model, J. Phys. Oceanogr., 6, 200–228, 1976.

Henderson, D. M. and Segur, H.: The role of dissipation in the evolution of ocean swell, J. Geophys. Res.-Oceans, 118, 5074–5091, https://doi.org/10.1002/jgrc.20324, 2013.

Hwang, P. A.: Duration and fetch-limited growth functions of wind-generated waves parameterized with three different scaling wind velocities, J. Geophys. Res., 111, 2156–2202, https://doi.org/10.1029/2005JC003180, 2006.

Hwang, P. A. and Wang, D. W.: Field measurements of duration-limited growth of wind-generated ocean surface waves at young stage of development, J. Phys. Oceanogr., 34, 2316–2326, 2004.

Jiang, H., Stopa, J. E., Wang, H., Husson, R., Mouche, A., Chapron, B., and Chen, G.: Tracking the attenuation and nonbreaking dissipation of swells using altimeters, J. Geophys. Res.-Oceans, 121, 2169–9291, https://doi.org/10.1002/2015JC011536, 2016.

Kahma, K. K. and Pettersson, H.: Wave growth in a narrow fetch geometry, Glob. Atmos. Ocean Syst., 2, 253–263, 1994.

Kantha, L.: A note on the decay rate of swell, OM, 11, 167–173, https://doi.org/10.1016/j.ocemod.2004.12.003, 2006.

Kats, A. V. and Kontorovich, V. M.: Drift stationary solutions in the weak turbulence theory, JETP Letters, 14, 265–267, 1971.

Kats, A. V. and Kontorovich, V. M.: Anisotropic turbulent distributions for waves with a non-decay dispersion law, Soviet Physics JETP, 38, 102–107, 1974.

Lavrenov, I., Resio, D., and Zakharov, V.: Numerical simulation of weak turbulent Kolmogorov spectrum in water surface waves, in: 7th International workshop on wave hindcasting and forecasting, 104–116, Banff, October 2002, available at: https://www.researchgate.net/publication/312210314_Lavr_7th_Workshop (last access: 30 May 2017), 2002.

Munk, W. H., Miller, G. R., Snodgrass, F. E., and Barber, N. F.: Directional Recording of Swell from Distant Storms, Philos. T. R. Soc. Lond., 255, 505–584, 1963.

Onorato, M., Osborne, A., Serio, M., Resio, D., Pushkarev, A., Brandini, C., and Zakharov, V. E.: Freely Decaying Weak Turbulence for Sea Surface Gravity Waves, Phys. Rev. Lett., 89, 144501, https://doi.org/10.1103/PhysRevLett.89.144501, 2002.

Pettersson, H.: Wave growth in a narrow bay, PhD thesis, University of Helsinki, available at: http://ethesis.helsinki.fi/julkaisut/mat/fysik/vk/pettersson/ (last access: 30 May 2017), [ISBN 951-53-2589-7 (Paperback) ISBN 952-10-1767-8 (PDF)], 2004.

Phillips, O. M.: On the dynamics of unsteady gravity waves of finite amplitude, J. Fluid Mech., 9, 193–217, 1960.

Pushkarev, A. and Zakharov, V.: On nonlinearity implications and wind forcing in Hasselmann equation, ArXiv e-prints, 2015.

Pushkarev, A. and Zakharov, V.: Limited fetch revisited: Comparison of wind input terms, in surface wave modeling, Ocean Mod-

ell., 103, 18–37, https://doi.org/10.1016/j.ocemod.2016.03.005, 2016.

Pushkarev, A. N., Resio, D., and Zakharov, V. E.: Weak turbulent approach to the wind-generated gravity sea waves, Phys. D: Nonlin. Phenom., 184, 29–63, 2003.

Pushkarev, A., Resio, D., and Zakharov, V.: Second generation diffusion model of interacting gravity waves on the surface of deep fluid, Nonlin. Processes Geophys., 11, 329–342, https://doi.org/10.5194/npg-11-329-2004, 2004.

Snodgrass, F. E., Groves, G. W., Hasselmann, K. F., Miller, G. R., Munk, W. H., and Powers, W. H.: Propagation of Ocean Swell across the Pacific, Philos. T. R. Soc. Lond., 259, 431–497, 1966.

Toba, Y.: Local balance in the air-sea boundary processes. Part I. On the growth process of wind waves, J. Oceanogr. Soc. Jpn., 28, 109–121, 1972.

Toffoli, A., Onorato, M., Bitner-Gregersen, E. M., and Monbaliu, J.: Development of a bimodal structure in ocean wave spectra, J. Geophys. Res., 115, 2156–2202, https://doi.org/10.1029/2009JC005495, 2010.

Tracy, B. and Resio, D.: Theory and calculation of the nonlinear energy transfer between sea waves in deep water, WES Rep. 11, US Army, Engineer Waterways Experiment Station, Vicksburg, MS, 1982.

Tsimring, L. S.: On the theory of swell decay under opposing wind, Meteorologiya i Gydrologiya, 6, 76–81, 1986 (in Russian).

Webb, D. J.: Non-linear transfers between sea waves, Deep-Sea Res., 25, 279–298, 1978.

Young, I. R.: Directional spectra of hurricane wind waves, J. Geophys. Res., 111, 2156–2202, https://doi.org/10.1029/2006JC003540, 2006.

Young, I. R., Hasselmann, S., and Hasselmann, K.: Computations of the response of a wave spectrum to a sudden change in wind direction, J. Phys. Oceanogr., 17, 1317–1338, 1987.

Young, I. R., Verhagen, L. A., and Banner, M. L.: A note on the bimodal directional spreading of fetch-limited wind waves, J. Geophys. Res., 100, 773–778, https://doi.org/10.1029/94JC02218, 1995.

Young, I. R., Babanin, A. V., and Zieger, S.: The Decay Rate of Ocean Swell Observed by Altimeter, J. Phys. Oceanogr., 43, 2322–2333, 2013.

Zakharov, V. E.: Statistical theory of gravity and capillary waves on the surface of a finite-depth fluid, Eur. J. Mech. B/Fluids, 18, 327–344, 1999.

Zakharov, V. E.: Theoretical interpretation of fetch limited wind-drivensea observations, Nonlin. Processes Geophys., 12, 1011–1020, https://doi.org/10.5194/npg-12-1011-2005, 2005.

Zakharov, V. E.: Energy balance in a wind-driven sea, Phys. Scr., T142, 014052, https://doi.org/10.1088/0031-8949/2010/T142/014052, 2010.

Zakharov, V. E. and Badulin, S. I.: On Energy Balance in Wind-Driven Seas, Doklady Earth Sciences, 440, 1440–1444, 2011.

Zakharov, V. E. and Filonenko, N. N.: Energy spectrum for stochastic oscillations of the surface of a fluid, Soviet Phys. Dokl., 160, 1292–1295, 1966.

Zakharov, V. E. and Pushkarev, A. N.: Diffusion model of interacting gravity waves on the surface of deep fluid, Nonlin. Processes Geophys., 6, 1–10, https://doi.org/10.5194/npg-6-1-1999, 1999.

Zakharov, V. E. and Zaslavsky, M. M.: Kinetic equation and Kolmogorov spectra in the weak-turbulence theory of wind waves, Izv. Atmos. Ocean. Phys., 18, 970–980, 1982.

Zakharov, V. E., Lvov, V. S., and Falkovich, G.: Kolmogorov spectra of turbulence, Part I, Springer, Berlin, 1992.

Zakharov, V. E., Korotkevich, A. O., Pushkarev, A. N., and Resio, D.: Coexistence of Weak and Strong Wave Turbulence in a Swell Propagation, Phys. Rev. Lett., 99, 164501, https://doi.org/10.1103/PhysRevLett.99.164501, 2007.

Zakharov, V. E., Badulin, S. I., Hwang, P. A., and Caulliez, G.: Universality of Sea Wave Growth and Its Physical Roots, J. Fluid Mech., 708, 503–535, https://doi.org/10.1017/jfm.2015.468, 2015.

Zaslavskii, M. M.: Nonlinear evolution of the spectrum of swell, Izv. Atmos. Ocean. Phys., 36, 253–260, 2000.

9

Quantifying the changes of soil surface microroughness due to rainfall impact on a smooth surface

Benjamin K. B. Abban[1], A. N. (Thanos) Papanicolaou[1,5], Christos P. Giannopoulos[1], Dimitrios C. Dermisis[2], Kenneth M. Wacha[3], Christopher G. Wilson[1], and Mohamed Elhakeem[4]

[1]Hydraulics and Sedimentation Lab, Department of Civil & Environmental Engineering, University of Tennessee – Knoxville, Knoxville, TN 37996, USA
[2]College of Engineering, Department of Chemical, Civil & Mechanical Engineering, McNeese State University, Lake Charles, LA 70605, USA
[3]USDA-ARS – National Laboratory for Agriculture and the Environment, Ames, IA 50011, USA
[4]Abu Dhabi University, Abu Dhabi, P.O. Box 59911, Abu Dhabi, United Arab Emirates
[5]Tennessee Water Resources Center, Knoxville, TN 37996, USA

Correspondence to: Athanasios Thanos N. Papanicolaou (tpapanic@utk.edu)

Abstract. This study examines the rainfall-induced change in soil microroughness of a bare smooth soil surface in an agricultural field. The majority of soil microroughness studies have focused on surface roughness on the order of ~ 5–$50\,\text{mm}$ and have reported a decay of soil surface roughness with rainfall. However, there is quantitative evidence from a few studies suggesting that surfaces with microroughness less than 5 mm may undergo an increase in roughness when subject to rainfall action. The focus herein is on initial microroughness length scales on the order of 2 mm, a low roughness condition observed seasonally in some landscapes under bare conditions and chosen to systematically examine the increasing roughness phenomenon. Three rainfall intensities of 30, 60, and 75 mm h^{-1} are applied to a smoothened bed surface in a field plot via a rainfall simulator. Soil surface microroughness is recorded via a surface-profile laser scanner. Several indices are utilized to quantify the soil surface microroughness, namely the random roughness (RR) index, the crossover length, the variance scale from the Markov–Gaussian model, and the limiting difference. Findings show a consistent increase in roughness under the action of rainfall, with an overall agreement between all indices in terms of trend and magnitude. Although this study is limited to a narrow range of rainfall and soil conditions, the results suggest that the outcome of the interaction between rainfall and a soil surface can be different for smooth and rough surfaces
and thus warrant the need for a better understanding of this interaction.

1 Introduction

Soil surface roughness influences many hydrologic processes such as flow partitioning between runoff and infiltration, flow unsteadiness, and soil mobilization and redeposition on scales ranging from a few millimeters to hillslope level (e.g., Huang and Bradford, 1990; Magunda et al., 1997; Zhang et al., 2014).

There are three distinct classes of microtopography surface roughness (Fig. 1a) for agricultural landscapes, each one of them depicting a representative length scale (Römkens and Wang, 1986; Potter, 1990). Following Oades and Waters (1991), the first class includes microrelief variations from individual soil grains to aggregates on the order of 0.053–2.0 mm. The second class consists of variations due to soil clods ranging between 2 and 100 mm. The third class of soil surface roughness is systematic elevation differences due to tillage, referred to as oriented roughness (OR), ranging between 100 and 300 mm.

From those outlined above, the first two classes are the so-called random roughness (RR), and constitute the main focus of the present research. RR is quantified on a sur-

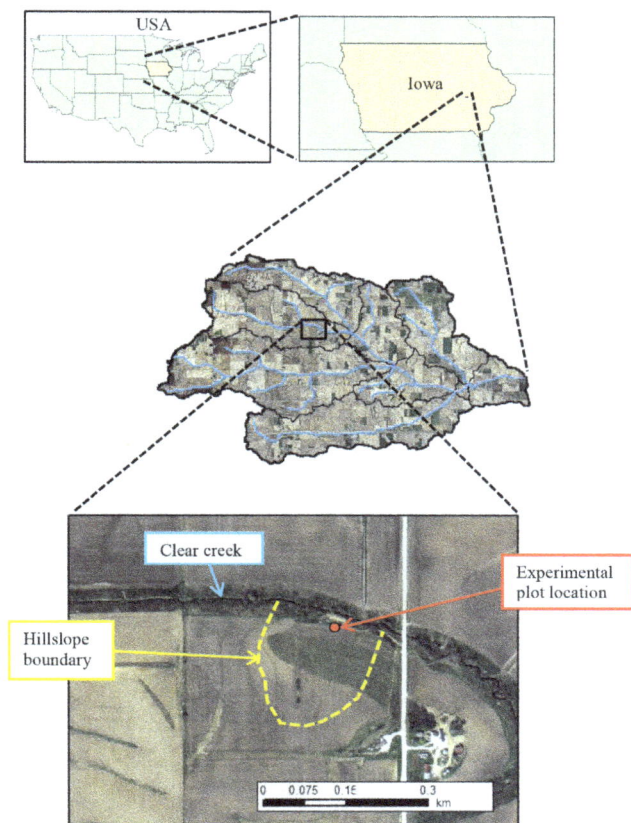

Figure 1. Location of experimental plot in the headwaters of Clear Creek, IA (41.74° N, −91.94° W).

ferred to as "rough". There are some quantitative indications that under bare smooth surface conditions, soil surface roughness may actually increase under the action of rainfall. Specifically, the study by Huang and Bradford (1992) calculated the semivariance with respect to length scale before and after rainfall, and an increase in roughness with rainfall was denoted using the Markov–Gaussian model for a surface with low initial roughness. Rosa et al. (2012) introduced an index (called the roughness index) estimated from the semivariogram to describe roughness, and an increase in the index with rainfall was observed under some conditions, and attributed to the fragmentation of aggregates and clods to smaller aggregates. Zheng et al. (2014) also reported an increase in values of the RR after the application of rainfall on smooth soil surfaces. However, none of the above studies acknowledged or related the increasing trend in surface microroughness to rainfall impact on smooth surfaces.

The main goal of this study is to examine changes in RR under rainfall impact for initial microroughness less than 2 mm, since this appears to be the lower limit of roughness scales examined in the literature. It is postulated that an increase in microroughness may occur under the action of rainfall on preexisting smooth surfaces due to the nature of the interaction between rainfall and the soil surface. To meet the goal, we employ four commonly used indices: the RR index, the crossover length, the variance scale from the Markov–Gaussian model, and the limiting difference. The last three indices are alternate methods and used here to supplement the RR index analysis for relative change in roughness.

2 Materials and methods

2.1 Experimental conditions

This study was conducted on an experimental plot of the US National Science Foundation Intensively Managed Landscapes Critical Zone Observatory in the headwaters of Clear Creek, IA (41.74° N, −91.94° W and an elevation of 250 m a.s.l. – above mean sea level; Figs. 1 and 2). The soil series at the plot where the experiments were conducted is Tama (fine-silty, mixed, superactive, Mesic Cumulic Endoaquoll) (http://criticalzone.org/iml/infrastructure/field-areas-iml/). It consists of 5 % sand, 26 % clay, 68 % silt, and an organic matter content of 4.4 %. The aggregate size distribution of the soil consists of 19 % of the soil size fraction less than 250 µm, 48 % between 250 µm and 2 mm, and 33 % greater than 2 mm. These soils contain both smectite and illite, with high cation exchange capacity between 15 and 30 cmol$_c$ kg^{-1}). The experimental plot was uniform in terms of downslope curvature, its gradient was 9 %, and the plot size was approximately 7 m long by 1.2 m wide.

The soil surface was prepared before each experiment by tamping using a plywood board to create a smoothened surface. This was done to ensure a consistency in surface rough-

face after correction for both slope and tillage marks. Contrary to OR, which changes seasonally and during crop rotations, RR changes on an event basis (Abaci and Papanicolaou, 2009). RR reflects the effects of rainfall action on the soil surface and inherently varies in space and time. As a result, RR affects key hydrologic processes at the soil scape and ultimately on the hillslope scale, e.g., infiltration, overland flow, etc. (Gómez and Nearing, 2005; Chi et al., 2012).

Several studies have been performed to characterize RR. Most have focused on initial microroughness length scales of 5–50 mm (e.g., Zobeck and Onstad, 1987; Gilley and Finkner, 1991). In these studies, a decay of roughness due to precipitation action is predicted, since rainfall impact and runoff "smoothen" the rough edges of soil grains, aggregates, and clods, especially in the absence of cover (Potter, 1990; Bertuzzi et al., 1990; Vázquez et al., 2008; Vermang et al., 2013). There are few studies that have examined surfaces with initial microroughness less than 5 mm, a low roughness condition observed seasonally in some landscapes under bare conditions (e.g., Kamphorst et al., 2000; Vázquez et al., 2008; Zheng et al., 2014). Hereafter, for shortness, tests with initial RR less than 5 mm will be referred to as "smooth", whereas tests with initial RR greater than 5 mm will be re-

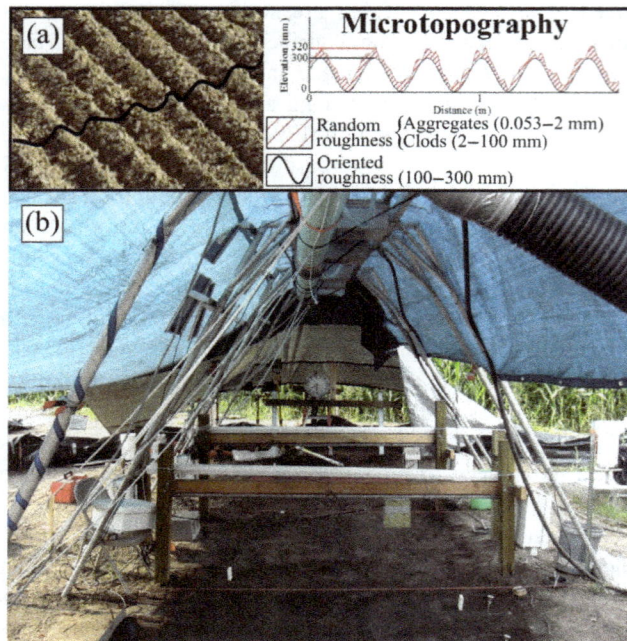

Figure 2. (a) Types of soil surface microroughness. (b) Experimental plot. The rainfall simulator is placed above the bare soil surface and a base made of wood is put into place to facilitate the movement of the surface-profile laser scanner.

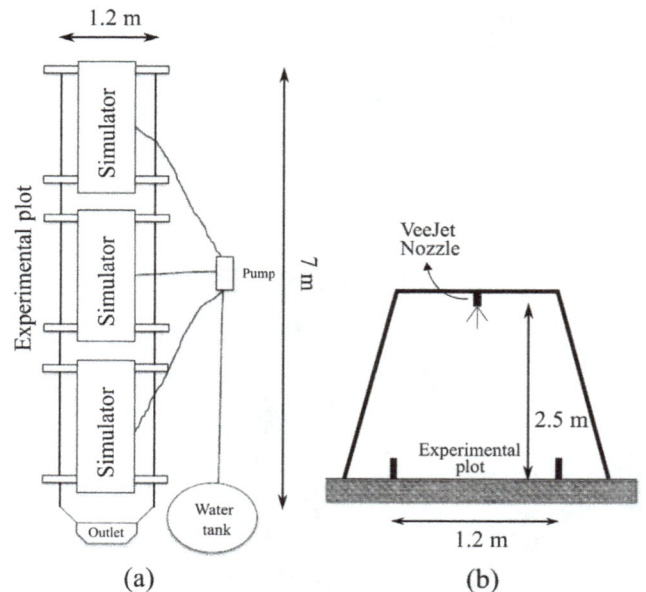

Figure 3. Setup of the experimental tests: (a) rainfall simulators are mounted in series and a pump provides them with water from a tank. (b) Rainfall simulators are placed and adjusted at a height of 2.5 m above the experimental plot surface to ensure drop terminal velocity is reached.

ness between the experiments, as well as to ensure that any potential bias introduced in the plot preparation would be also be consistent, if not minimal. This was confirmed by the observed roughness of the experiment replicates. Rainfall was applied to the plot using Norton ladder multiple intensity rainfall simulators designed by the USDA-ARS National Soil Erosion Research Laboratory, IN. Figure 3 shows the setup for all the experimental runs considered in the present study. For each test, three rainfall simulators were mounted in series over the experimental plot (Fig. 3a) and approximately 2.5 m atop the plot surface (Fig. 3b) in order to ensure that raindrop terminal velocity was reached. Water was continuously pumped from a water tank under controlled pressure, and uniform rainfall was applied through oscillating VeeJet nozzles which provided spherical drops with median diameters between 2.25 and 2.75 mm and a terminal velocity between 6.8 and 7.7 m s^{-1} depending on the rainfall intensity. The distribution of raindrop sizes generated by the rainfall simulators was calibrated using a disdrometer and followed a Marshall–Palmer distribution (Elhakeem and Papanicolaou, 2009), which is a widely accepted distribution for natural raindrop sizes in the US Midwest, where the study was performed (Marshall and Palmer, 1948). The calibration of the raindrop sizes was achieved by adjusting the pressure and swing frequency of the VeeJet nozzles. This level of attention was taken to minimize any potential biases compared to natural rainfall with respect to raindrop size distribution,

and, thus, render the rainfall simulation experiments scalable to other regions experiencing the same type of soil, bare surface, roughness conditions, and natural rainfall characteristics.

Surface elevations were obtained prior to and after the completion of the experiments via an instantaneous digital surface-profile laser scanner (Darboux and Huang, 2003), developed by the USDA-ARS National Soil Erosion Research Laboratory, IN (Fig. 4a). Laser scanner measurements before the runs confirmed that the overall microrelief was less than 2 mm. Horizontal and vertical accuracies of the laser are 0.5 mm. Thus, microroughness features less than 0.5 mm may not have been captured in the analysis. Points were measured every 1 mm. The system consists of two laser diodes mounted 40 cm apart to project a laser plane over the targeted surface. The beam is captured by an 8 bit, high-resolution progressive scan charge-couple device camera with 1030 rows × 1300 columns and a 9 mm lens. The camera and lasers are mounted on a 5 m long carriage assembly, and their movement on the carriage is controlled by software that regulates the travel distance based on a user-specified distance (Fig. 4a). Information captured by the camera is recorded with an attached computer. The information from each scan is converted into a set of (x, y, z) coordinates using a calibration file and the software developed from the USDA-ARS National Soil Erosion Research Laboratory for data transformation as explained by Darboux and Huang (2003). The set of (x, y, z) coordinates obtained for

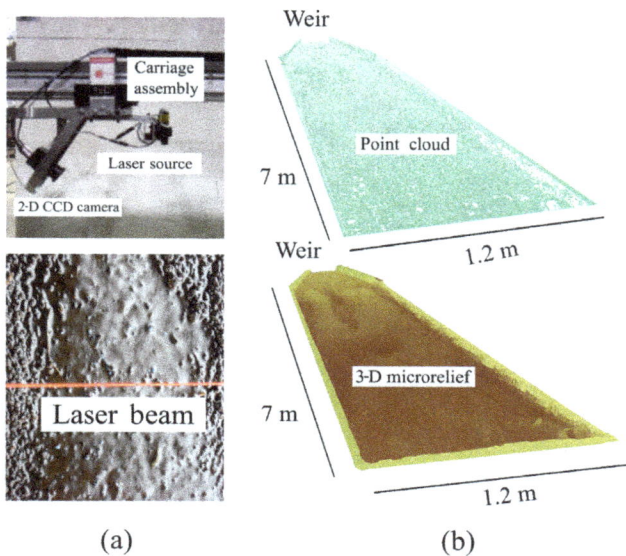

Figure 4. (a) Instantaneous digital surface-profile laser scanner used in the experimental runs and laser beam projected on the soil surface. (b) Cloud of (x, y, z) data acquired from the laser scanner for an experimental test along with the associated 3-D representation of the soil surface microrelief through inverse distance weighted interpolation.

each experiment are imported into ArcGIS 10.3.1 in order to create the corresponding digital elevation models (DEMs) through inverse distance weighting interpolation and thereby visualize or analyze the surfaces (Fig. 4b). The resulting DEMs have a horizontal resolution of 1 mm and an accuracy of 0.5 mm in the vertical.

Three tests of varying rainfall intensity were conducted on the experimental plot. Rainfall intensities were 30, 60, and 75 mm h^{-1} for experiments 1–3, respectively. These simulated intensities represent typical storms observed in the region of South Amana where the plot is located (Huff and Angel, 1992). Three replicates of each rainfall intensity case were performed until steady-state conditions were achieved, and repeatability was confirmed by evaluation of changes in RR at specific cross sections in the rain-splash-dominated zone. It was found that on average, the relative error of the RR ratios between replicates did not exceed 7 %. The volumetric water content was recorded via six 5TE soil moisture sensors manufactured by Decagon Devices, Inc. and placed along the plot to a depth of 10 mm. The initial volumetric water content was found to be similar for each experiment and approximately equal to 35 % at the whole plot, where the field capacity of the specific soil is 38 %. Each experiment was run for nearly 5 h, sufficiently long to reach steady-state conditions, as confirmed by weir readings and discrete samples taken at the outlet of the plot. The infiltration rate was estimated during all rainfall simulation runs by subtracting the measured runoff rates from the constant rainfall rates.

This approach has been commonly used in plot experiments and provides a good estimate of the spatially averaged infiltration rates (e.g., Mohamoud et al., 1990; Wainwright et al., 2000). Averaged saturated hydraulic conductivity values ranged from 3.20 to 4.56 mm h^{-1}, which are in agreement with the averaged saturated hydraulic conductivity value of 4.3 mm h^{-1} measured by Papanicolaou et al. (2015a) using semiautomated double-ring infiltrometers at the field where the study was performed. Although the average saturated hydraulic conductivity values were low with respect to the applied rainfall rates, minimal ponding was observed on the experimental plot, owing to the smooth bare conditions and the high plot gradient of 9 %, which led to low depressional storage.

The initial microroughness length scale in Experiment 1 (1.17 mm) was greater than that of Experiment 2 (0.42 mm) and Experiment 3 (0.32 mm; see Table 1). This is attributed to the different timing of the experiment runs with respect to tillage. Experiment 1 was performed in early August, soon after harvest, so the soil surface had recently been disturbed. However, for Experiments 2 and 3, which were performed in late September, the soil presented less surface disturbance due to the cumulative action of runoff from upslope areas on the plots arising from natural rainfall within that period (Papanicolaou et al., 2015b). Therefore, despite tamping with plywood, remnants of tillage effects remained in Experiment 1, yielding different initial microroughness length scales to those in Experiments 2 and 3. This, however, is not an issue since all the results are presented herein in a dimensionless form (see Sect. 2.2 below on the index ratios). All cases, nonetheless, exhibited initial microroughness length of less than 2 mm, corresponding to smooth surface bed conditions, as confirmed with the laser scanner. Dry soil bulk density was 1.25 g cm^{-3} for Experiment 1, and about 6 % higher for Experiments 2 and 3 due to self-weight consolidation of soil.

Figure 5a provides an example of the experimental plot at prerainfall and postrainfall conditions. Since the focus of this research is only on plot regions where raindrop detachment is dominant over runoff, we are using the scanned profiles that correspond only to these upslope locations, which are shown in Fig. 5b. Rill formation was not observed in these regions throughout the experiments. Visual observations confirmed that raindrop detachment was dominant and the main driver of the change in soil surface roughness. For scanned profiles within the region of interest (ROI; i.e., a selected 200 mm × 200 mm window size), we extracted the data for further statistical and geostatistical analyses by utilizing the public-domain R software (https://www.r-project.org/). The geostatistics ("gstat") and spatial analysis ("sp") libraries were imported to create sample semivariograms.

Table 1. Summary of the rainfall-induced change in the RR index in the experimental tests of this study, as well as in experiments reported in the literature. Smooth conditions refer to initial microroughness less than 5 mm. Cumulative rainfall amounts are also provided.

Rainfall intensity $(mm\,h^{-1})$	Cumulative rainfall (mm)	Soil type	Prerainfall RR (mm)	Postrainfall RR (mm)	RR ratio
			Present study		
30	150	silty clay loam	1.17	1.57	1.34
60	300	silty clay loam	0.42	1.48	3.55
75	375	silty clay loam	0.32	1.46	4.56
			Vázquez et al. (2008)*		
30	85	silt loam	3.39	3.70	1.09
30	50	silt loam	3.00	2.13	0.71
65	195	silt loam	4.72	5.10	1.08
			Zheng et al. (2014)		
40	~60	silty clay loam	2.01	2.35	1.17
90	~135	silty clay loam	2.40	2.68	1.12

* The Vázquez et al. (2008) study looked at RR evolution under successive rainfall events, unlike the other two studies. Postrainfall RR data presented for Vázquez et al. (2008) are those that were determined on completion of the last rainfall succession in each experiment.

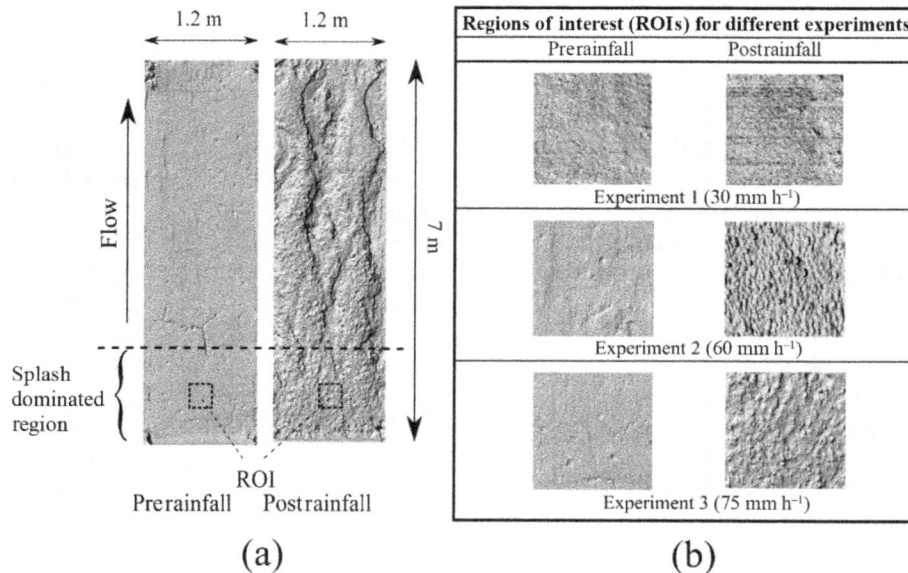

Figure 5. (a) Experimental plot under pre- and postrainfall conditions for an experimental test. The dashed boxes indicate the extent of the region of interest (ROI), where raindrop detachment is dominant over runoff. **(b)** Scanned profiles extracted from the laser-scanned areas of the three experimental tests considered, under both pre- and postrainfall conditions.

2.2 Soil surface roughness quantification

According to Paz-Ferreiro et al. (2008), the RR index, which was first proposed by Allmaras et al. (1966), is the most widely used statistical microrelief index for the evaluation of soil surface roughness. The RR index was initially calculated per Allmaras et al. (1966) as the standard deviation of the log-transformed residual point elevation data. In this study, it is calculated according to Currence and Lovely (1970) as the standard deviation of bed surface elevation data around the mean elevation, after correction for slope using the best-fit plane and removal of tillage effects in the individual height readings:

$$RR = \sqrt{\frac{\sum\limits_{i=1}^{n} (Z_i - \overline{Z})^2}{n}}, \qquad (1)$$

where Z_i and $\overline{Z}$ are individual elevation height readings and their mean, respectively, and n is the total number of readings. The RR index calculated from Eq. (1) is the principal method to quantify soil surface roughness due to its frequent and widespread use in various studies and landscape models as a descriptor of microroughness. The RR index, however, requires that there is no spatial correlation between the surface elevations (Huang and Bradford, 1992). Hence, special care must be taken in adopting the RR index. If correlation exists within a certain spatial scale, the RR index will likely change with the changing window size of observed data (Paz-Ferreiro et al., 2008) and may be dependent on the resolution of the measurement device (Huang and Bradford, 1992). Thus, alternative scale-independent methods that consider spatial correlation have been developed by other researchers in order to address this issue. These methods include first-order variogram analysis (Linden and Van Doren, 1986; Paz-Ferreiro et al., 2008), semivariogram analysis (Vázquez et al., 2005; Oleschko et al., 2008; Rosa et al., 2012; Vermang et al., 2013), fractal models based on fractional Brownian motion (Burrough, 1983; Vázquez et al., 2005; Papanico-laou et al., 2012; Vermang et al., 2013), multifractal analysis (Lovejoy and Schertzer, 2007; Vázquez et al., 2008), Markov–Gaussian models (Huang and Bradford, 1992; Vermang et al., 2013), and two-dimensional Fourier transform models (Cheng et al., 2012), among others. We herein employ additional indices derived from the first-order variogram and the semivariogram as alternatives to the RR index, which is also utilized accounting for its limitations. These include the crossover length, the Markov–Gaussian variance length scale, and the limiting difference.

The crossover length derived from semivariogram analysis is an index that is commonly used in most recent soil microrelief studies to describe surface microroughness. It has the advantage of its quantification being scale-independent through the consideration of the spatial correlation between surface elevations (Vidal Vázquez et al., 2007; Paz-Ferreiro et al., 2008; Tarquis et al., 2008). The semivariogram is calculated from the following equation:

$$\gamma(h) = \frac{1}{2n(h)} \sum_{i=1}^{n(h)} [Z(x_i + h) - Z(x_i)]^2, \qquad (2)$$

where $\gamma(h)$ is the semivariance, h is the lag distance between data points, $Z(x)$ is the elevation height value at location x after correction for both slope and tillage marks, and $n(h)$ is the total number of pairs separated by lag distance h considered in the calculation. The semivariogram is the plot of the semivariance with respect to the lag distance.

Key indices for describing soil surface roughness can be derived from the semivariogram. Assuming a fractional Brownian motion model for describing soil surface roughness (as proposed in the pioneering work of Mandelbrot and Van Ness, 1968), the following expression for $\gamma(h)$ that incorporates the generalized Hurst exponent, H is obtained (Huang and Bradford, 1992; Vidal Vázquez et al., 2007; Paz-Ferreiro et al., 2008; Tarquis et al., 2008):

$$\gamma(h) = l^{2-2H} h^{2H}, \qquad (3)$$

where H is a measure of the degree of correlation between the surface elevations at lag distance h with $0 < H < 1$, and l is the crossover length. The crossover length is a measure of the vertical variability of soil surface roughness on the particular scale where the fractal dimension is estimated, and hence greater roughness is associated with larger crossover length values and vice versa (Huang and Bradford, 1992). The generalized Hurst exponent is a less sensitive descriptor of soil surface evolution as influenced by rainfall (Vázquez et al., 2005), and hence attention is mostly centered on the crossover length. Given the semivariogram plot calculated using Eq. (2), H and l can be extracted by fitting a power law relationship in the form of $y = Ax^B$ to the semivariance-lag distance data, where $y = \gamma(h)$ and $x = h$. According to Eq. (3), the B regression variable gives the generalized Hurst exponent value and the A regression variable yields the crossover length.

The Markov–Gaussian model is a random process that has been adopted for the quantification of soil surface roughness (Huang and Bradford, 1992; Vermang et al., 2013). In that case, the semivariogram is written as an exponential-type function with the following form:

$$\gamma(h) = \sigma^2 \left(1 - e^{-h/L}\right), \qquad (4)$$

where σ is the variance length scale, representing the roughness of a surface on the large scale, and L is the correlation length scale, which is a measure of the rate at which small-scale roughness variations approach the constant value of σ. These indices are obtained by fitting the exponential-type function of Eq. (4) to the semivariogram obtained from Eq. (2).

Finally, the limiting difference (LD) index is another index adopted to quantify soil surface roughness. It is calculated from the first-order variogram with elevation data corrected for both slope and tillage marks (Linden and Van Doren, 1986; Paz-Ferreiro et al., 2008), which is written in the following form:

$$\Delta Z(h) = \frac{1}{n(h)} \sum_{i=1}^{n(h)} |Z(x_i + h) - Z(x_i)|. \qquad (5)$$

Then, a linear relationship is fitted between $1/\Delta Z(h)$ and $1/h$:

$$1/\Delta Z(h) = a + b/h. \tag{6}$$

The LD index is then calculated as $LD = 1/a$. LD has units of length and represents the value of the first-order variance at large lag distances. It is considered to be an indicator of soil surface roughness, and is thus adopted in the present study as an additional roughness index.

In order to negate the effects of the differences that existed in the initial microrelief amongst the three runs due to the different timing of the experiments (see Sect. 2.1) and to compare rainfall-induced changes in relative terms, the results from the rainfall experiments are presented in the form of ratios of the roughness indices. More precisely, the RR ratio, defined as the ratio of the postrainfall RR index over the RR index prior to the rainfall (RR_{post}/RR_{pre}), is calculated for each experiment. Semivariograms are plotted under pre- and postrainfall conditions at the ROI to assess the spatial correlation of surface elevations. Along the same lines, ratios between pre- and postrainfall conditions are calculated for the crossover length, the variance length scale of the Markov–Gaussian model, and the limiting difference to assess changes in microroughness along with the RR ratio.

3   Results

3.1   Changes in the RR index

Based on visual inspection of the DEMs in Fig. 5b, it is evident that microroughness in the splash-dominated region increases with rainfall. Table 1 summarizes the results of this study along with results from other studies focused on smooth surfaces, documenting the RR index values before and after the rainfall events, the cumulative rainfall, and the associated RR ratio. The present study, along with Vázquez et al. (2008) and Zheng et al. (2014) generally report an increase in RR with rainfall under the conditions examined. The Vázquez et al. (2008) study, however, differs from the present study and that of Zheng et al. (2014) in that it examined roughness evolution under successive rainfall events per run. Only the RR data collected on completion of the last rainfall succession in each run conducted by Vázquez et al. (2008) are presented in Table 1. The final RR values after the last rainfall succession were selected for being the more closely comparable to the steady-state conditions examined herein. Although both Vázquez et al. (2008) and Zheng et al. (2014) recorded an increase in RR with rainfall, they had significantly lower values of the RR ratio than the present study. This could be due to several factors including, but not limited to, lower applied rainfall intensity and amount, the initial surface microroughness, and different soil conditions.

Other studies not included in Table 1 have also shown increasing trends of roughness with rainfall, as quantified with the use of different indices. For instance, Huang and Bradford (1992) calculated the semivariograms for different sur-faces and used fractal and Markov–Gaussian parameters to quantify the roughness. Markov–Gaussian analysis showed a relative increase in the roughness parameter for a surface of low initial roughness. Finally, Rosa et al. (2012) introduced the roughness index, which is estimated from the semivariogram sill (i.e., the upper value where the semi-variance levels out), in order to quantify roughness, and observed an increase with rainfall under low initial roughness conditions. That increase was attributed to the fragmentation of aggregates and clods to smaller aggregates but was not linked to smooth bare soil surface conditions. Overall, the experimental evidence suggests that the interaction between rainfall and smooth soil surfaces can lead to an increase in microroughness.

The results outlined above for the use of the RR index as a descriptor of change in microroughness have been based on the assumption that there is no statistically significant spatial correlation in elevation readings between neighboring locations at the ROI. This condition was indeed not violated due to the choice in ROI. The following subsection outlines and discusses the results of the semivariogram analysis and additional indices used to confirm the validity of the assumption and their comparison with the RR index method.

3.2   Changes in alternative roughness indices

Semivariograms and first-order variograms were obtained from geostatistical analysis and plotted at four different angles – 0, 45, 90, and 135° – with respect to the downslope direction. Since the action of rainfall is isotropic and adds no systematic trend along any direction, no significant differences were expected between semivariograms. A nonparametric test for spatial isotropy was performed per Guan et al. (2004) using the public domain R statistical package with the "spTest" library. The spatial isotropy hypothesis was confirmed ($p < 0.05$). Thus, no bias was determined in taking any direction to calculate the semivariograms and the associated crossover lengths.

The semivariograms calculated at the ROI were chosen to be in the downslope direction at an angle of 0° and are presented for each experiment in Fig. 6. The vertical dashed lines designate the lag distances above which the spatial autocorrelation of the elevations is not statistically significant. These lag distances are approximately 10 mm, so the selected 200 mm window size of the ROI is almost 20 times greater than the spatial autocorrelation range. This implies that the window size of the ROI falls on the scale of the semivariogram sill (which is defined as the near-constant value of semivariance at large lag distances where the semivariogram levels out – see horizontal dashed lines in Fig. 6). RR is directly related to the semivariogram sill (e.g., Vázquez et al., 2005; Vermang et al., 2013); therefore it can be considered independent of the selected window size, given that the latter far exceeds the spatial autocorrelation range.

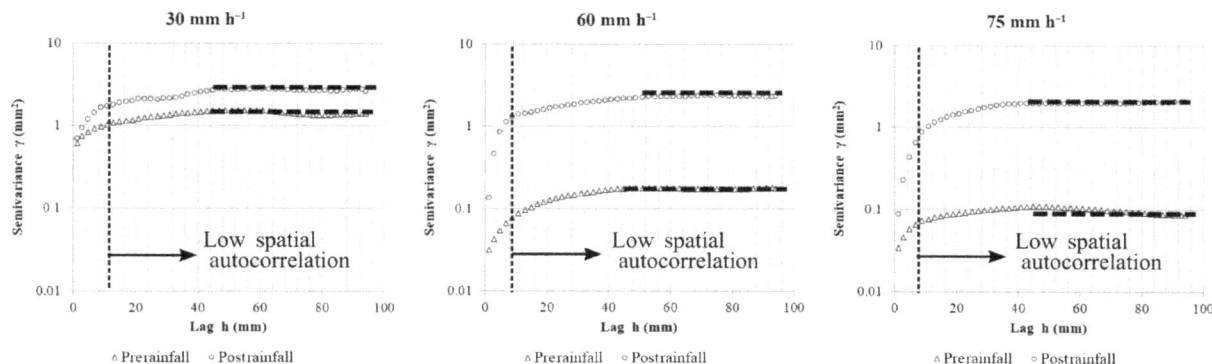

Figure 6. Semivariograms at the region of interest for the three experimental tests, under pre- and postrainfall conditions. Horizontal dashed lines indicate the semivariogram sills and vertical dashed lines indicate the lag distance above which the spatial autocorrelation of the elevations is negligible.

Figure 6 shows that the postrainfall sills are greater than their corresponding prerainfall values. Also, the difference in sills between pre- and postrainfall conditions for the 30 mm h^{-1} precipitation intensity is much lower than those of the 60 and 75 mm h^{-1} events. These observations are in accordance with visual inspection of the surfaces as well as with the results noted earlier for the RR ratio (see Table 1). Complete agreement between the trends of the RR index, the semivariogram sill, and visual inspection of the surfaces justifies the use of the RR index as a representative and unbiased descriptor of microroughness.

Table 2 lists the crossover length, the Markov–Gaussian variance length scale and the limiting difference indices for the three experimental tests, and their relative change after the rainfall. These indices show an increase with rainfall that is of the same magnitude and trend as the RR index and crossover length and provide a supplemental analysis about the role of rainfall intensities on the relative increase in roughness. Our findings were compared against those reported in the literature. Huang and Bradford (1992) studied the evolution of soil surface roughness with the Markov–Gaussian variance length scale, and saw an increase of 6 % in roughness for a surface of low initial roughness. Moreover, Paz-Ferreiro et al. (2008), who used the LD index to quantify soil surface roughness, also recorded a 10 % increase in the LD index for a low roughness conventional tillage soil surface. The higher relative increase in roughness seen in our study (Table 2) compared to other studies is attributed to the lower initial roughness conditions in addition to different soil types and management.

Overall, the results provided suggest that all the indices employed in this study may be used interchangeably to characterize rainfall-induced changes in soil surface roughness and can capture an increase in soil surface roughness, especially for smooth soil surfaces. For these microroughness scales, the relative increase in roughness is also shown to in-

Table 2. Summary of the rainfall-induced change in the crossover length, the Markov–Gaussian variance length scale and limiting difference indices for the experimental tests of this study.

Rainfall intensity (mm h^{-1})	Cumulative rainfall (mm)	Pre-rainfall value	Post-rainfall value	Index ratio
		l (mm)		
30	150	0.71	0.73	1.03
60	300	0.09	0.20	2.13
75	375	0.15	0.39	2.56
		σ (mm)		
30	150	1.19	1.63	1.37
60	300	0.42	1.52	3.62
75	375	0.31	1.43	4.56
		LD (mm)		
30	150	0.79	0.87	1.10
60	300	0.26	0.87	3.39
75	375	0.15	0.71	4.84

crease with rainfall intensity under the conditions examined herein.

4 Discussion and conclusions

Many studies have examined the response of rough surfaces to rainfall and have reported a decay of roughness. Few studies have assessed microscale variation of smooth surfaces in response to rainfall under controlled conditions. The experiments presented herein were designed to help us decipher the role of rain splash on RR for smooth surfaces with initial microroughness on the order of 2 mm by isolating the role of other factors such as runoff, variable water content, bare soil surface, and soil texture, among others. Our results show a consistent increase in roughness under the action of rainfall, with an overall agreement between all the roughness indices examined herein in terms of trend and magnitude. Our findings are consistent with findings of other studies that have examined length scales less than 5 mm and suggest the possible existence of a characteristic roughness threshold below which RR is expected to increase due to the action of rainfall. The value of this threshold may depend on the specific soil and rainfall conditions. A caveat of our study is that due to the limited range of conditions examined herein more experiments are needed to further solidify the conditions under which RR is expected to increase under rainfall action. An outcome of this study is the awareness that within landscape regions where smooth surfaces are present, an increase in RR may occur during the early part of the storm where rain splash action is more important than runoff.

This study suggests that the effects of the interaction between rainfall and a soil surface can be different for smooth and rough surfaces, and highlights the need for a better understanding of the interaction due to its potential impact on hydrologic response. This potential impact is demonstrated with the following established pedotransfer function for the effects of soil crusting, roughness, and rainfall kinetic energy on the bare hydraulic conductivity, K_{br} (Risse et al., 1995):

$$K_{br} = K_b \left[CF + (1 - CF)e^{-C \cdot E_a (1 - RR_t / RR_{t-max})} \right], \qquad (7)$$

where K_b is the baseline hydraulic conductivity, CF is the crust factor, C is the soil stability factor, E_a is the cumulative rainfall kinetic energy since the last tillage, RR_t is random roughness height, and RR_{t-max} is the maximum random roughness height. Using the following typical values for the study site based on the literature (Flanagan and Nearing, 1995; Chang, 2010), $E_a = 10\,000\,\text{J m}^{-2}$, $C = 0.0002\,\text{m}^2\,\text{J}^{-1}$, and $RR_{t-max} = 40\,\text{mm}$, the percentage change in bare hydraulic conductivity for increasing roughness can be estimated for an initial RR_t value of 2 mm and minimal CF factor. Performing the analysis for the range of random roughness ratios observed in this study (~ 1.3–4.5), the percentage increase in hydraulic conductivity is found to range between 5 and 42 %, which will have a significant impact on rainfall–runoff partitioning.

It is recognized that the soil preparation method in our study could have introduced some bias to the soil properties such as aggregate size distribution, compaction, and aggregate stability. Nonetheless, with regard to the purpose for which this study was designed, this preparation method ensured consistency in the initial and final roughness states, as confirmed by replications of our experimental runs. It is also recognized that drier, silty-type soils may not exhibit the increase in RR shown here. Further, the role of sealing may be important on roughness development under bare soil conditions and needs further examination. Soil water retention characteristics of the soils under sealing and its implication to RR must be considered (Saxton and Rawls, 2006). Finally,

the role of successive storm events on changing roughness for smooth surfaces is not covered in this study and needs to be examined.

The exact mechanisms leading to increase in roughness remain unknown and are not the focus of this study. However, changes in roughness during a storm event have been attributed to compression and drag forces from the raindrop impact on the soil, angular displacement due to rain splash, aggregate fragmentation, and differential swelling (Al-Durrah and Bradford, 1982; Warrington et al., 2009; Rosa et al., 2012; Fu et al., 2016). Regions exhibiting different median raindrop diameters may experience different soil surface roughness evolution due to different aggregate fragmentation and rain splash effects (Warrington et al., 2009; Rosa et al., 2012; Fu et al., 2016). Future research should explore these mechanisms.

Competing interests. The authors declare that they have no conflict of interest.

Acknowledgements. The present study was in part supported by the National Science Foundation grant EAR 1331906 for the Critical Zone Observatory for intensively managed landscapes (IML-CZO), which comprises a multi-institutional collaborative effort. The authors, especially the corresponding author, would like to acknowledge the help provided by Chi-Hua Huang from the USDA-ARS National Soil Erosion Research Lab, West Lafayette, IN, regarding the purchase of the laser system used in this research to map the RR. The fifth author was partially supported by the University of Iowa NSF IGERT program, Geoinformatics for Environmental and Energy Modeling and Prediction. This research was supported by the NASA EPSCoR Program (grant no. NNX10AN28A) and the Iowa Space Grant Consortium (grant no. NNX10AK63H). The first author during part of this analysis has been supported by the USDA-AFRI grant. Finally, we would like to thank the anonymous reviewers, whose insightful comments and suggestions led to an improved paper.

Edited by: Daniel Schertzer

References

Abaci, O. and Papanicolaou, A. N.: Long-term effects of management practices on water-driven soil erosion in an intense agricultural sub-watershed: monitoring and modelling, Hydrol. Process., 23, 2818–2837, https://doi.org/10.1002/hyp.7380, 2009.

Al-Durrah, M. M. and Bradford, J. M.: The mechanism of raindrop splash on soil surfaces, Soil Sci. Soc. Am. J., 46, 1086, https://doi.org/10.2136/sssaj1982.03615995004600050040x, 1982.

Allmaras, R. R., Burwell, R. E., Larson, W. E., and Holt, R. F.: Total porosity and random roughness of the interrow zone as influenced by tillage, USDA Conservation Re. Rep. 7, USDA, Washington, D.C., 16 pp., 1966.

Bertuzzi, P., Rauws, G., and Courault, D.: Testing roughness indices to estimate soil surface roughness changes due to simulated rainfall, Soil Till. Res., 17, 87–99, https://doi.org/10.1016/0167-1987(90)90008-2, 1990.

Burrough, P. A.: Multiscale sources of spatial variation in soil. I. The application of fractal concepts to nested levels of soil variation, J. Soil Sci., 34, 577–597, https://doi.org/10.1111/j.1365-2389.1983.tb01057.x, 1983.

Chang, Y.: Predictions of saturated hydraulic conductivity dynamics in a midwestern agriculture watershed, Iowa, MS Thesis, The University of Iowa, Iowa City, IA, USA, 2010.

Cheng, Q., Sun, Y., Lin, J., Damerow, L., Schulze Lammers, P., and Hueging, H.: Applying two-dimensional Fourier Transform to investigate soil surface porosity by laser-scanned data, Soil Till. Res., 124, 183–189, https://doi.org/10.1016/j.still.2012.06.016, 2012.

Chi, Y., Yang, J., Bogart, D., and Chu, X.: Fractal Analysis of Surface Microtopography and its Application in Understanding Hydrologic Processes, T. ASABE, 55, 1781–1792, https://doi.org/10.13031/2013.42370, 2012.

Currence, H. D. and Lovely, W. G.: The analysis of soil surface roughness, T. ASAE, 13, 710–714, 1970.

Darboux, F. and Huang, C.: An instantaneous-profile laser scanner to measure soil surface microtopography, Soil Sci. Soc. Am. J., 67, 92–99, https://doi.org/10.2136/sssaj2003.9200, 2003.

Elhakeem, M. and Papanicolaou, A. N.: Estimation of the runoff curve number via direct rainfall simulator measurements in the State of Iowa, USA, Water Resour. Manage., 23, 2455–2473, https://doi.org/10.1007/s11269-008-9390-1, 2009.

Flanagan, D. C. and Nearing, M. A. (Eds.): USDA Water Erosion Prediction Project: Hillslope Profile and Watershed Model Documentation, NSERL Report No. 10, USDA-ARS National Soil Erosion Research Laboratory, West Lafayette, IN, USA, 1995.

Fu, Y., Li, G., Zheng, T., Li, B., and Zhang, T.: Impact of raindrop characteristics on the selective detachment and transport of aggregate fragments in the Loess Plateau of China, Soil Sci. Soc. Am. J., 80, 1071, https://doi.org/10.2136/sssaj2016.03.0084, 2016.

Gilley, J. E. and Finkner, S. C.: Hydraulic roughness coefficients as affected by random roughness, T. ASAE, 34, 897–903, https://doi.org/10.13031/2013.31746, 1991.

Gómez, J. A. and Nearing, M. A.: Runoff and sediment losses from rough and smooth soil surfaces in a laboratory experiment, Catena, 59, 253–266, https://doi.org/10.1016/j.catena.2004.09.008, 2005.

Guan, Y., Sherman, M., and Calvin, J. A.: A nonparametric test for spatial isotropy using subsampling, J. Am. Stat. Assoc., 99, 810–821, https://doi.org/10.1198/016214504000001150, 2004.

Huang, C. and Bradford, J. M.: Depressional storage for Markov–Gaussian surfaces, Water Resour. Res., 26, 2235–2242, https://doi.org/10.1029/WR026i009p02235, 1990.

Huang, C. and Bradford, J. M.: Applications of a laser scanner to quantify soil microtopography, Soil Sci. Soc. Am. J., 56, 14–21, https://doi.org/10.2136/sssaj1992.03615995005600010002x, 1992.

Huff, F. A. and Angel, J. R.: Rainfall Frequency Atlas of the Midwest, Midwestern Climate Center Research Report 92-03, Midwestern Climate Center Research, Champaign, IL, 1992.

Kamphorst, E. C., Jetten, V., Guérif, J., Pitkanen, J., Iversen, B. V., Douglas, J. T., and Paz, A.: Predicting depressional storage from soil surface roughness, Soil Sci. Soc. Am. J., 64, 1749, https://doi.org/10.2136/sssaj2000.6451749x, 2000.

Linden, D. R. and Van Doren, D. M.: Parameters for characterizing tillage-induced soil surface roughness, Soil Sci. Soc. Am. J., 50, 1560, https://doi.org/10.2136/sssaj1986.03615995005000060035x, 1986.

Lovejoy, S. and Schertzer, D.: Scaling and multifractal fields in the solid earth and topography, Nonlin. Processes Geophys., 14, 465–502, https://doi.org/10.5194/npg-14-465-2007, 2007.

Magunda, M. K., Larson, W. E., Linden, D. R., and Nater, E. A.: Changes in microrelief and their effects on infiltration and erosion during simulated rainfall, Soil Technol., 10, 57–67, https://doi.org/10.1016/0933-3630(95)00039-9, 1997.

Mandelbrot, B. B. and Van Ness, J. W.: Fractional Brownian motions, fractional noises and applications, SIAM Review, 10, 422–437, 1968.

Marshall, J. S. and Palmer, W. M. K.: The distribution of raindrops with size, J. Meteorol., 5, 165–166, https://doi.org/10.1175/1520-0469(1948)005<0165:TDORWS>2.0.CO;2, 1948.

Mohamoud, Y. M., Ewing, L. K., and Boast, C. W.: Small plot hydrology: I. Rainfall infiltration and depression storage determination, T. ASAE, 33, 1121–1131, https://doi.org/10.13031/2013.31448, 1990.

Oades, J. and Waters, A.: Aggregate hierarchy in soils, Aust. J. Soil Res., 29, 815–828, https://doi.org/10.1071/SR9910815, 1991.

Oleschko, K., Korvin, G., Muñoz, A., Velazquez, J., Miranda, M. E., Carreon, D., Flores, L., Martínez, M., Velásquez-Valle, M., Brambila, F., Parrot, J. F., and Ronquillo, G.: Mapping soil fractal dimension in agricultural fields with GPR, Nonlin. Processes Geophys., 15, 711–725, https://doi.org/10.5194/npg-15-711-2008, 2008.

Papanicolaou, A. N., Tsakiris, A. G., and Strom, K.: The use of fractals to quantify the morphology of cluster microform, Geomorphology, 139–140, 91–108, https://doi.org/10.1016/j.geomorph.2011.10.007, 2012.

Papanicolaou, A. N., Elhakeem, M., Wilson, C. G., Burras, C. L., West, L. T., Lin, H., Clark, B., and Oneal, B. E.: Spatial variability of saturated hydraulic conductivity at the hillslope scale: Understanding the role of land management and erosional effect, Geoderma, 243–244, 58–68, https://doi.org/10.1016/j.geoderma.2014.12.010, 2015a.

Papanicolaou, A. N., Wacha, K. M., Abban, B. K., Wilson, C. G., Hatfield, J. L., Stanier, C. O., and Filley, T. R.: From soilscapes to landscapes: A landscape-oriented approach to simulate soil organic carbon dynamics in intensively managed landscapes, J. Geophys. Res.-Biogeo., 120, 2375–2401, https://doi.org/10.1002/2015JG003078, 2015b.

Paz-Ferreiro, J., Bertol, I., and Vázquez, E. V.: Quantification of tillage, plant cover, and cumulative rainfall effects on soil surface microrelief by statistical, geostatistical and fractal indices, Nonlin. Processes Geophys., 15, 575–590, https://doi.org/10.5194/npg-15-575-2008, 2008.

Potter, K. N.: Soil properties effect on random roughness decay by rainfall, T. ASAE, 33, 1889–1892, 1990.

Risse, L. M., Liu, B. Y., and Nearing, M. A.: Using curve numbers to determine base-line values of Green-Ampt effective hydraulic conductivities, Water Resour. Bull., 31, 147–158, 1995.

Römkens, M. J. and Wang, J. Y.: Effect of tillage on surface roughness, T. ASAE, 29, 429–433, https://doi.org/10.13031/2013.30167, 1986.

Rosa, J. D., Cooper, M., Darboux, F., and Medeiros, J. C.: Soil roughness evolution in different tillage systems under simulated rainfall using a semivariogram-based index, Soil Till. Res., 124, 226–232, https://doi.org/10.1016/j.still.2012.06.001, 2012.

Saxton, K. E. and Rawls, W. J.: Soil water characteristic estimates by texture and organic matter for hydrologic solutions, Soil Sci. Soc. Am. J., 70, 1569, https://doi.org/10.2136/sssaj2005.0117, 2006.

Tarquis, A. M., Heck, R. J., Grau, J. B., Fabregat, J., Sanchez, M. E., and Antón, J. M.: Influence of thresholding in mass and entropy dimension of 3-D soil images, Nonlin. Processes Geophys., 15, 881–891, https://doi.org/10.5194/npg-15-881-2008, 2008.

Vázquez, E. V., Miranda, J. G. V., and González, A. P.: Characterizing anisotropy and heterogeneity of soil surface microtopography using fractal models, Ecol. Model., 182, 337–353, https://doi.org/10.1016/j.ecolmodel.2004.04.012, 2005.

Vázquez, E. V., Moreno, R. G., Miranda, J. G. V., Díaz, M. C., Requejo, A. S., Paz-Ferreiro, J., and Tarquis, A. M.: Assessing soil surface roughness decay during simulated rainfall by multifractal analysis, Nonlin. Processes Geophys., 15, 457–468, https://doi.org/10.5194/npg-15-457-2008, 2008.

Vermang, J., Norton, L. D., Baetens, J. M., Huang, C., Cornelis, W. M., and Gabriels, D.: Quantification of soil surface roughness evolution under simulated rainfall, T. ASABE, 56, 505–514, https://doi.org/10.13031/2013.42670, 2013.

Vidal Vázquez, E., Miranda, J. G. V., and Paz González, A.: Describing soil surface microrelief by crossover length and fractal dimension, Nonlin. Processes Geophys., 14, 223–235, https://doi.org/10.5194/npg-14-223-2007, 2007.

Wainwright, J., Parsons, A. J., and Abrahams. A. D.: Plot-scale studies of vegetation, overland flow and erosion interactions: case studies from Arizona and New Mexico, Hydrol. Process., 14, 2921–2943, https://doi.org/10.1002/1099-1085(200011/12)14:16/17<2921::AID-HYP127>3.0.CO;2-7, 2000.

Warrington, D. N., Mamedov, A. I., Bhardwaj, A. K., and Levy, G. J.: Primary particle size distribution of eroded material affected by degree of aggregate slaking and seal development, Eur. J. Soil Sci., 60, 84–93, https://doi.org/10.1111/j.1365-2389.2008.01090.x, 2009.

Zhang, X., Yu, G. Q., Li, Z. B., and Li, P.: Experimental study on slope runoff, erosion and sediment under different vegetation types, Water Resour. Manage., 28, 2415–2433, 2014.

Zheng, Z. C., He, S. Q., and Wu, F.: Changes of soil surface roughness under water erosion process: Soil surface roughness under water erosion, Hydrol. Process., 28, 3919–3929, https://doi.org/10.1002/hyp.9939, 2014.

Zobeck, T. M. and Onstad, C. A.: Tillage and rainfall effects on random roughness: A review, Soil Till. Res., 9, 1–20, https://doi.org/10.1016/0167-1987(87)90047-X, 1987.

10

Non-linear effects of pore pressure increase on seismic event generation in a multi-degree-of-freedom rate-and-state model of tectonic fault sliding

Sergey B. Turuntaev[1,2,3] **and Vasily Y. Riga**[2]

[1]Institute of Geosphere Dynamics, Russian Academy of Sciences, Moscow, 119334, Russian Federation
[2]Moscow Institute of Physics and Technology, Moscow, 141701, Russian Federation
[3]All-Russian Research Institute of Automatics, Moscow, 127055, Russian Federation

Correspondence to: Sergey B. Turuntaev (s.turuntaev@gmail.com)

Abstract. The influence of fluid injection on tectonic fault sliding and seismic event generations was studied by a multi-degree-of-freedom rate-and-state friction model with a two-parametric friction law. A system of blocks (up to 25 blocks) elastically connected to each other and connected by elastic springs to a constant-velocity moving driver was considered. Variation of the pore pressure due to fluid injection led to variation of effective stress between the first block and the substrate. Initially the block system was in a steady-sliding state; then, its state was changed by the pore pressure increase. The influence of the model parameters (number of blocks, spring stiffness, velocity weakening parameter) on the seismicity variations was considered. Various slip patterns were obtained and analysed.

1 Introduction

Despite the fact that the rate-and-state model of friction was proposed in the second half of the previous century, the interest in it has increased in recent years. The rate-and-state model (Gu et al., 1984; Dieterich, 1992; Abe and Kato, 2013) was adopted as a quite appropriate basis for describing seismic processes in the Earth's crust and for modelling relevant geophysical systems. Currently, it is believed that this model describes the seismic process most adequately.

Brace and Byerlee (1966) proposed considering unstable frictional sliding along tectonic faults as a model of earthquakes. The model included a suggestion that a cohesion existing in some parts of tectonic fault prevents free slipping along it and leads to an accumulation of a shear stress to a critical level, after which the slip and the earthquake occur.

Peculiarities of the friction force dependence on the duration of the stationary state of the contact and on the velocity of the motion along the fault were examined by Dieterich (Dieterich, 1992). Gu et al. (1984) experimentally investigated various modes of the frictional movements and determined empirical constants whose values are used in many modern variants of the rate-and-state equation.

The rate-and-state equation was considered by Hobbs (Hobbs, 1990) by means of non-linear dynamics methods. Change in the friction was studied as a function of displacement and velocity at a variation of the stiffness coefficient in the rate-and-state equation. A similar approach was implemented by Erickson et al. (2008); they examined an appearance of chaotic solutions in the one-parameter velocity-dependent friction equation.

Abe and Kato (Abe and Kato, 2013, 2014) examined two- and three-degree-of-freedom spring-block models with a one-parameter rate-and-state friction law and obtained different slip patterns for such system. By varying stiffness parameters, they obtain periodic recurrence of the seismic and aseismic events and several types of seismicity chaotic behaviour.

Turuntaev et al. (2012) showed using the Grassberger–Procaccia method (Grassberger and Procaccia, 1983) that the man-made impact underground leads to an increase in the "regularity" of the seismic regime. To explain the increase

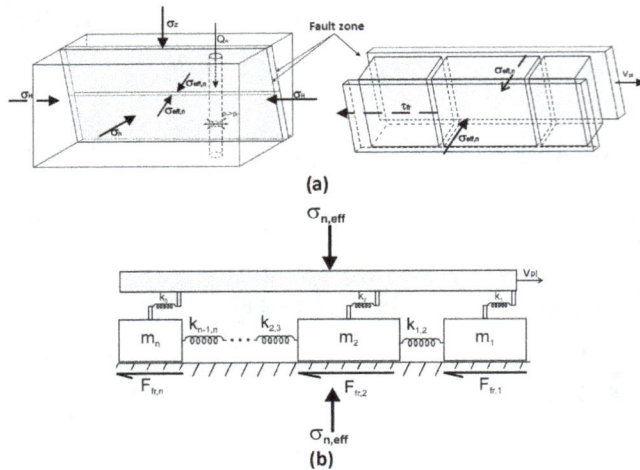

Figure 1. (a) The block model of an active tectonic fault; **(b)** schematic diagram of a multi-degree-of-freedom spring-block model.

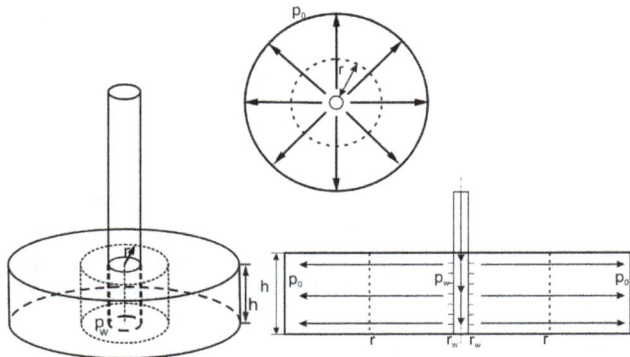

Figure 2. Radial flow in a homogeneous reservoir.

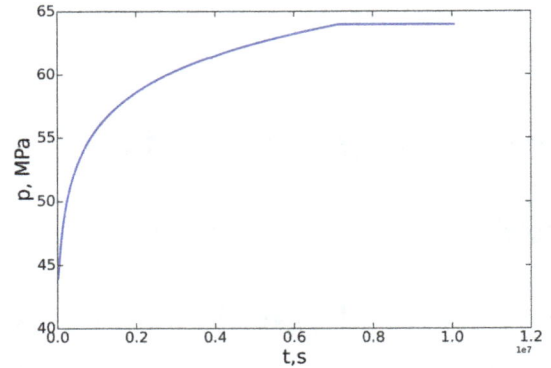

Figure 3. Pore pressure change at the boundary between the first block and substrate.

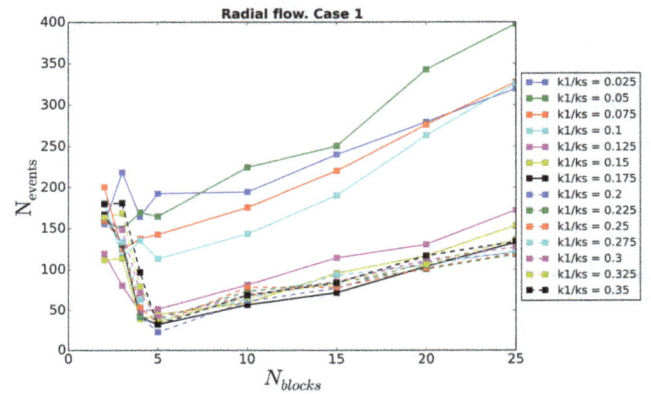

Figure 4. Cumulative number of events vs. number of blocks.

in the seismic regime regularity, a model of the fault motion defined by the two-parameter velocity dependent friction law was considered.

In the presented paper, we consider the two-parameter type of the friction law in a multi-degree-of-freedom spring-block model and change the value of critical shear stress in the rate-and-state equation, suggesting that this is the value varied by human impact (by fluid injection and corresponding pore pressure change). Here we use the classical pore-elastic model of radial filtration of injected fluid to calculate the typical pore pressure change.

2 The model description

2.1 Spring-block model

The tectonic fault model proposed by Burridge and Knopov (Burridge and Knopov, 1966) looks like a system of blocks elastically connected to each other (Fig. 1a, b). Each block

moves under the net action of elastic forces from adjacent blocks and driver and friction force from the stationary substrate. Here, the multi-degree-of-freedom system is investigated. Every block of mass m_i is connected by a spring of stiffness k_l to the driver moving at a rate v_{pl}, and linked with each other by springs of stiffness $k_{n-1,n}$. The motion equation may be written as Eq. (1):

$$\begin{cases} m_1 \ddot{x}_1 = k_1 \left(v_{pl} t - x_1 \right) - k_{12} \left(x_1 - x_2 \right) - F_{fr1}, \\ m_2 \ddot{x}_2 = k_2 \left(v_{pl} t - x_2 \right) + k_{12} \left(x_1 - x_2 \right) - k_{23} \left(x_2 - x_3 \right) - F_{fr2}, \\ \dots \end{cases} \quad (1)$$

where $F_{\text{fri}} = S_i \tau_i$ is the force of friction between the block number i and the substrate, S_i is the area of the block surface, τ_i is the shear stress, t is time and x_i is the displacements of the blocks relative to the driver.

We assume that the friction shear stress at the block boundary obeys the following two-parameter friction law:

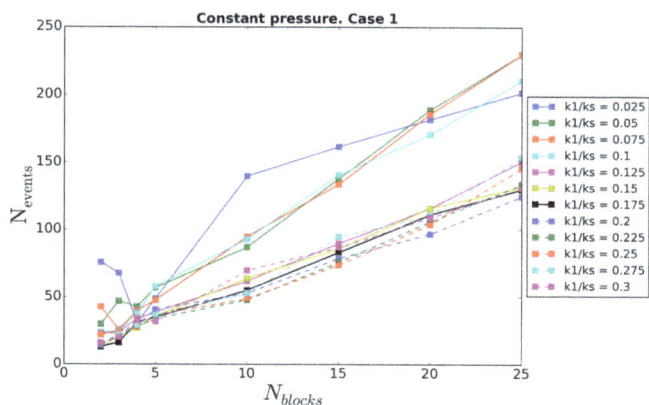

Figure 5. Cumulative number of events vs. number of blocks.

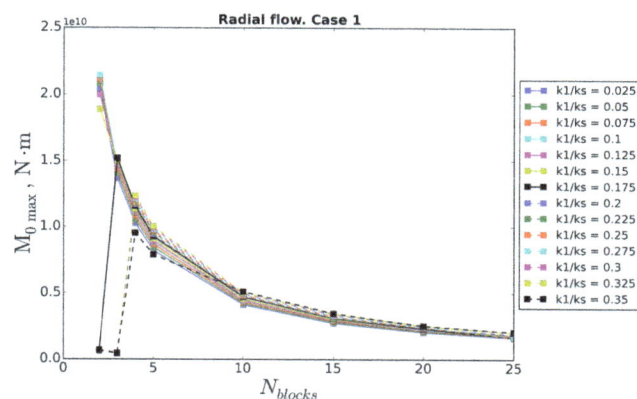

Figure 8. Maximum seismic moment of event vs. number of blocks.

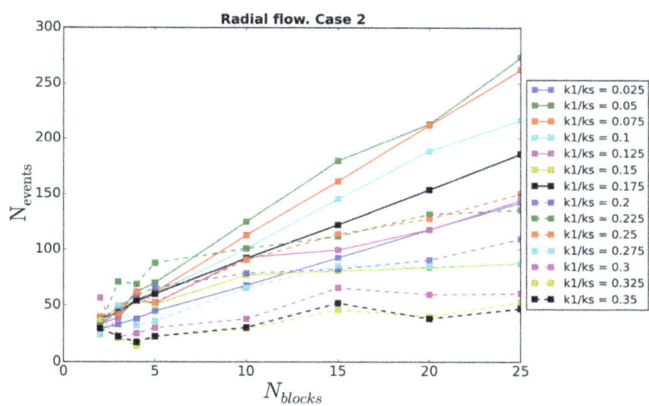

Figure 6. Cumulative number of events vs. number of blocks.

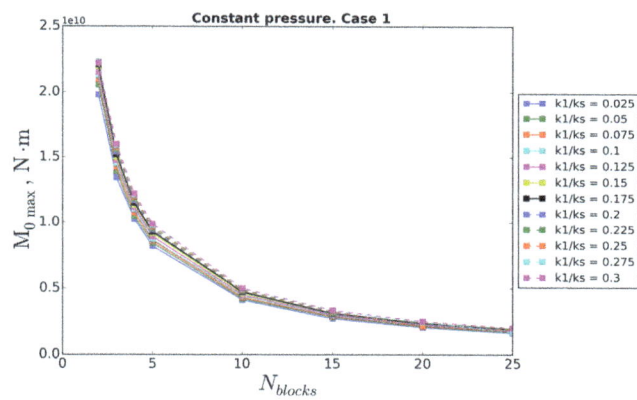

Figure 9. Maximum seismic moment of event vs. number of blocks.

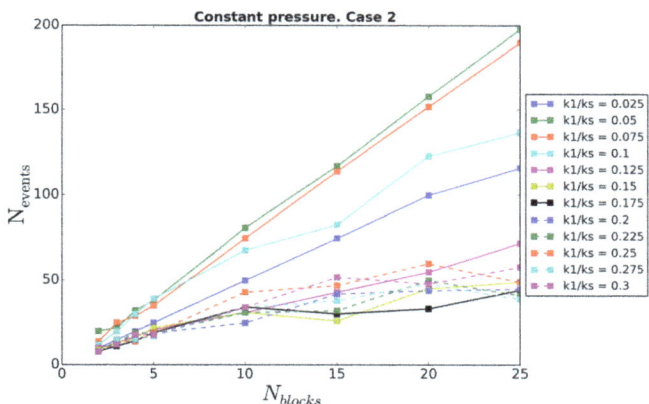

Figure 7. Cumulative number of events vs. number of blocks.

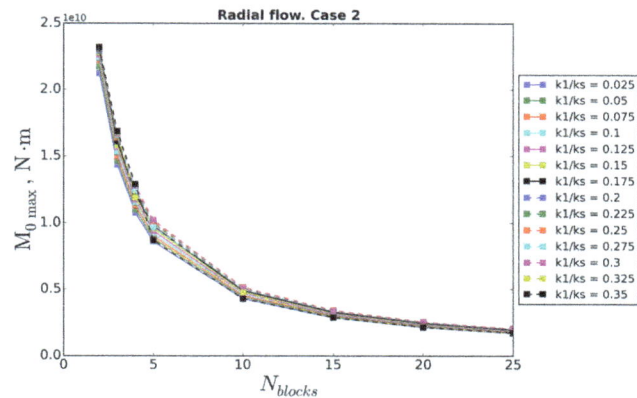

Figure 10. Maximum seismic moment vs. number of blocks.

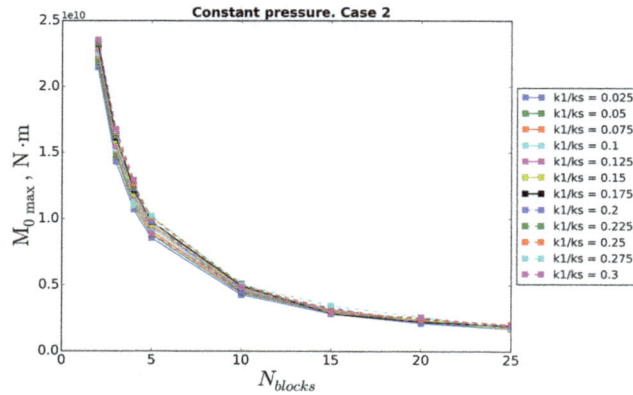

Figure 11. Maximum seismic moment vs. number of blocks.

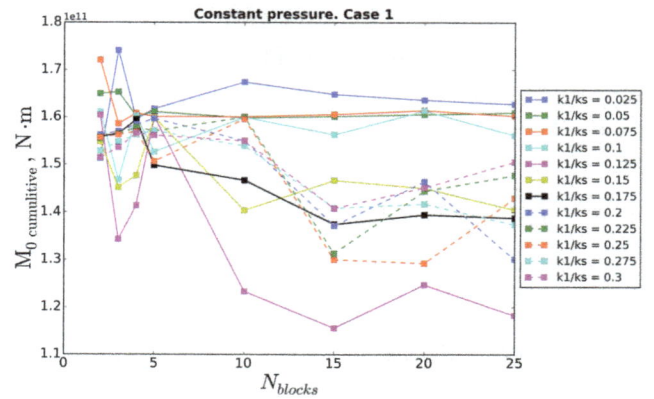

Figure 13. Cumulative seismic moment vs. number of blocks.

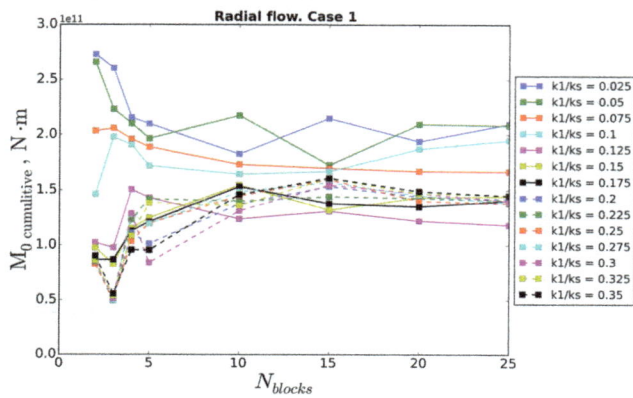

Figure 12. Cumulative seismic moment vs. number of blocks.

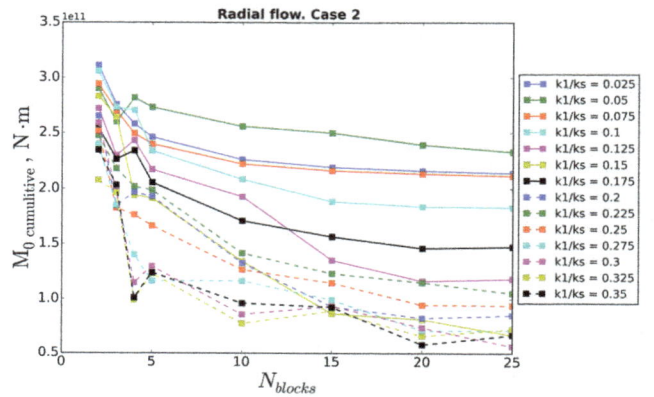

Figure 14. Cumulative seismic moment vs. number of blocks.

$$\tau = \tau^* + A \ln\left(\frac{v}{v^*}\right) + \theta_1 + \theta_2, \tag{2}$$

$$\tau^* = \tau_0 + \mu(\sigma_n - p), \tag{3}$$

$$\dot{\theta}_i = -\frac{v}{L_i}\left[\theta_i + B_i ln(v/v^*)\right], \tag{4}$$

where θ_1, θ_2 are the state parameters, A and B_1, B_2 are constants that represent the rate and the time dependences of the friction, respectively, L_1, L_2 are characteristic slip distances, $v*$ is a reference velocity, σ_n is a normal stress, p is fluid pore pressure, τ_0 is a cohesion, μ is the Coulomb friction coefficient, and $\tau*$ is a critical shear stress. Here, the values of constants A, B_i, and L_i were taken from the experiments of Gu et al. (1984).

As was shown by Gu et al. (1984), if $A - B_1 - B_2 < 0$, the friction shows velocity weakening, which can lead to stick–slip motion; otherwise, if $A - B_1 - B_2 \geq 0$, the friction shows velocity strengthening. For the single-degree-of-freedom spring-block model with the spring stiffness k, the so-called critical stiffness k_{cr} (per unit area of block surface) is defined by Eq. (5):

$$k_{cr} = \frac{2A}{L_1 + L_2}\left[(\beta_1 - 1) + \rho^2(\beta_2 - 1) + 2\rho(\beta_1 + \beta_1 - 1)\right.$$
$$\left. + \sqrt{\left\{\left[(\beta_1 - 1) + \rho^2(\beta_2 - 1)\right]^2 + 4\rho^2(\beta_1 + \beta_1 - 1)\right\}}\right]/(4\rho) \tag{5}$$

where $\beta_1 = \frac{B_1}{A}, \beta_2 = \frac{B_2}{A}$, and $\rho = \frac{L_1}{L_2}$.

If $k < k_{cr}$ and $A - B_1 - B_2 < 0$, the stick–slip occurs. Let us suppose that all the blocks have the same friction parameters and stiffness, and that these parameters satisfy the conditions for stick–slip. Initially, all blocks are moving with the velocities equal to the driver velocity. To study the difference between the injection-induced seismicity and the natural seismicity, two sets of numerical calculations were conducted (Set 1 and Set 2). In the first set ("natural" seismicity case), a perturbation in the form of an instant increase in the first block velocity was introduced equal to the velocity of the driver (as considered by Hobs, 1990). In the second set, the pore pressure in the boundary between the first block and the substrate was increased with time in accordance with the pore-elastic equation solution.

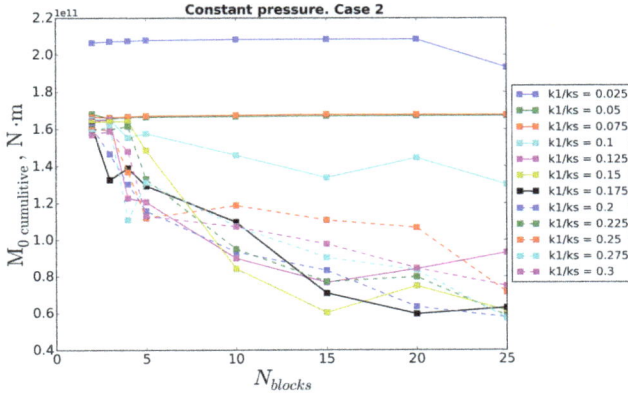

Figure 15. Cumulative seismic moment vs. number of blocks.

Table 1. Values of coefficients A and k_{cr} used in calculations.

Case	A	k_{cr}
1	3.3×10^4 Pa	1.06×10^{10} Pa m^{-1}
2	3.2×10^4 Pa	1.11×10^{10} Pa m^{-1}

The parameters in all the simulations were the following: $B_1 = 3.3 \cdot 10^4$ Pa, $B_2 = 2.772 \cdot 10^4$ Pa, $L_1 = 2.5 \cdot 10^{-7}$ m, $L_2 = 5.2 \cdot 10^{-6}$ m, $v_{pl} = 10^{-9}$ m s^{-1} (3.2 cm yr^{-1}), $k_s = 9.04 \cdot 10^9$ Pa m^{-1} (stiffness per unit area of the block), $\tau^* = 99$ MPa; the block mass obeyed the condition $\frac{mv_{pl}^2}{AS} \ll 1$; S was the area of the block contact with the substrate. By using such a small mass, we can neglect the inertness of the system and Eq. (5) will be relevant for our system. For both sets of the calculations, two cases were considered, which differed by the values of A and k_{cr} (Table 1). It was shown (Gu et al., 1984; Hobbs, 1990) that the one-block system will move chaotically in Case 1 and periodically in Case 2.

2.2 Pore pressure change

To estimate the pore pressure change, we considered radial flow of fluid in an infinite homogeneous reservoir of constant thickness from the injection well with a negligibly small radius (Fig. 2). The initial reservoir pressure was assumed to be the same everywhere and equal to p_0. The volumetric flow rate from the well was constant and equal to Q_0. The assumptions were the following: the permeability was constant (independent of the pressure), and the fluid had small and constant compressibility. To express the condition for constant flow rate u_r at the well bore, Darcy's law was used:

$$u_r = -\frac{k}{\mu}\frac{dp}{dr}, \tag{6}$$

$$Q_0 = 2\pi h r \cdot u_r = -\frac{2\pi h k}{\mu} r \frac{dp}{dr}. \tag{7}$$

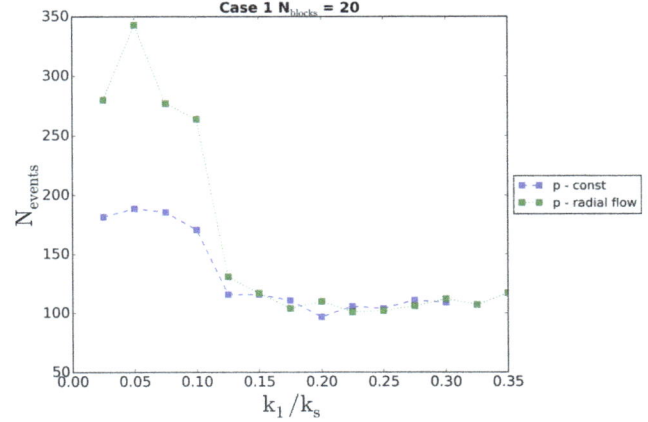

Figure 16. Event cumulative number dependence on the stiffness of the interblock link.

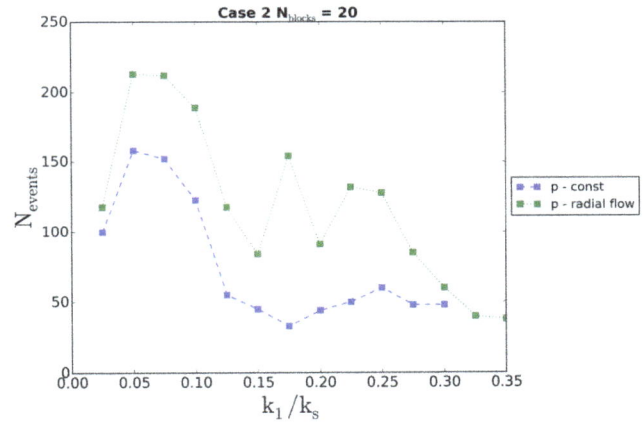

Figure 17. Event cumulative number dependence on the stiffness of the interblock link.

So we got the standard diffusivity equation, where $D = \frac{k}{\varphi\mu c}$ is the hydraulic diffusivity (Matthews and Russel, 1967):

$$\begin{cases} \dfrac{\partial p(r,t)}{\partial t} = D\left(\dfrac{\partial^2 p(r,t)}{\partial r^2} + \dfrac{1}{r}\dfrac{\partial p(r,t)}{\partial r}\right), \\ Q_0 = -\dfrac{2\pi h k}{\mu} r \dfrac{dp}{dr}\Big|_{r=r_w}, \ (r_w \to 0), \\ p(+\infty,t) = p_0, \\ p(r,0) = p_0. \end{cases} \tag{8}$$

The solution of this equation with the above initial and boundary conditions reads as

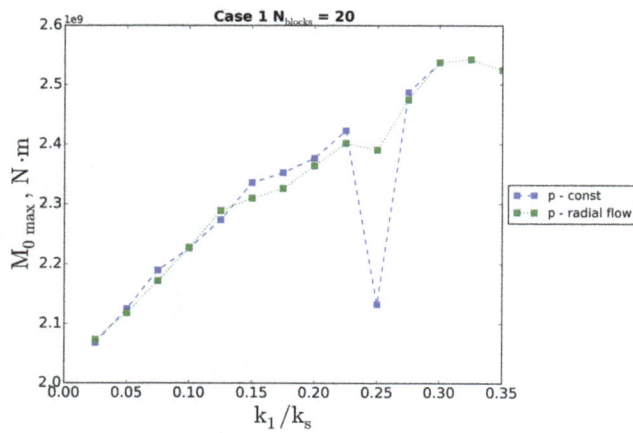

Figure 18. The event maximum seismic moment dependence on the stiffness of the interblock link.

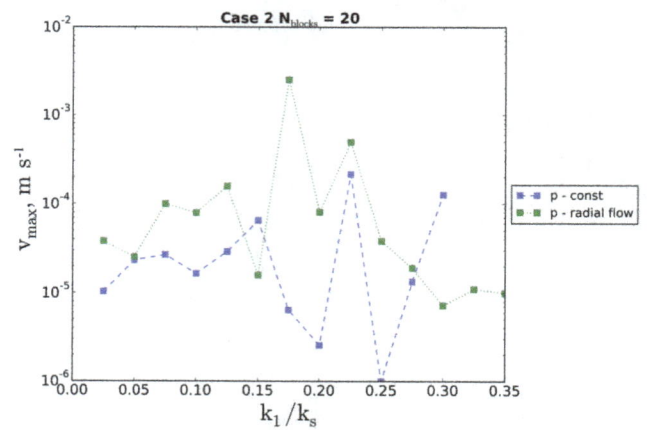

Figure 21. Maximum velocity dependence on the stiffness of the interblock link.

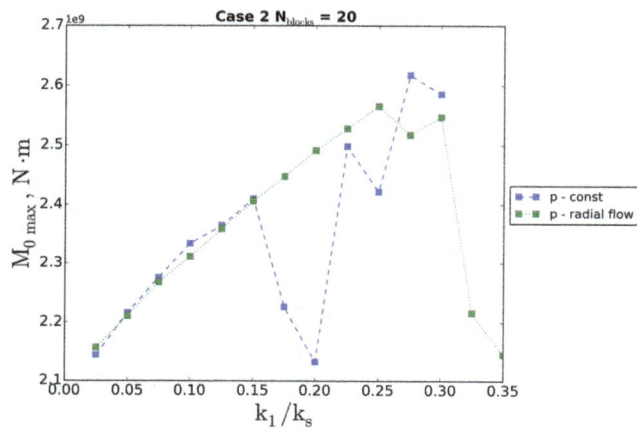

Figure 19. The event maximum seismic moment dependence on the stiffness of the interblock link.

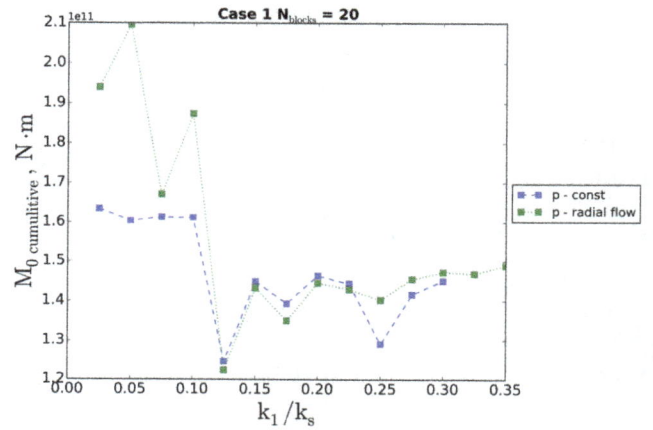

Figure 22. The block system cumulative seismic moment dependence on the stiffness of the interblock link.

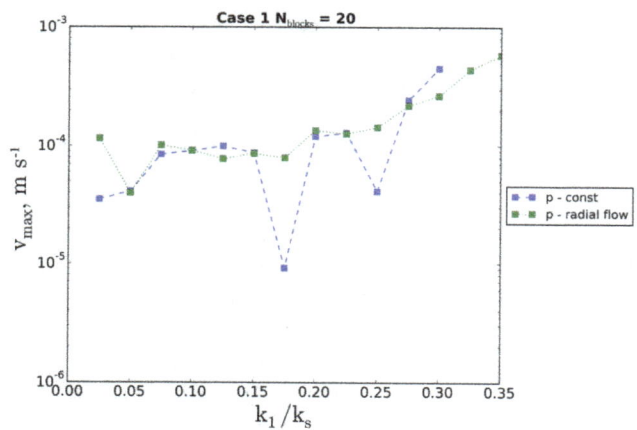

Figure 20. Maximum velocity dependence on the stiffness of the interblock link.

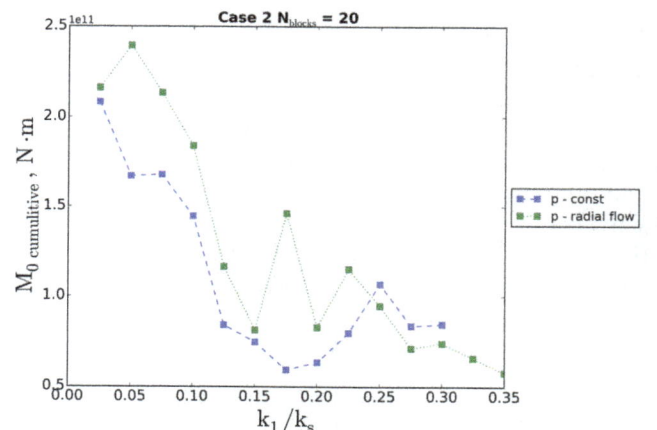

Figure 23. The block system cumulative seismic moment dependence on the stiffness of the interblock link.

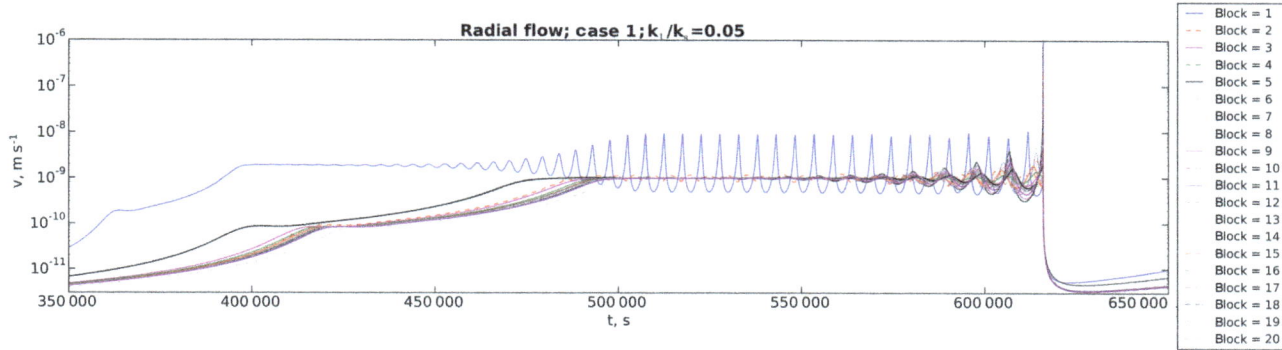

Figure 24. Block velocity variations in time for the system consisting of 20 blocks in Case 1.

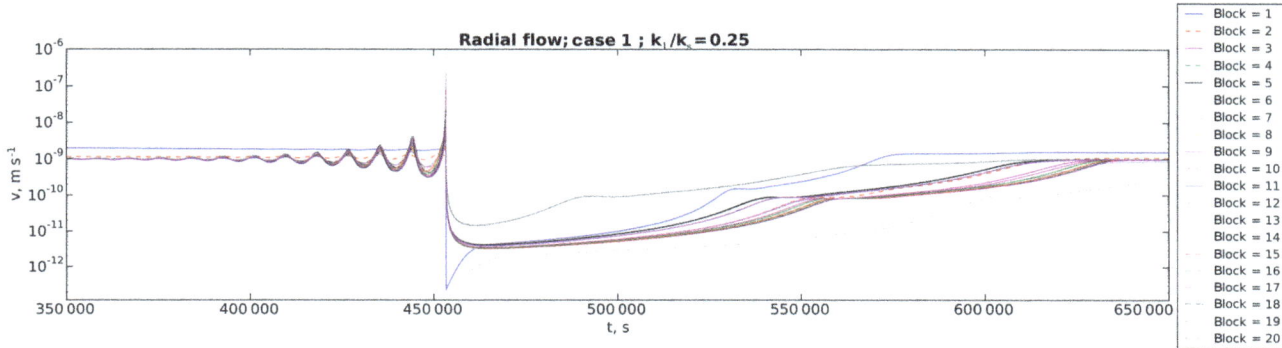

Figure 25. Block velocity variations in time for the system consisting of 20 blocks in Case 1.

$$p = \frac{Q_0 \mu}{4\pi k h} Ei\left(\frac{r^2}{4Dt}\right) + p_0, \tag{9}$$

$$Ei(t) = \int_x^\infty \frac{e^{-t}}{t} dt. \tag{10}$$

The values of parameters used in the calculations were close to the parameters of the Basel project (Häring et al., 2008): $r = 100$ m, $Q_0 = 1.5$ m^3 min^{-1}, $p_0 = 44$ MPa, $\mu = 0.284$ Pa s, $h = 46$ m, $k = 4$ mD, $D = 0.065$ m^2 s^{-1} (Dinske, 2010). We stopped the pressure growth at the first block boundary when it exceeded the value 64 MPa (the corresponding time is approximately $7.13 \cdot 10^6$ s, Fig. 3). Instead of exponential integral Ei(9) we used its approximation (Abramovitz and Stigan, 1979):

$$Ei(x) = \begin{cases} -\ln(\gamma_1 x) \quad 0 < x \leq 0,01 \\ -\ln(\gamma_1 x) + a_1 x + a_2 x^2 + a_3 x^3 + a_4 x^4 + a_5 x^5, \\ \quad 0,01 < x \leq 1 \\ \left(\frac{x^2 + b_1 x + b_2}{x^2 + c_1 x + c_2}\right) \frac{e^{-x}}{x} \quad 1 < x < +\infty \end{cases} \tag{11}$$

where $a_1 = 0.99999193$; $a_2 = -0.24991055$; $a_3 = 0.05519968$; $a_4 = -0.00976004$; $a_5 = 0.00107857$; $b_1 =$

2.334733; $b_2 = 0.250621$; $c_1 = 3.330657$; $c_2 = 1.681534$; $\gamma_1 = 1.7810$.

3　Results

To study the influence of the number of blocks in the multi-degree-of-freedom spring-block system on characteristics of the simulated seismicity (the total number of events, the maximum and cumulative seismic moments) for the "natural" and "induced" cases, the calculations for 2, 3, 4, 5, 10, 15, 20 and 25 blocks were made for the same motion durations – 1 million seconds. The time restriction was related to the computational complexity of 25 block simulations. During this time, the pressure changed significantly in Set 2 (near 11 MPa). The number of events, the maximum seismic moment of one event and the cumulative seismic moment of all events and all blocks are shown in Figs. 4–15. The calculations were made for different ratios of stiffness of the springs between the blocks k_1 to stiffness of links between the driver and the blocks k_s.

It can be seen that if the pore pressure did not change (Set 1, Figs. 5, 7, 9, and 11), the number of events grows almost linearly with the increase in the number of blocks for all values of stiffness of springs between the blocks in both cases

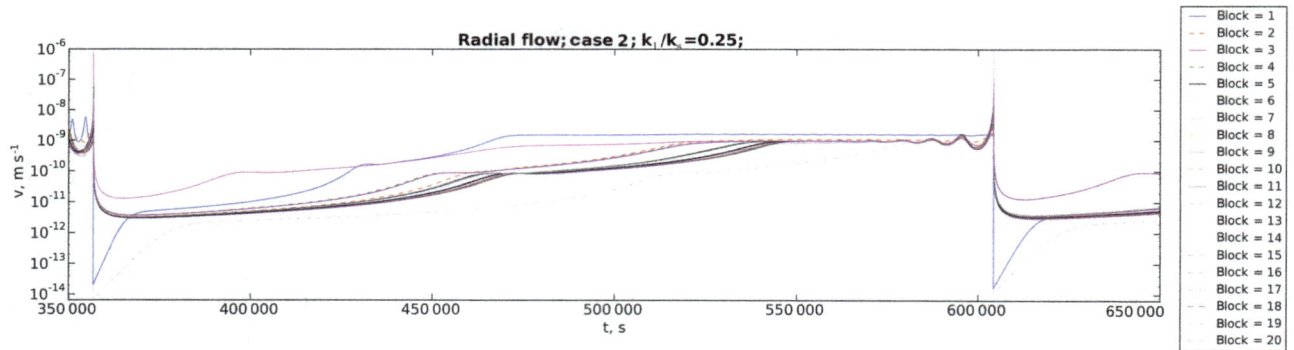

Figure 26. Block velocity variations in time for the system consisting of 20 blocks in Case 2.

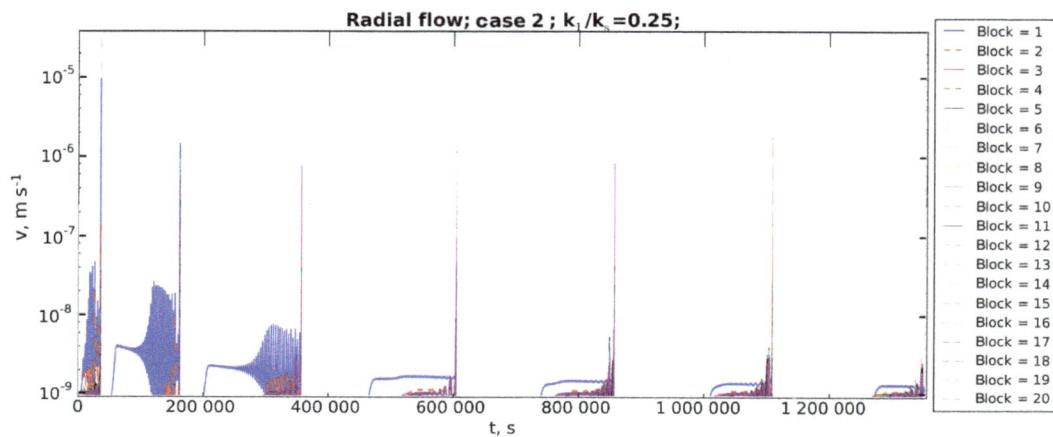

Figure 27. Seismic event occurrences in time for the system consisting of 20 blocks in Case 2.

(1 and 2); the maximum seismic moment of the events decreases with the increase in the number of blocks.

However, for small values of k_1 (equal to $0.025k_s$) the total seismic moment does not depend on the number of blocks for both cases. In Case 1 and $k_1 > 0.1k_s$ for $N_{blocks} \leq 10$ the cumulative seismic moment slightly decreases; for $N_{blocks} > 10$ it almost does not change. In Case 2 the cumulative seismic moment decreases with an increase in the number of blocks. For Set 2, when the pore pressure increases (Figs. 4, 6, 8, and 10), the dependence is more complicated: in Case 1 for $k_1 \leq 0.1k_s$ the number of events also grows linearly with the increase in the number of blocks, but for $0.1k_s < k_1 \leq 0.35k_s$ the number of events decreases with the increase in the number of blocks up to 5, and only then does it start to increase linearly. The maximum seismic moment decreases in both cases; the deviation of one point in Set 2 ("induced" seismicity simulation) from the main trend is caused by insufficient calculating time. The total seismic moment almost does not change in Case 1, and gradually decreases in Case 2.

Now, let us consider the change in the behaviour of the system consisting of 20 blocks with the change in the stiffness of the link between the blocks k_1. In Case 1 (both sets, Figs. 16,

18, and 20), the total number of events initially decreases with the increase in k_1 and then stabilizes at a value around 100, while the maximum seismic moment and the maximum block velocity increase almost monotonically. These results can be explained by the following. Case 1 corresponds to the chaotic behaviour of the one-block system; the characteristic feature of that behaviour is the quick changes in the block velocity. If there are many blocks, the interaction of one block with its neighbours prevents a significant increase in the block velocity. At low values of k_1 every neighbouring block reacts with time lag to movement of the first block, and all blocks move asynchronously and disturb each other. The same effect causes a large number of events. With an increase in k_1, the first block perturbation transfers faster to other blocks; the system starts moving more synchronously, which leads to the increase in the block velocities and in the event seismic moments. At the same time, the total number of perturbations experienced by each block decreases, which leads to the decrease in the number of events. All these features are illustrated in Figs. 24 and 25. For convenience, we consider a short period of time and truncate the maximum value of the velocity.

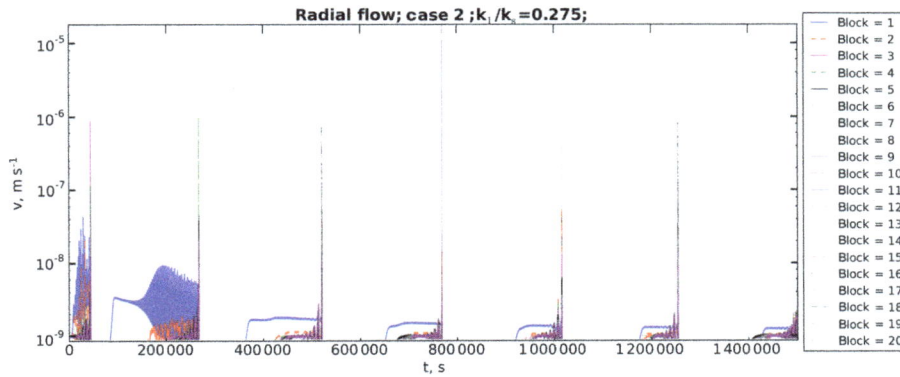

Figure 28. Seismic event occurrences in time for the system consisting of 20 blocks in Case 2.

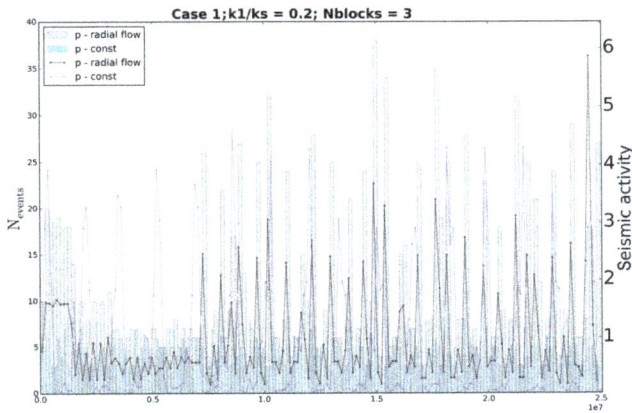

Figure 29. Time variation of seismic activity.

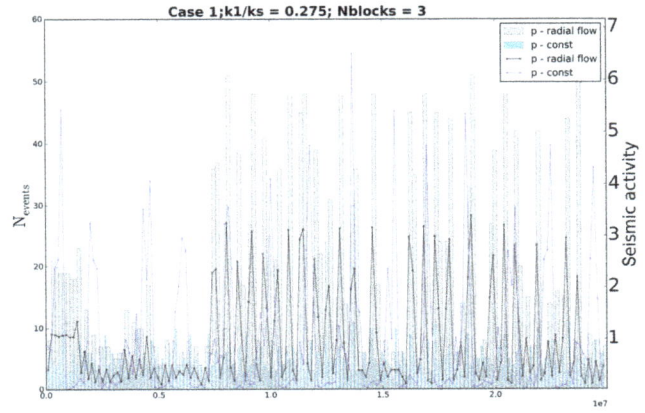

Figure 31. Time variation of seismic activity.

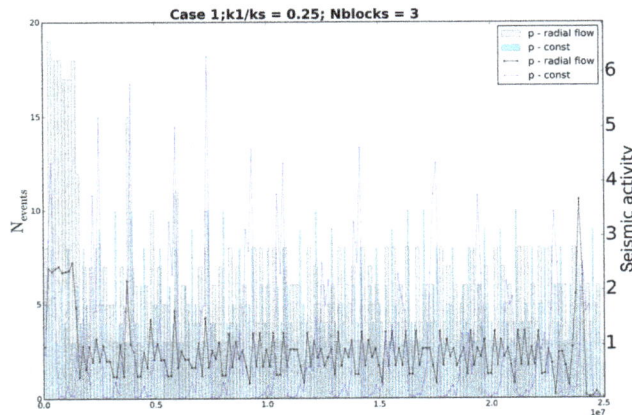

Figure 30. Time variation of seismic activity.

Case 2 is characterized by slower changes in the velocity with time than Case 1 (Fig. 26). That is why there is no clear dependence of the number of events and the block maximum velocity on the interblock link stiffness. Such behaviour becomes more evident with a decrease in parameter A.

Our model demonstrates that the influence of the interblock link stiffness on the behaviour of the studied systems is very strong. By changing the stiffness, we may get periodic or chaotic motion of the system and occurrence of the first strong seismic event almost immediately after the injection start or after a relatively long time (compare Figs. 27 and 28); furthermore, the main seismic activity may occur at the moment of the injection start, when the pressure gradient is the highest, or in the post-injection phase. In Figs. 27–30 the seismic activity variations in the form of the number of events per 10 days (left vertical axis) and the ratio of the cumulative seismic moment of events to the average cumulative seismic moment per 10 days (right vertical axis) are shown for both "natural" (Set 1) and "induced" (Set 2) seismicity. The "natural" seismic activity variations have almost the same amplitudes during all considered time intervals, while the "induced" seismic activity variations depend on interblock link stiffness: in the case of small stiffness the amplitude of the seismic activity during injection is almost the same as in the post-injection period (Fig. 29). When the stiffness becomes higher, the seismicity during injection becomes twice greater than the post-injection activity (Fig. 30);

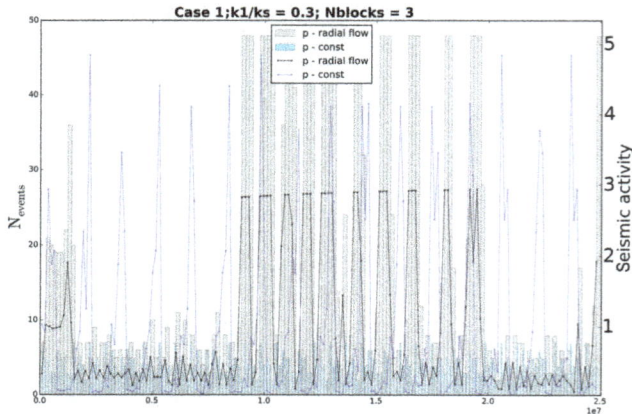

Figure 32. Time variation of seismic activity.

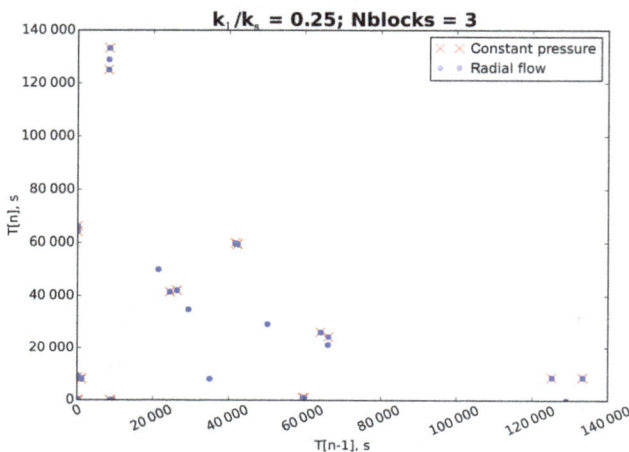

Figure 33. Iteration map of recurrence intervals of seismic events, T_n denotes the time interval between the nth and $(n+1)$th events. The map includes events that occurred at time $t \geq 8 \times 10^6$ s.

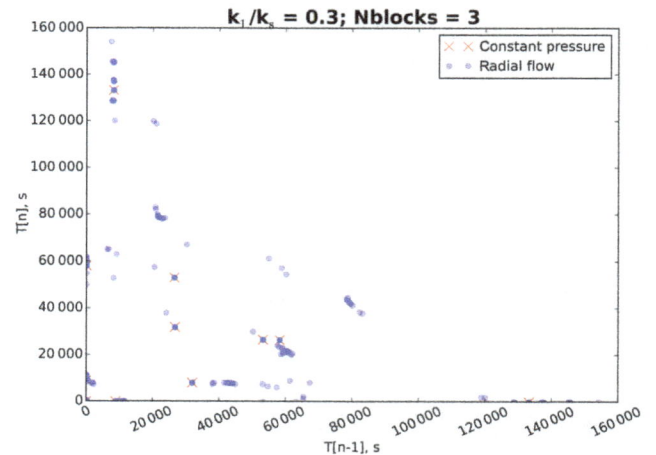

Figure 34. Iteration map of recurrence intervals of seismic events; T_n denotes the time interval between the nth and $(n+1)$th events. The map includes events that occurred at time $t \geq 8 \times 10^6$ s.

tions of a system of blocks (consisting of up to 25 blocks) elastically connected to each other and connected by elastic springs to a constant-velocity moving driver (the multi-degree-of-freedom spring-block model). The rate-and-state friction model with the two-parametric friction law was adopted for description of the friction between the blocks and the substrate. Initially the block system was in steady-sliding state; then, its state was disturbed by the pore pressure increase. The influences of the model parameters (the number of the blocks, the spring stiffness, the velocity weakening parameter) on the process of the model seismicity variations were considered.

It was shown that the considered spring-block system could exhibit different types of motion with different patterns. The motion could be periodic or chaotic; the magnitude of the seismic events depends on fragmentation of the fault system (the number of blocks in the considered model) and may have different values. The analysis shows that the stiffness of the link between the blocks affects significantly the behaviour of the model and the resulting seismicity, so the main seismic activity could appear directly after the start of the fluid injection or in the post-injection phase. Such influence of the injection on seismicity could be observed in the real cases. However, the parameters in the rate-and-state model are known only from laboratory experiments, and it is hard to believe that one should use the same values to describe the real-scale phenomena. Yet our study showed that it is possible to select more suitable parameters that will allow one to match results of calculations and data of real observations. It can be concluded that the considered model has the potential to be used for the estimations of the possible fluid-induced seismicity activity variations.

Competing interests. The authors declare that they have no conflict of interest.

the further increase in the interblock link stiffness leads to a significant increase in the post-injection activity (Figs. 31 and 32).

The recurrence maps of the seismic event sequences are shown in Figs. 33 and 34. It could be seen that for $k_1/k_s = 0.25$ the time intervals between two events converge to several points for both "induced" and "natural" seismicity (only post-injection seismic activity is considered). For $k_1/k_s = 0.3$, the "natural" seismicity shows periodic variations, while the "induced" seismicity has more complicated chaotic behaviour. For other values of k_1 both "induced" and "natural" seismicity shows the chaotic variations.

4 Discussion and conclusions

The problem of the influence of the fluid injection on the tectonic fault sliding and generation of the seismic events was studied by numerical calculations of the peculiarities of mo-

Acknowledgements. The financial support of the Russian Foundation for Basic Research (project no. 16-05-00869) is acknowledged.

Edited by: A. Dyskin

References

Abe, Y. and Kato, N.: Complex Earthquake Cycle Simulations Using a Two-Degree-of-Freedom Spring-Block Model with a Rate- and State-Friction Law, Pure Appl. Geophys. 170, 745–765, doi:10.1007/s00024-011-0450-8, 2013.

Abe, Y. and Kato, N.: Intermittency of earthquake cycles in a model of a three-degree-of-freedom spring-block system, Nonlin. Processes Geophys., 21, 841–853, doi:10.5194/npg-21-841-2014, 2014.

Abramovitz, M. and Stigan, I.: Handbook of special functions, M., Nauka, 56–59, 1979 (in Russian).

Brace, W. F. and Byerlee, J. D.: Stick-slip as mechanism for earthquakes, Science, 153, 990–992, doi:10.1126/science.153.3739.990, 1966.

Burridge, R. and Knopov, L.: Model and theoretical seismicity, B. Seismol. Soc. Am., 67, 341–371, 1967.

Dieterich, J. H.: Earthquake nucleation on faults with rate and state-dependent friction, Tectonophysics, 211, 115–134, doi:10.1029/96JB00529, 1992.

Dinske, C.: Interpretation of fluid-induced seismicity at geothermal and Hydrocarbon Reservoir of Basel and Cotton Valley, PhD Thesis, 151 pp., 2010.

Erickson, B., Birnir, B., and Lavallée, D.: A model for aperiodicity in earthquakes, Nonlin. Processes Geophys., 15, 1–12, doi:10.5194/npg-15-1-2008, 2008.

Grassberger, P. and Procaccia, I.: Measuring the strangeness of strange attractors, Physica D: Nonlinear Phenomena, 9, 189–208, doi:10.1016/0167-2789(83)90298-1, 1983.

Gu, J. C., Rice, J. R., Ruina, A. L., and Tse, S. T.: Slip motion and stability of a single degree of freedom elastic system with rate-and-state dependent friction, J. Mech. Phys. Solids, 32, 167–196, doi:10.1016/0022-5096(84)90007-3, 1984.

Häring, M. O., Schanz, U., Ladner, F., and Dyer, B. C.: Characterisation of the Basel 1 enhanced geothermal system, Geothermics, 37, 469–495, doi:10.1016/j.geothermics.2008.06.002, 2008.

Hobbs, B. E.: Chaotic behaviour of frictional shear instabilities, Rockbursts and Seismicity in Mine, Mineapolis, 87–91, 1990.

Matthews, C. S. and Russel, D. G.: Pressure buildup and flow tests in wells, Monograph, Society of Petroleum Engineers of AIME, Dallas, TX, 167, 1967.

Turuntaev, S. B., Vorohobina, S. V., and Melchaeva, O. Y.: Identification of anthropogenic changes of seismic regime using methods of nonlinear dynamics, Phys. Earth, 3, 52–65, 2012.

11

The fully nonlinear stratified geostrophic adjustment problem

Aaron Coutino and Marek Stastna

Department of Applied Mathematics, University of Waterloo, Waterloo, Canada

Correspondence to: Aaron Coutino (acoutino@uwaterloo.ca)

Abstract. The study of the adjustment to equilibrium by a stratified fluid in a rotating reference frame is a classical problem in geophysical fluid dynamics. We consider the fully nonlinear, stratified adjustment problem from a numerical point of view. We present results of smoothed dam break simulations based on experiments in the published literature, with a focus on both the wave trains that propagate away from the nascent geostrophic state and the geostrophic state itself. We demonstrate that for Rossby numbers in excess of roughly 2 the wave train cannot be interpreted in terms of linear theory. This wave train consists of a leading solitary-like packet and a trailing tail of dispersive waves. However, it is found that the leading wave packet never completely separates from the trailing tail. Somewhat surprisingly, the inertial oscillations associated with the geostrophic state exhibit evidence of nonlinearity even when the Rossby number falls below 1. We vary the width of the initial disturbance and the rotation rate so as to keep the Rossby number fixed, and find that while the qualitative response remains consistent, the Froude number varies, and these variations are manifested in the form of the emanating wave train. For wider initial disturbances we find clear evidence of a wave train that initially propagates toward the near wall, reflects, and propagates away from the geostrophic state behind the leading wave train. We compare kinetic energy inside and outside of the geostrophic state, finding that for long times a Rossby number of around one-quarter yields an equal split between the two, with lower (higher) Rossby numbers yielding more energy in the geostrophic state (wave train). Finally we compare the energetics of the geostrophic state as the Rossby number varies, finding long-lived inertial oscillations in the majority of the cases and a general agreement with the past literature that employed either hydrostatic, shallow-water equation-based theory or stratified Navier–Stokes equations with a linear stratification.

1 Introduction

Geostrophic balance, namely the balance between the pressure gradient and the Coriolis pseudoforce, is observed to hold to a good approximation for many large-scale motions in the atmosphere and the ocean. The process through which some disturbed state reaches this balance is called geostrophic adjustment. The linear problem was first considered by Rossby (1937). Using conservation of momentum and mass, he derived the geostrophic steady state corresponding to an initial perturbation. In the original publication, Rossby noted that the final state of the system possessed less energy than the initial state. The cause of this difference was identified by Cahn (1945), who showed that the end state is reached via inertial oscillations, which disperse energy through waves. Since then, numerous papers have used a variety of methods such as asymptotic expansions and numerical integration to solve this linear problem. There has been a great deal of published work on the linear problem (Ou, 1984; Gill, 1976; Middleton, 1987; Washington, 1964; Mihaljan, 1963), but little on the fully nonlinear one. This is partly because nonlinear problems rarely yield analytical solutions in closed form, and partly because numerical methods applied to the problem must accurately resolve multiple length scales.

Kuo and Polvani (1999), a key paper in the study of the nonlinear problem, considered the adjustment problem in the context of the single-layer shallow-water equations in one dimension. The authors built on the results of Killworth (1992) and performed a numerical analysis of the fully nonlinear problem with "dam break" initial conditions (see Gill, 1982). The authors found that the nonlinearity and rotation led to bore generation, with the bores dissipating energy as they propagated away from the geostrophic state. Since the nonlinear shallow-water equations neglect non-hydrostatic dispersion, these bores manifested as shock-like fronts. This is in contrast to the non-rotating stratified adjustment problem which leads to the generation of either a rank-ordered train of internal solitary waves or an undular bore. Indeed, in this dispersive system, for the majority of parameter space, breaking is not observed. The authors also found that the inertial oscillations within the geostrophic state can persist for long times and are highly dependent on the initial conditions. In their analysis of the energy within the geostrophic state, the authors found that the ratio of change in kinetic energy (ΔKE) to change in potential energy (ΔPE) tended to $\frac{1}{3}$, which is the theoretically predicted value for both the linear and nonlinear problems (see Boss and Thompson, 1995). However, Kuo and Polvani (1999) showed that, even for late times, the energy ratio fluctuated by up to 30 % around the $\frac{1}{3}$ value.

While primarily viewed from a theoretical framework, rotation-modified adjustment has been shown to arise naturally in the ocean. Examples of this are upwelling fronts which can create the initial density anomaly that then must adjust; see Chia et al. (1982) for a more complete discussion. More recently, Ledwell et al. (2004) and Oakey and Greenan (2004), as part of the Coastal Mixing and Optics experiment, showed the presence of patches of well-mixed regions (density anomalies) throughout a background of stable fluid along the New England shelf. The specifics of this adjustment were investigated in Lelong and Sundermeyer (2005). The authors performed fully 3-D numerical simulations of moderate resolution, using the nonhydrostatic equations under the Boussinesq approximation, of the adjustment process resulting from one of these density anomaly patches. To allow for more reasonable computation, the authors followed the procedure outlined in Lelong and Dunkerton (1998), and reduced the physical ratio for $\frac{N}{f}$ (by varying f and holding N constant), where N is the buoyancy frequency and f is the Coriolis parameter. For analysis, they separated the energy into kinetic and potential, and separated the domain into two regions, an inner region associated with the geostrophic state and an outer region associated with the waves. Since the initial conditions are static, the initial energy of the system is contained solely as potential energy. By comparing the energy within different areas of the simulation to the initial energy, the authors found that initial conditions with $\frac{Ro_r}{L} = 1$, where Ro_r is the Rossby radius of deformation, and L is the half-width of the initial state, were the most effective at generating kinetic energy in the geostrophic state. This is in contrast to cases with $\frac{Ro_r}{L} < 1$, where rotation effects dominate and little potential energy is converted to kinetic, or cases with $\frac{Ro_r}{L} > 1$, where the potential energy is primarily converted to wave energy. In all cases considered by these authors the radiated wave train was weak, composed of long waves and well approximated by linear theory.

The nonlinear effects on the rotating adjustment problem have been investigated analytically using multiple-scale perturbation analysis of the shallow-water and fully stratified equations. In part one of a two-part paper series, Zeitlin et al. (2003a) perturb the rotating shallow-water equations using the Rossby number as their small parameter. The authors proceed to confirm that a slow–fast splitting is possible, with the slow state largely remaining in geostrophic balance and largely unaffected by the fast state. In the waves that are generated, Zeitlin et al. (2003a) observe shock formation and present a semi-quantitative criterion for this, based on the initial conditions. In the second paper, Zeitlin et al. (2003b), the authors generalize their results to the case of continuous stratification and also consider two-layer and quasi-two-layer stratifications. Zeitlin et al. (2003b) perform a number of asymptotic expansions for different initial isopycnal deviation regimes. They conclude that for large deviations the model strongly depends on the ratio of the layer depths and that the waves produced from the initialization obey a Schrödinger-type modulation equation. For small deviations the waves generated are not impacted by the geostrophic state, which is left to evolve according to the standard quasi-geostrophic (QG) equations.

Rotation-influenced nonlinear waves have also been considered using a model nonlinear wave equation, in this case a member of the Korteweg–de Vries (KdV) family of equations. The KdV equation is the simplest model equation that allows for a balance between nonlinear and dispersive effects, with a rich mathematical structure which makes predictions of the evolution of an initial state that are remarkably robust in both laboratory and field settings (see Johnson, 1997). A rotation-modified version of the KdV equation was first derived by Ostrovsky (see Grimshaw et al., 2012 for an in-depth discussion of the equation properties and references to the Russian literature). This new equation was subsequently analysed both through theoretical solutions found by asymptotic expansions and through numerical solutions. Investigation of the model equations revealed that the precise balance between nonlinearity and dispersion that leads to the traditional soliton solution of the KdV equation is destroyed by the addition of rotation, and that over time the soliton breaks down into a nonlinear wave packet (Grimshaw and Helfrich, 2008). This hypothesis was later supported by experimental results (Grimshaw et al., 2013). From a theoretical point of view, Grimshaw and Helfrich (2008) also found that the extended nonlinear Schrödinger (NLS) equation provides a good qualitative description of the wave packet. While the mathematical developments of the rotation-modified theory are substantial, it is also true that this theory has a number of pathologies not observed in the non-rotating KdV-based theory, which in itself has been shown to misrepresent aspects of large amplitude solitary waves (Lamb, 1997, is one of many papers to discuss some of these discrepancies).

Work has also been performed using models with higher-order nonlinearity (Helfrich, 2007) but weak nonhydrostatic effects, as well as with the full set of stratified Euler equations (Stastna et al., 2009). Both of these studies considered the breakdown of an initial solitary wave in the presence of rotation. Helfrich suggested that the initial solitary wave breaks down into a coherent leading wave packet with a trailing tail of waves. Stastna et al. (2009) suggested that for large amplitude, exact internal solitary waves that are solutions to the Dubreil–Jacotin–Long (DJL) equation, Helfrich's result was observed for artificially high rotation rates, while rotation rates typical of mid-latitudes led to a disturbance that never fully separated from the trailing tail. Despite differences in details, the qualitative features observed in both studies were quite similar. Additionally, they also performed collision experiments, finding that the packets that emerge from the initial solitary waves can merge during collisions and hence do not interact as classical solitons. Finally, Stastna et al. (2009) also found that by increasing the width of a flat-crested wave, more energy was deposited into the tail. It remains to reconcile the two sets of results in detail, likely by systematically reducing the solitary wave amplitude used as an initial condition.

In this paper, we present the results of high-resolution simulations of the geostrophic adjustment of a stratified fluid with a single pycnocline on an experimental scale. Our simulations consider the full set of stratified Euler equations using a pseudo-spectral collocation method. We begin by providing and reviewing the non-rotating case and the changes that arise when the polarity of the initial condition is changed. Next we present the general evolution of the rotating case using classical theory and two "base" cases, one of which is comparable to one of the cases presented in Grimshaw et al. (2013). We subsequently identify the manner in which nonlinearity is exhibited in the problem, focusing on both the wave train and the geostrophic state and its inertial oscillations. We are able to clearly show the generation of a leftward propagating wave from the initial condition (especially evident for wider initial perturbations) and its subsequent reflection from the left wall. This wave train interacts with the geostrophic state, before and after reflecting off the left wall of the domain, and then continues to propagate rightward across the tank. This is of potential interest to future experiments. We then focus on the geostrophic state in detail, specifically examining the change in kinetic energy and the change in potential energy for different initial widths, as well as the changes in the kinetic energy in the geostrophic state and the propagating wave train as the Rossby number varies. These results make the closest contact with the work of Lelong and Sundermeyer (2005). Finally we draw a number of conclusions based on our findings and identify directions for future work.

2 Methods

For the following numerical simulations, the full set of stratified Navier–Stokes equations for an incompressible fluid were used, though no span-wise variations were consid-

ered. Rotation was incorporated using an f plane approximation and the non-traditional Coriolis terms were dropped. For a review of the effects of the non-traditional Coriolis terms, see Gerkema et al. (2008). The x axis is taken as parallel to the flat bottom with the z axis pointing upward ($\hat{k}$ is the upward directed unit vector). The origin is placed in the bottom left corner so that both axes are positive. The incompressible Navier–Stokes equations for velocity $\boldsymbol{u} = [u(x, z, t), v(x, z, t), w(x, z, t)]$, density $\rho(x, z, t)$, and pressure $P(x, z, t)$ are

$$\frac{D\boldsymbol{u}}{Dt} + (-fv, fu, 0) = -\nabla P - \rho' g \hat{k} + \nu \nabla^2 \boldsymbol{u}, \tag{1}$$

$$\nabla \times \boldsymbol{u} = 0, \tag{2}$$

$$\frac{D\rho}{Dt} = 0, \tag{3}$$

where f is the constant Coriolis parameter, g is acceleration due to gravity, and ν is the kinematic viscosity. In accordance with convention, we have divided the momentum equation by the constant reference density ρ_0 and absorbed the hydrostatic pressure into the pressure P. We make the Boussinesq approximation for density and write $\rho = \rho_0(1 + \rho'(x, z, t))$, where ρ' is considered a small perturbation. Due to our interest in the wave dynamics in the main water column, as opposed to details of the boundary-layer dynamics, we impose free slip boundary conditions at the top and bottom of our domain. This will also ensure that the boundary layer does not play a significant role in the simulations on which we report. The walls allow us to mimic a lock–release set-up that is used to create waves in many laboratory set-ups (Carr and Davies, 2006; Grue et al., 2000; Helfrich and Melville, 2006). We have chosen to neglect the span-wise dimension (y), as the lab results in Grimshaw et al. (2013) were performed away from any side boundaries and the authors elected to neglect any curvature from the waves created. Another change from Grimshaw et al. (2013) is that we have a rigid lid as opposed to their free surface; this is due to the computational difficulty of a moving boundary.

In the following set of experiments the dominant dimensionless number is the Rossby number. This number is defined as $Ro = \frac{U}{fL}$, where U is the typical wave speed, L is the typical length scale, and f is the Coriolis parameter. This reflects a ratio of the inertia term to the Coriolis pseudoforce term (henceforth just force). When the Coriolis force dominates, the fluid can reach a balance between the rotation and pressure terms, i.e., geostrophic balance. Since the full equations contain the diffusion term $\nu\nabla^2$, it can be used to form the dimensionless Reynolds number which is given by $Re = \frac{UL}{\nu}$. U and L are the same as for Ro and ν is the kinematic viscosity. The other relevant number considered is the Froude number $Fr = \frac{U}{c}$, which compares the typical wave speed U to the theoretical wave speed c. In addition to these traditional dimensionless numbers we also define a nonlinearity parameter α. Following from Kuo and Polvani (1997) this is defined as $\alpha = \frac{\eta}{H_1}$, where η is the height of the displacement in isopycnals and H_1 is the height of the undisturbed fluid interface. This parameter can be used to modify the strength of the nonlinearity, and is well suited for shallow-water equations. However, for the full set of incompressible Navier–Stokes equations some ambiguity is introduced by the vertical structure of the stratification and the initial perturbation. Nevertheless, we have found α to be a useful parameter, likely since the disturbances in our simulations are dominated by mode-1 waves.

The numerical simulations presented here were performed using an incompressible Navier–Stokes equation solver which implements a pseudo-spectral collocation method (SPINS), presented in Subich et al. (2013). The solver uses spectral methods resulting in the order of accuracy scaling with the number of grid points. To deal with the build-up of energy in the high wavenumbers, an exponential filter is used after a specific wavenumber cut-off.

We computed a series of 2-D lab-scale numerical simulations on a similar scale to the physical experiments presented in Grimshaw et al. (2013), which were performed using the 13 m diameter rotating platform at the LEGI-Coriolis Laboratory in Grenoble. Motivated by the results presented in Stastna et al. (2009), a domain 4 times longer than the physical tank ($Lx = 52$ m) was used, as 13 m is an insufficient length when considering lower (closer to physical) rotation rates. In addition to this it was decided to change the tank depth to a more evenly divisible 0.4 m (from a laboratory value of 0.36 m). The density difference was set to 1 % to match Grimshaw and Helfrich (2008). The different physical parameters related to the initial set-up are illustrated in Fig. 1.

In total, 8192 grid points were used to resolve the 52 m length of the tank and 192 points were used for the 0.4 m height, providing a 0.006 m horizontal resolution and a 0.002 m vertical resolution. To easily compare these numerical results to the experimental values in Grimshaw et al. (2013), our Coriolis parameter was based on their lowest presented value of f, which had a value of 0.105 s^{-1}. It was also decided to base the initial perturbation width w_0 on twice the Rossby radius of deformation ($Ro_r = \frac{U}{f}$) so as to allow for a neater examination of the parameter space. We used the same change in density as Grimshaw et al. (2013), 1 % between the upper and lower fluids. In each of the simulations, the initial conditions were given by a quiescent fluid, and a density field defined via the isopycnal displacement η,

$$\rho'(x, z, t = 0) = -0.005 \tanh\left(\frac{z - \eta - 0.3}{0.01}\right), \qquad (4)$$

$$\eta = \pm 0.05 \exp\left[-\left(\frac{x}{w}\right)^8\right], \qquad (5)$$

where w is the half-width of the initial perturbation and the sign changes correspond to changes in perturbation polarity.

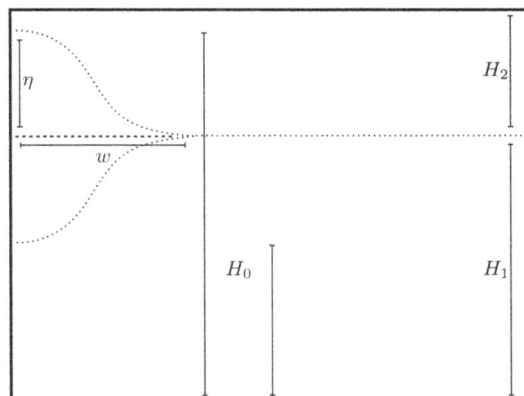

Figure 1. A schematic of the tank simulation set-up which illustrates the different parameters. The dotted line represents the isopycnal found at the centre of the pycnocline on the far right of the domain. The largest deflection (both polarities are shown in the figure) occurs at the left end point of the domain. H_1 and H_2 represent the depth of fluid below and above the centre of the undisturbed pycnocline respectively. H_0 is the maximum or minimum height of the pycnocline created by the initial conditions. η is the isopycnal displacement. w is the width of the initial condition defined from the left-hand wall to where the pycnocline reaches within 1 % of the undisturbed height.

3 Results

In this section we present the results of multiple numerical simulations. Parameters were primarily modified by changing either the initial width of the perturbation or by changing the rotation rate. Using the initial width as the typical length scale, $L = w$, we are thus varying the Rossby number. The resulting values of Ro are shown in Table 1. Across all the cases, the depth does not change, and hence does neither does the two-layer linear longwave speed $U = \sqrt{g \frac{\Delta\rho}{\rho_0} \frac{H_1 H_2}{H_1 + H_2}} = 0.0858$ m s^{-1}, where $g = 9.81$ m s^{-2}, $\Delta\rho = 10$ kg m^{-3}, $\rho_0 = 1000$ kg m^{-3}, $H_1 = 0.3$ m, and $H_2 = 0.1$ m. Using the same wave speed and length scales as for the Rossby number, the corresponding Reynolds numbers can be calculated; however, since we are primarily concerned with internal waves, viscosity is negligible until the waves disperse to scales where viscosity is dominant Vallis (2006). The kinematic viscosity was the same for all simulations, $\nu = 1 \times 10^{-6}$ m^2 s^{-1}. Several additional experiments were carried out by changing the initial wave amplitude, which results in different "nonlinearity" parameters. For the initial amplitude of $\eta = 0.05$ m and undisturbed pycnocline height $H_1 = 0.3$ m we have $\alpha = 0.1667$. For the cases where amplitude is halved and quartered, corresponding alpha values are $\alpha = 0.0833$ and $\alpha = 0.0416$. The initial value of $\alpha = 0.1667$ allows for an easy comparison to many of the figures in Kuo and Polvani (1999) which are based on a value of 0.1. In addition to the simulations above, another set of simulations was performed

Table 1. Rossby number of each simulation, where $f_0 = 0.105 \, \text{s}^{-1}$ and $w_0 = 1.63 \, \text{m}$.

Ro	$\frac{1}{4}w_0$	$\frac{1}{2}w_0$	w_0	$2w_0$	$4w_0$
f_0	2	1	$\frac{1}{2}$	$\frac{1}{4}$	$\frac{1}{8}$
$\frac{1}{2}f_0$	4	2	1	$\frac{1}{2}$	$\frac{1}{4}$
$\frac{1}{4}f_0$	8	4	2	1	$\frac{1}{2}$

using the opposite polarity of the initial disturbance. These opposite polarity simulations correspond exactly to the cases seen in Table 1, the only difference being the sign in the isopycnal displacement used in the initial conditions.

Several simulations were also performed on an extra-long tank to investigate the long-time results of adjustment. For these simulations the tank length was $L = 260 \, \text{m}$ and the number of horizontal grid points was increased to $16\,384$, providing a $0.0158 \, \text{m}$ resolution. The vertical height and grid points were kept the same from the smaller case.

Unless otherwise stated the following scaling is used for all figures: $T = \frac{1}{f}$, $L_z = Lz$, and $L_x = Ro_r$, with $Lz = 0.4 \, \text{m}$ corresponding to the depth of the tank. For kinetic energy, we scale by the maximum kinetic energy for all spaces and times *to show relative changes*.

3.1 The non-rotating case

We begin by reproducing the results of the adjustment problem without rotation. The solution to this problem is well known, though we are not aware of any references that present the result in detail. We thus state the result, with a numerical example, and briefly outline the weakly nonlinear theory behind it. Non-rotating adjustment yields either a rank-ordered train of solitary waves or an undular bore forming from the initial disturbance, depending upon the polarity of the initial disturbance. Examples of these two cases are shown in Fig. 2. Since there is no rotation, the advective timescale was chosen, $T = \frac{L}{U}$, to nondimensionalize time, with the initial width $w = \frac{1}{2}w_0$ chosen for the typical length scale. The stark difference between these cases is readily apparent in both types of plots.

The result may be understood in terms of KdV theory. Using the notation of Lamb (1997), separation of variables is applied to the streamfunction so that

$$\psi(x, z, t) = B(x, t)\phi(z). \qquad (6)$$

The vertical structure is determined from a linear eigenvalue problem, while to first order in amplitude and aspect ratio $B(x, t)$ is governed by a KdV equation for waves propagating in each direction. The KdV equation corresponding to rightward propagating waves reads as

$$B_t = -cB_x + 2cr_{10}BB_x + r_{01}B_{xxx}, \qquad (7)$$

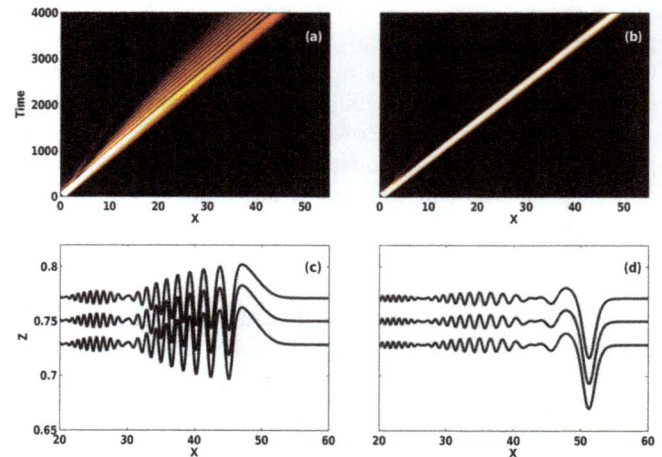

Figure 2. A space–time filled contour plot of vertically integrated kinetic energy and density isocontours at $t = 4275$ showing the differences between the positive and negative polarity initial conditions with $w = \frac{1}{2}w_0$. Panels (**a**) and (**c**) correspond to the positive polarity case, and panels (**b**) and (**d**) to the negative case.

where c is the linear longwave speed, r_{10} is the nonlinearity coefficient, and r_{01} is the dispersive coefficient. The numerical value for c is computed from the linear longwave eigenvalue problem (Lamb's Eq. 8a), while r_{10} and r_{01} are computed from the integral expressions involving the eigenfunctions of the same problem (Lamb's Eqs. 10a and b). The dispersive coefficient, r_{01}, is always negative, while the nonlinear coefficient, r_{10}, switches sign depending on the functional form of the stratification. In the case of a two-layer flow exact expressions can be derived. Solitary wave solutions of Eq. (7) are of the classical sech^2 form. The propagation speed equals the linear longwave speed to leading order, with a nonlinear correction that is proportional to amplitude and r_{10} (Lamb's Eq. 17). Thus the sign of r_{10} also determines solitary wave polarity. In the absence of background shear currents this implies that stratifications centered above (below) the mid-depth yield solitary waves of depression (elevation). All numerical experiments performed with exact solitary waves computed using the DJL equation that we are aware of match the predictions of the KdV theory presented above, as far as solitary wave polarity is concerned. Of course, KdV theory is not necessarily a quantitatively accurate predictor of the structure of large solitary waves (Lamb, 1997, is one of many papers to discuss some of the discrepancies).

3.2 Rotation-modified evolution

As discussed in the introduction, a variety of model equations have been derived that account for the effects of rotation, with Grimshaw and Helfrich (2008) providing a relatively recent summary. The essential aspects of the role of rotation can be gleaned from linear theory. In this case the

streamfunction is governed by

$$\left(\mu^2\psi_{xx}+\psi_{zz}\right)_{tt}+\frac{1}{Ro^2}\psi_{zz}+\frac{1}{Fr^2}N(z)^2\psi_{xx}=0, \qquad (8)$$

where $\mu=\frac{H}{L}$ is the aspect ratio. When the assumption of a linear stratification is made, the vertical structure of ψ is sinusoidal, so that for the first vertical mode a separation of variables like Eq. (6) yields

$$\left(\mu^2 B_{xx}-\frac{\pi^2}{H^2}B\right)_{tt}-\frac{\pi^2}{H^2}\frac{1}{Ro^2}B+\frac{1}{Fr^2}B_{xx}=0. \qquad (9)$$

The well-known dispersion relation of rotation-modified internal waves in a channel is readily recovered by assuming a travelling wave solution (Vallis, 2006). In the hydrostatic limit this equation reduces to the simplest example of a partial differential equation that is both hyperbolic and dispersive, the classical Klein–Gordon equation of mathematical physics,

$$B_{tt}+\frac{1}{Ro^2}B=\frac{H^2}{\pi^2 Fr^2}B_{xx}. \qquad (10)$$

By using the plane wave ansatz, it can immediately be seen that the dispersion relation yields a lower bound on frequency in the longwave limit, and hence the phase speed is unbounded in this same limit. This is the central problem that model equation theories such as the Ostrovsky equation face when implemented numerically. It is also readily apparent from Eq. (10) that a non-trivial, time independent state is possible, and this state corresponds to the geostrophic state of classical geophysical fluid dynamics (Vallis, 2006). Furthermore, it is clear that a spatially independent inertial oscillation is a possible solution to the equation. However, for a given initial condition it is not immediately obvious what the precise split is between the portion of the initial state that propagates away and the portion left behind. While the case in which the disturbance that emanates from the initial condition is small enough to be well described by linear wave theory has been studied in detail by Lelong and Sundermeyer (Lelong and Sundermeyer, 2005), the significant amount of literature on the combined effects of nonlinearity, dispersion, and rotation, and especially the experimental results in Grimshaw et al. (2013), suggest that the initial value problem should be reconsidered without a priori approximations.

Using our definition of the Rossby number, we find that the experiments presented in Grimshaw et al. (2013), with a Coriolis parameter of $f=0.105\,\text{s}^{-1}$, had a corresponding Rossby number of 0.667. Therefore, we consider the $f=f_0$ and $w=w_0$ case ($Ro=0.5$) as our baseline. We also pick a negative initial condition to match their configuration. With this in mind, Fig. 3 compares this baseline case with one where the only difference is that the rotation rate has been quartered ($f=\frac{1}{4}f_0$, $Ro=2$). Figure 3a and c correspond

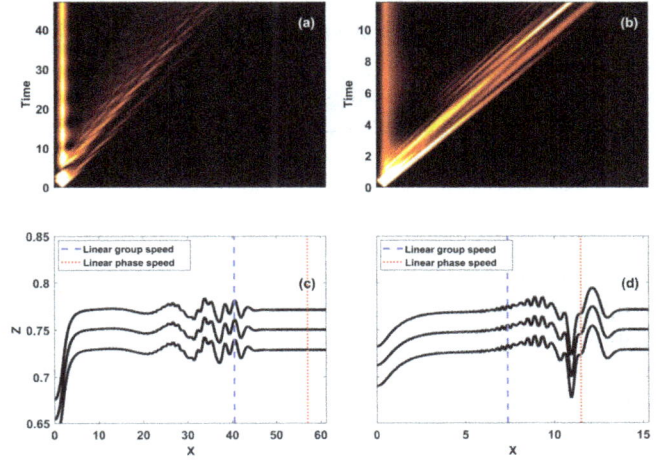

Figure 3. Panels (a) and (b) show a space–time plot of vertically integrated kinetic energy, while panels (c) and (d) show three density isolines at $t=47.25$ and $t=11.81$ respectively. Panels (a) and (c) correspond to the $Ro=0.5$, $f=f_0$, and $w=w_0$ case, and panels (b) and (d) to the $Ro=2$, $f=\frac{1}{4}f_0$, and $w=w_0$ case. The vertical lines in panels (c) and (d) represent the distance the waves would have travelled according to the linear phase and group speed.

to the baseline case, while Fig. 3b and d correspond to the reduced rotation case. Figure 3a and b show vertically integrated kinetic energy space–time plots, while Fig. 3c and d show density isolines at $t=47.25$ and $t=11.81$ respectively. The vertical lines correspond to the locations of the waves that emanate from the initial disturbance as described by linear theory. We computed the spectrum of the horizontal velocities to extract the dominant wavenumbers ($k\approx0.84$ and $k\approx0.48$ respectively) and used the algorithm outlined in Stastna and Rowe (2007) to calculate the speeds. We have presented both the linear phase and linear group speeds.

Comparing the results seen in Fig. 3 with Fig. 2b and c (since they both began with a negative polarity initial condition), there are immediate differences in both styles of plots. The most striking of these differences are the retention of energy in the geostrophic state, and the spreading of the ejected waves. The geostrophic state is visible in all plots along the left-hand side of the tank (near the wall). Comparing the two columns in Fig. 3, the case with a stronger rotation rate traps more energy in the geostrophic state. We will investigate differences in the geostrophic state in Sect. 3.4. The wave spreading is visible in the space–time plots as the waves propagate and within the new structure of the density isolines. At both rotation rates, the solitary wave, which is produced in the non-rotating case, has broken down into a series of smaller waves. A transition also appears to occur in the wave speed as the rotation rate changes. In Fig. 3d the wave front roughly corresponds to the linear phase speed (which is, in turn, a good approximation of the solitary wave propagation speed), while in Fig. 3c the front appears to have shifted to the linear group speed. This suggests that the low rotation case develops a wave train that can be interpreted as a rotation-modified solitary wave (at least on the timescales considered), while the high rotation case develops a wave train that can be interpreted as a wave packet. The importance of nonlinearity for both of these cases, and indeed for the geostrophic state, remains to be assessed.

Observing the structure that appears throughout the figure, we argue that these waves closely resemble a modulated wave packet as presented in Grimshaw et al. (1998), instead of the rotation-modified bore seen in Kuo and Polvani (1997). When comparing to the work done by Kuo and Polvani (1997), we first note that for our simulations the nonlinear parameter is quite small at $\alpha \approx 0.166$. However, their work suggests that, even for this small value and smooth initial condition, breaking will still occur. The present simulations were carried out with the full set of incompressible Navier–Stokes equations. As such, the dispersion that is neglected in the shallow-water equations, used by Kuo and Polvani (1997), becomes important when the wave front steepens. Dispersion breaks the front down into a train of smaller waves and eliminates shock formation. In addition to the change in steepening dynamics, the initially localized waves disperse over time, yet are observed to remain bound together (corresponding to the width of the packet envelope). For this reason we find that the modulated wave packet is a better description for these dynamics, though we note that in all our simulations the wave packet never completely separates from the trailing waves. This description is also supported by the shift in propagation speed to the linear group speed, as this is the first-order estimate of the speed which a wave packet would propagate at.

3.3 Nonlinear and polarity effects

Since the majority of the classical literature on the geostrophic adjustment problem considers the linear problem, it is important to clearly identify those aspects of our simulations that are nonlinear in nature. One way to investigate the nonlinear effects in the evolution, shown in Fig. 3, is to consider how the spectrum evolves in time, since (in the absence of dissipation) linear dispersive waves maintain the spectrum of the initial conditions for all times. Figure 4 shows the spectrum of the horizontal velocity profile at the surface (the results at other depths, and indeed for other fields, yielded qualitatively unchanged results) at various times for both cases shown in Fig. 3. Respectively, these correspond to $t = 15.75$ and $t = 3.94$, $t = 31.50$ and $t = 7.87$, $t = 47.25$ and $t = 11.81$, and $t = 63$ and $t = 15.75$. The spectral power density was scaled by the maximum power for all profiles shown in order to highlight the differences. It is readily apparent from Fig. 4a that in the $Ro = 0.5$, $f = f_0$, and $w = w_0$ case there is little change in the spectrum as time evolves. As time increases there appears to be a slow decay in the power at the excited wavenumbers. There

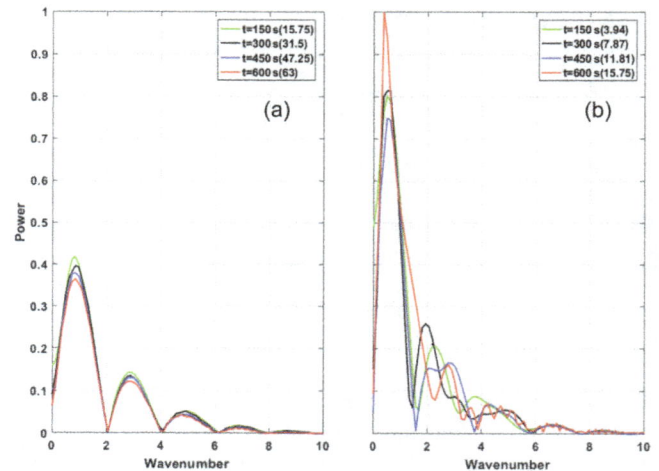

Figure 4. A comparison of how the horizontal velocity spectra for the $Ro = 0.5$, $f = f_0$, and $w = w_0$ case (**a**) and the $Ro = 2$, $f = \frac{1}{4} f_0$, and $w = w_0$ case (**b**) change over time. The spectral power has been scaled by the maximum between the cases to highlight the differences. The green line corresponds to $t = 15.75$ and $t = 3.94$, the black line to $t = 31.50$ and $t = 7.87$, the blue line to $t = 47.25$ and $t = 11.81$, and the red line to $t = 63$ and $t = 15.75$. There are clear changes in the spectra for the weakly rotating case as the waves evolve. In the strong rotation case the only temporal differences are a slow decay. The differences seen in the weakly rotating case highlight the nonlinear effects which are present in this regime.

is no shift in wavenumber for the various peaks, or indeed any other major change evident in the spectrum. This is not the case in Fig. 4b for the $Ro = 2$, $f = \frac{1}{4} f_0$, and $w = w_0$ case. The spectrum in this case contains large fluctuations (more than 25 % for the peak value) in power and shifts in the excited wavenumbers. These changes in the spectra as time evolves are hallmark effects of nonlinearity, and hence indicate that, while the emanating wavetrain in the $Ro = 0.5$ case appears to be well described by linear theory, if we consider weaker rotation effects such as in the $Ro = 2$ case, linear theory is no longer a useful description.

In order to investigate these effects in a more systematic manner, we started from the case with $w = \frac{1}{2} w_0$ (which was the smallest width that still produced a solitary wave) and ran several simulations where we varied the amplitude (by halving it), varied the rotation rate (which was quartered to show clear differences), compared the different polarities, and considered an extremely small amplitude "nearly linear" case. To compare with known nonlinear wave results we computed the non-rotating version of a number of the cases. A comparison of the 1-D-averaged KE for a number of the cases is presented in Fig. 5, where we have scaled any reduced amplitude cases so that, were linear theory to apply, the curves would collapse onto a single profile. Kinetic energy was chosen as the variable shown, since it provides information about both

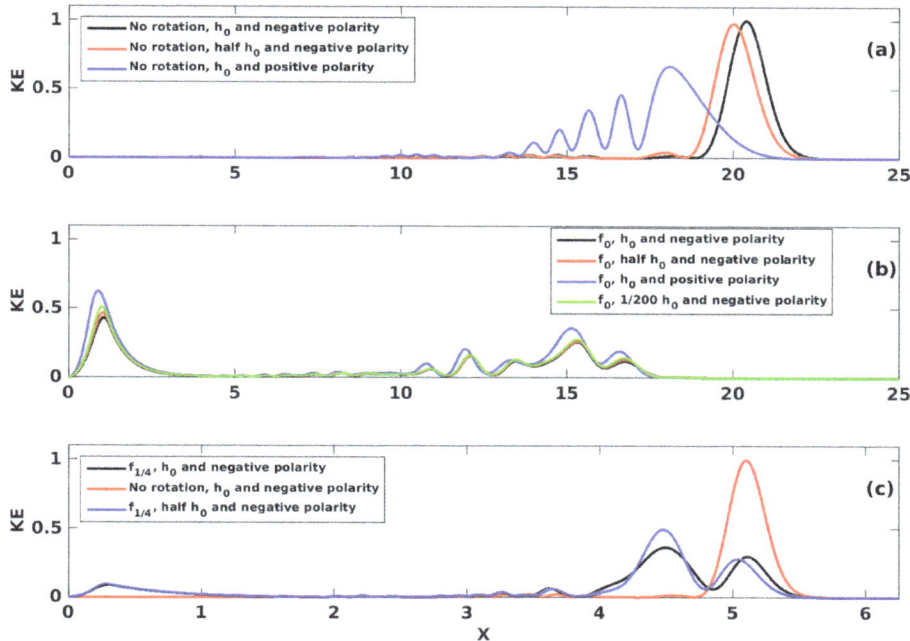

Figure 5. A comparison of the 1-D KE for several different cases to outline the effects of nonlinearity. All plots are taken at $t = 1140$ for the non-rotating case, $t = 25.2$ for the f_0 case ($w = \frac{1}{2}w_0$, $Ro = 1$), and $t = 6.3$ for the $\frac{1}{4}f_0$ case ($w = \frac{1}{2}w_0$, $Ro = 4$). The reduced amplitude nearly linear cases have been scaled by the change in amplitude squared. The kinetic energy has also been normalized by the maximum of the non-rotating negative polarity case. Panel **(a)** highlights the differences that arise from halving the initial amplitude and by changing the polarity of the non-rotating case. Panel **(b)** compares the same changes as the panel above; however, we have included rotation, and also consider a "nearly linear" case with an initial amplitude of $\frac{1}{200}\eta_0$. Panel **(c)** is a comparison of different negative polarity cases, one with no rotation and the standard amplitude, and two others at $\frac{1}{4}f_0$ with standard and half amplitude.

the structure of the dynamics and the magnitude of the velocities. Figure 5a shows how the non-rotating adjustment yields waves that are profoundly affected by changes in polarity and to a lesser degree by changes in amplitude. A solitary wave train is observed for the negative polarity case and an undular bore for the positive polarity case. The change in amplitude results in a phase shift; however, the amplitude of the solitary wave remains nearly constant. These results are a clear indication of nonlinear behaviour for the non-rotating case. In Fig. 5b we compare several cases with rotation following a similar methodology to Fig. 5a; however, we have included our "nearly linear" case where the amplitude has been reduced by a factor of 200. The change in polarity does not significantly change the dynamics of the ejected waves, with the largest change between these cases being that the positive polarity case has a higher amplitude both within the wave packet and in the geostrophic state. The effect of changing the amplitude does not significantly change the wave packet, since the wave packet is quite small in this case, and hence to leading order can be understood from the point of view of linear dispersive wave theory, similarly to what was observed based on the spectrum in the discussion above (Fig. 3 and the related discussion). For the geostrophic state, the changes in the initial disturbance amplitude result in changes to the am-

plitude and the location of the peak in kinetic energy, with the reduction in amplitude yielding a greater than linear response in the amplitude of the geostrophic state. Again, changing polarity yields the most significant changes. Linear theory, as exemplified by the green curve, provides a reasonable prediction, though details are amplitude dependent. For Fig. 5c we kept the polarity of the initial disturbance negative and compared the change in amplitude for a smaller rotation rate as well as the non-rotating case. The lower rotation rate allows for more energy to be deposited into the wave train. The primary change for the reduction in amplitude is that the individual waves within the wave packet of the scaled reduced case (in blue) appear to be larger in amplitude compared to the base case (in black). There also appears to be a slight phase shift between the cases (consistent with a packet that travels at a slightly different speed). Comparing these two cases to the non-rotating case shows that, while the peak in energy has been shifted back, the wave front of the solitary wave and the wave packets are at roughly the same location. For this lower rotation case there is very little difference in the geostrophic state as a result of amplitude reduction, implying that for low rotation rates the geostrophic state can be well described by linear theory.

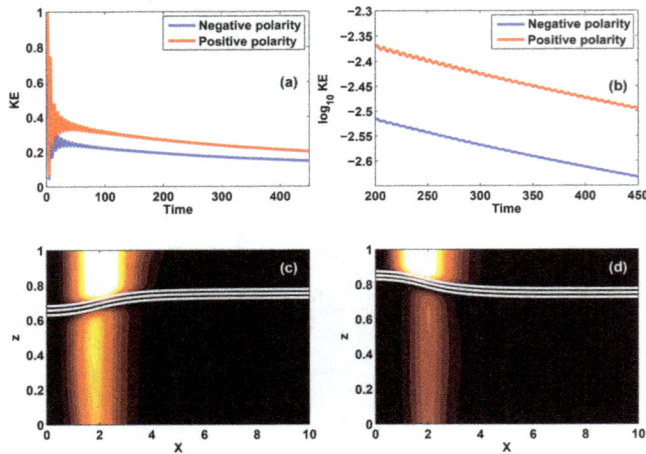

Figure 6. Long-time simulations comparing the differences in the geostrophic state for negative and positive initializations. For both simulations $f = 2f_0$, $w = \frac{1}{2}w_0$ and $Ro = \frac{1}{2}$. Panel **(a)** presents the long-time time series of vertically integrated kinetic energy in the geostrophic state for both cases. Panel **(b)** shows the base-10 logarithm of the geostrophic state kinetic energy for both cases after the packet has been ejected. Panels **(c)** and **(d)** show the shaded distribution of kinetic energy in the geostrophic state along with contours of constant density, for a negative and positive initial polarity respectively.

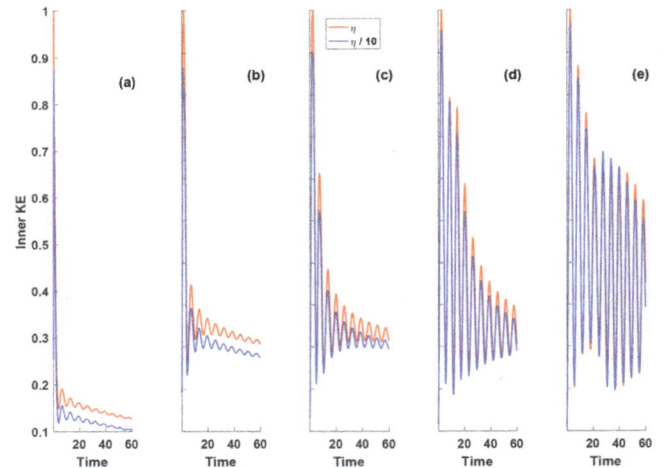

Figure 7. The difference in total inner kinetic energy between our original amplitude cases and cases where the amplitude has been reduced by a factor of 10. The reduced cases have then been linearly scaled to account for this amplitude change. The original energies are shown in red, while the reduced ones are in blue. Discrepancies between the two cases are due to nonlinear effects. **(a)** $w = \frac{1}{4}w_0$ $(Ro = 2)$, **(b)** $w = \frac{1}{2}w_0$ $(Ro = 1)$, **(c)** $w = w_0$ $(Ro = 0.5)$, **(d)** $w = 2w_0$ $(Ro = 0.25)$, and **(e)** $w = 4w_0$ $(Ro = 0.125)$.

To investigate the nonlinear effects that arise from changes in polarity in the geostrophic state (Fig. 5b), long-time simulations with a rotation rate set to $f = 2f_0$ and an initial width of $w = \frac{1}{2}w_0$ were computed, resulting in a Rossby number of $\frac{1}{2}$. These results are presented in Fig. 6. Figure 6a and b show the vertically integrated kinetic energy at the location of the maximum induced by the geostrophic state. Figure 6c and d show the total spatial distribution of the normalized kinetic energy in the region around the geostrophic state at 720 s (151.2), along with three contours of constant density, for a negative and positive initial polarity respectively. Figure 6a clearly shows the energy difference that was seen in Fig. 6b, indicating that the positive polarity case appears more efficient at keeping energy in the geostrophic state. Figure 6b shows the time series of the logarithm of kinetic energy after the packet has been ejected. If we ignore the inertial oscillations, which appear to be rapid on the timescale shown, we can see a clear decay. From this panel it is also possible to note that the oscillations appear to persist significantly longer in the positive case. By computing the logarithm of the time series (Fig. 6b) we are able to show that the decay is nearly exponential, with the positive polarity case decaying roughly 5 % faster. The decay rate decreases over time. From the bottom two panels, Fig. 6c and d, it is clear that the polarity of the geostrophic state strongly modifies the vertical distribution of the kinetic energy. Note in particular the difference in strength of kinetic energy below the pycnocline and the tilt of the high kinetic region that follows the deformed pycnocline.

To quantify the nonlinear behaviour of the geostrophic state and the inertial oscillations that accompany it, Fig. 7 shows the differences in total kinetic energy within the geostrophic state between our original amplitude cases and cases with a 10-fold reduction in amplitude. As in Fig. 5 we have scaled up the reduced amplitude cases, and thus for a purely linear problem there should be no differences between the two curves shown. Figure 7 shows five cases of differing initial width and the same Coriolis parameter $f = f_0$, Fig. 7a to $w = \frac{1}{4}w_0$ $(Ro = 2)$, Fig. 7b to $w = \frac{1}{2}w_0$ $(Ro = 1)$, Fig. 7c to $w = w_0$ $(Ro = 0.5)$, Fig. 7d to $w = 2w_0$ $(Ro = 0.25)$, and Fig. 7e to $w = 4w_0$ $(Ro = 0.125)$. Though the nearly linear case does behave in a qualitatively similar manner to the original cases, there are key differences. For one, in Fig. 7a–c there are clear magnitude differences between the two cases. For almost all times the scaled kinetic energy curve for the amplitude reduced case lies below the corresponding curve for the original cases. This remains true for the later cases (Fig. 7d and e), though not to the same extent. Second, closely examining the inertial oscillations reveals that they appear to decay more rapidly for the amplitude reduced cases compared to the original cases; this can especially be seen in Fig. 7c and d. Only for the two rightmost panels $(Ro \leq 0.25)$ could it be said that the two curves are nearly coincident. Thus even the geostrophic state exhibits clear evidence of nonlinearity.

The primary dynamic variable for these simulations is the Rossby number since both changes to rotation rate and changes to the initial width are both just modifications to

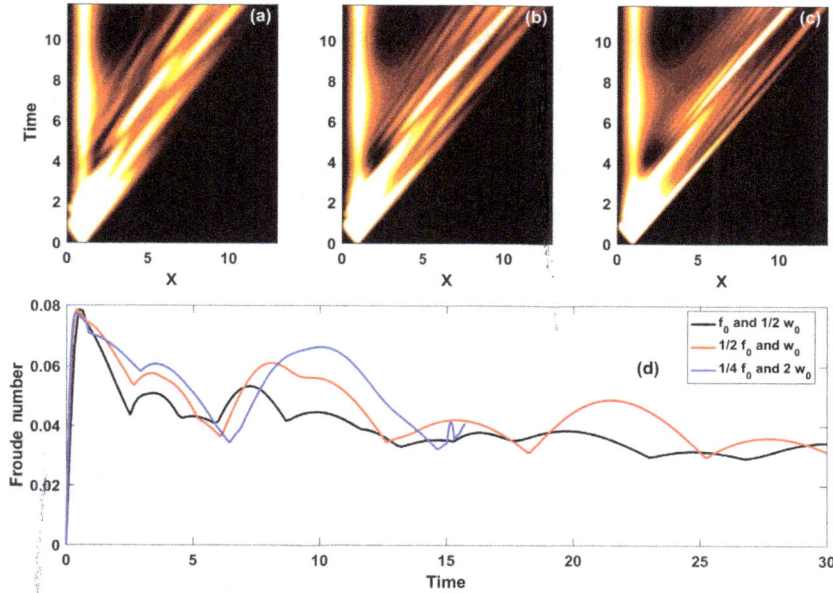

Figure 8. A space–time plot of kinetic energy. The different columns correspond to different combinations of f and w used to form a value of $Ro = 1$. **(a)** $f = f_0$ and $w = \frac{1}{2}w_0$, **(b)** $f = \frac{1}{2}f_0$ and $w = w_0$, and **(c)** $f = \frac{1}{4}f_0$ and $w = 2w_0$. The second row corresponds to the same case as the first columns, but the aspect ratio has been scaled by the change in rotation rate.

this dimensionless parameter. A different manner in which the effects of nonlinearity may be investigated is by asking whether the dynamics collapse onto a single case for the same Rossby number; Fig. 8a–c show the space–time plots of vertically integrated kinetic energy for three cases with the same Rossby number but different combinations of parameters. The Froude number, $Fr = \frac{U}{c}$, is computed dynamically, with U set by the maximum horizontal velocity from the simulation at a given time and c given by the linear group speed calculated using the algorithm outlined in Stastna and Rowe (2007) using the dominant wavenumbers ($k \approx 0.84$ and $k \approx 0.48$ respectively) for each rotation rate. Fig. 8a corresponds to $f = f_0$ and $w = \frac{1}{2}w_0$, Fig. 8b to $f = \frac{1}{2}f_0$ and $w = w_0$, and Fig. 8c to $f = \frac{1}{4}f_0$ and $w = 2w_0$. The axis of Fig. 8b and c has been scaled by the corresponding change in Coriolis parameter ($\frac{1}{2}$ and $\frac{1}{4}$ respectively). Figure 8d shows the time series of the Froude number, $Fr = \frac{U}{c}$. While the oscillations of the geostrophic state near $x = 1$ in the space–time plots are quantitatively similar, the number and shape of the waves that are produced in the wave train are slightly different. The reason for these differences is highlighted in Fig. 8d, where the Froude numbers match for early times, but begin to drift rapidly. These results again illustrate the importance of nonlinear effects within this system and the necessity to include such effects when modeling the system.

During the analysis of the numerical experiments that varied the width of the initial condition, an interesting observation about multiple wave trains was made. The initial condition yields both rightward and leftward propagating waves. For narrow initial conditions the leftward propagating waves reflect from the left wall early in the simulation and are difficult to disentangle from the initially rightward propagating wave train. However, for wider initial conditions the leftward propagating waves must travel a longer distance before reflecting off the wall, allowing for them to appear separate from rightward propagating waves. This interaction is shown in Fig. 9 using the potential energy field because the amount of span-wise velocity created in the geostrophic state is so great that it drowns out this reflection signal in the kinetic energy field. Both cases maintain the same Coriolis parameter $f = f_0$, but Fig. 9a corresponds to $w = 0.5w_0$, while Fig. 9b corresponds to $w = 2w_0$. In Fig. 9a it is difficult to distinguish the two wave trains (though we have superimposed coloured arrows in order to accentuate the pattern for the reader). This distinction is much clearer in Fig. 9b, where the leftward travelling wave takes roughly twice as long to reach the left wall. In this case it is possible to distinguish the wave trains within the pattern of waves that are produced. Thus a natural method of generation of waves in a tank will create waves in both directions which must be accounted for in the interpretation of physical experiments.

3.4 The geostrophic state

In the rotation-modified adjustment problem there are two dominant features, the geostrophic state that is left over from the initial conditions and the train of Poincaré waves that carries energy away from it. For this section we will focus on the dynamics, and changes, of the geostrophic state. We will primarily be comparing our results with

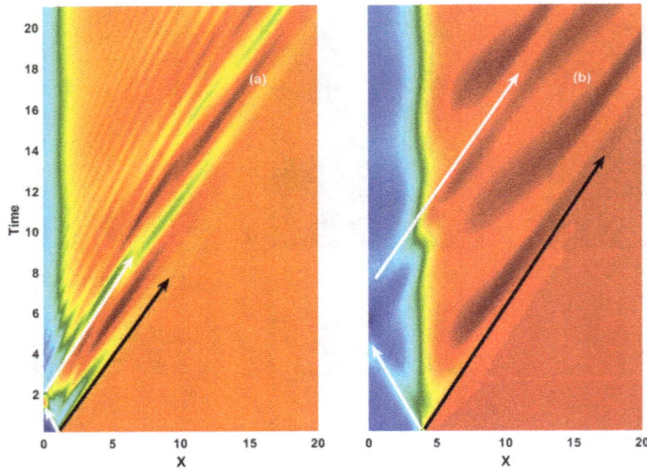

Figure 9. Space–time pseudocolour plots of the change in potential energy for the $Ro = 1$, $f = f_0$, and $w = 0.5w_0$ case (a), and the $Ro = 0.25$, $f = f_0$, and $w = 2w_0$ case (b). As for the kinetic energy, it has been scaled to the maximum value. Visible in both cases (though significantly easier to see in panel b), there are both rightward (black arrow) and leftward (white arrow) propagating wave trains created by the initial conditions. The leftward travelling wave train will eventually reflect off the close left-hand wall and propagate rightwards. If the initial conditions are quite narrow, the leftward propagating wave train reflects quickly off the wall and is difficult to disentangle from the rightward propagating wave train.

Table 2. A comparison of the notation in our work and that of Lelong and Sundermeyer (2005).

Our notation	Lelong and Sundermeyer (2005)
Ro_r	R
Ro	R/L
w	L
Geostrophic state KE / PE	KE_v / PE_v
Outside geostrophic state KE / PE	KE_w / PE_w

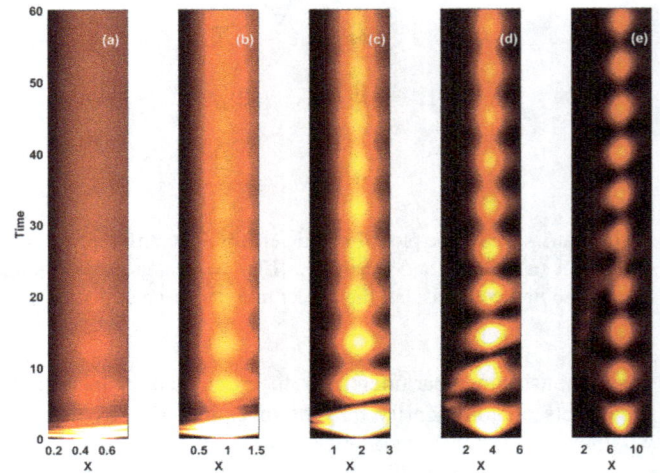

Figure 10. A space–time plot of vertically integrated kinetic energy, in the geostrophic state, for different values of w while f is held constant at $f = f_0$. (a) $w = \frac{1}{4}w_0$ ($Ro = 2$), (b) $w = \frac{1}{2}w_0$ ($Ro = 1$), (c) $w = w_0$ ($Ro = 0.5$), (d) $w = 2w_0$ ($Ro = 0.25$), and (e) $w = 4w_0$ ($Ro = 0.125$).

those from Lelong and Sundermeyer (2005), who performed 3-D numerical simulations of the adjustment problem. Lelong and Sundermeyer (2005) focused on the energetics of the geostrophic state that are generated by a well-mixed region of intermediate density fluid. Though not exactly the same case, Lelong and Sundermeyer (2005) performed their simulations using the full set of equations and thus provide an apt point of comparison. To facilitate this comparison between our work and theirs, Table 2 provides a translation of our notation to that of Lelong and Sundermeyer (2005).

A major difference between the two sets of experiments is the background stratification and density anomaly. As given explicitly in Sect. 2, we have a two-layer stratification given by the tanh function with an anomaly also given by the tanh function. Lelong and Sundermeyer (2005) use a localized anomaly diffusivity to create a two-lobed axisymmetric lens density perturbation, with a linear background stratification.

Figure 10 shows space–time plots of vertically integrated kinetic energy within the geostrophic state for five cases of different initial widths where the rotation rate has been held constant at $f = f_0$. These cases have Rossby numbers 2, 1, $\frac{1}{2}$, $\frac{1}{4}$, and $\frac{1}{8}$ for Fig. 10a, b, c, d, and e respectively. The figure has been saturated by the maximum kinetic energy across all cases. Once $Ro \geq 1$ (Fig. 10b–e), the geostrophic state shows clear oscillations within the kinetic energy. These spikes in

kinetic energy occur during the vertical oscillations of the geostrophic state. It is also possible to see from this figure that all the cases initially spike with roughly the same magnitude of kinetic energy, but then differ greatly depending on the Rossby number. In Fig. 10d and e it is possible to identify the reflected wave interfering with the oscillations of the geostrophic state, matching the features seen in Fig. 9.

To compare the energy within the geostrophic state between cases, and with the published literature, we horizontally integrate the geostrophic state (the region shown in Fig. 10) to produce a time series of both the kinetic and potential energies. Following what was done in Kuo and Polvani (1997), we compute the difference in these energies compared to the initial state. The results of this are shown in Fig. 11. The extent of the geostrophic state is defined as twice the distance from the left-hand wall to the maximum in kinetic energy. Due to the nature of our initial conditions, namely that we start with a smooth transition and still fluid, the ratio $\Delta KE/\Delta PE$, which is used in

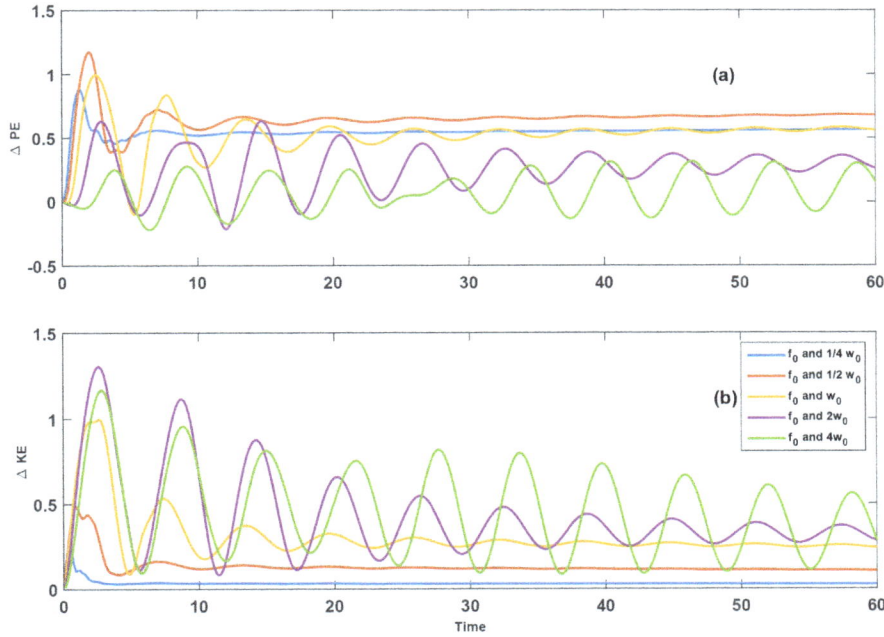

Figure 11. The changes in potential and kinetic energy, compared to the initialization, for the cases in Fig. 6. Panel **(a)** corresponds to the change in potential energy, and panel **(b)** to the changes in kinetic energy. The energy in both plots has been scaled by the base case ($f = f_0$, $w = w_0$).

Kuo and Polvani (1997), is difficult to use since the change in potential energy may be zero if it reaches its initial state during oscillation. While it would be possible to use the reciprocal of the ratio, we have chosen to present the differences separately (Fig. 11a for ΔPE and Fig. 11b for ΔKE). The energy in both plots has been scaled by the base case of $f = f_0$ and $w = w_0$. As in the results presented in Fig. 8 of Lelong and Sundermeyer (2005), for larger initial widths (smaller Rossby numbers) there are correspondingly larger oscillations in both potential and kinetic energy. Our simulations show that these oscillations persist for long times, in agreement with the results of Kuo and Polvani (1997). In a similar manner to what is seen in Lelong and Sundermeyer (2005), the case with $Ro = 1$ (f_0 and $\frac{1}{2} w_0$) appears to retain the maximum amount of potential energy (as opposed to kinetic energy for Lelong and Sundermeyer, 2005). However, we have verified that the results seen in their figure can be generated by scaling the kinetic energy by the initial energy (not shown here). We also computed the linear kinetic energy for the geostrophic state following Boss and Thompson (1995)'s Eq. (9), using the parameter set for our base case. We then compared this to the maximum kinetic energy in the geostrophic state. We calculated the linear KE to be 4.06327×10^{-5}, while our KE was 3.63771×10^{-5}, which is roughly an 11 % difference.

We next consider how the time evolution of the total kinetic energy inside the geostrophic state compares to that outside; this is shown in Fig. 12. From Fig. 9a we can see

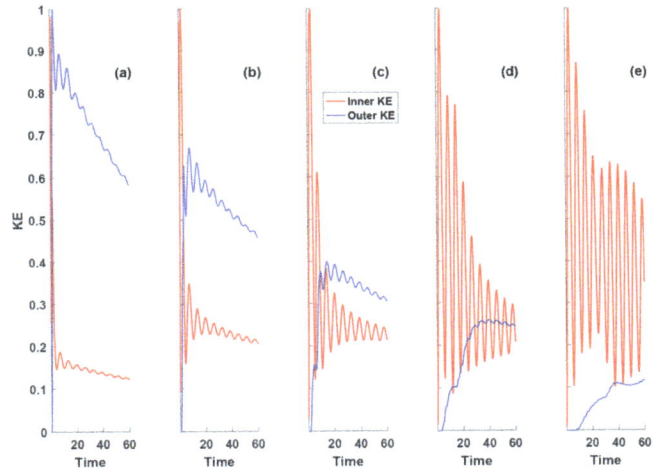

Figure 12. The total kinetic energy located inside (blue) and outside (red) of the geostrophic state. In this figure we have the same cases as in Figs. 10 and 11, separated into their own panels. Panel **(a)** corresponds to $\frac{1}{4} w_0$ ($Ro = 2$), **(b)** corresponds to $\frac{1}{2} w_0$ ($Ro = 1$), **(c)** corresponds to w_0 ($Ro = 0.5$), **(d)** corresponds to $2w_0$ ($Ro = 0.25$), and **(e)** corresponds to $4w_0$ ($Ro = 0.125$). The red line (inner) is the energy inside the geostrophic state, while the blue line (outer) corresponds to that outside.

that for low rotation rates there is much less energy retained within the geostrophic state, and that the oscillations of this remaining kinetic energy are very small. We can compare this panel with Fig. 13 of Lelong and Sundermeyer (2005) and see similar results, namely the dominance of the kinetic energy outside the geostrophic state. We should also note that there is a steady decrease in kinetic energy in the outer kinetic energy which is due to dissipation of the waves throughout the numerical domain. As in Fig. 6 of Lelong and Sundermeyer (2005) we see a similar separation between the energy of the inner and outer regions for the $Ro = 1$ case (our Fig. 12b). The inner kinetic energy oscillations of Fig. 12d and e do not appear to have reached a steady state by the end of our simulation, but have reached a sufficiently close value to interpret. In Fig. 12d there is an equal amount of energy in the final inner and outer regions. These results seen in Fig. 12d and e are consistent with the results from Lelong and Sundermeyer (2005)'s Fig. 14, namely significantly more energy being retained inside the geostrophic state, resulting in much larger amplitude oscillations.

Motivated by the results shown in Fig. 12, we considered the vertical structure of the inertial oscillations. We confirmed that for all cases the gradient Richardson number (including the v component of shear) never dips below 0.25 in the stratified region, thereby suggesting the inertial waves are not strong enough to induce shear instability. Indeed, the isopycnal displacements associated with the inertial oscillations were never larger than about 2.5 % of the total depth. For early times and low Rossby numbers ($Ro \leq 0.125$), the spatiotemporal (in z and t) structure of the kinetic energy field induced by the inertial waves followed a separable structure. For the $Ro = 0.25$ case evidence of a nonseparable structure was clear for $t > 100$. In comparison, for the $Ro = 0.5$ case a non-separable structure was evident by $t = 60$. However, since the inertial oscillations are smaller in this case at later times ($t > 150$), the signature of the inertial waves is masked by that of the geostrophic state.

4 Conclusions

In this paper we have taken a systematic approach to the classical rotation-modified stratified adjustment problem. Building on results based on shallow-water theory presented in Kuo and Polvani (1997), we have shown that by using the fully nonlinear incompressible Navier–Stokes equations, under the Boussinesq approximation, the waves that are ejected from the geostrophic state do not steepen to a shock. Once the wave front steepens sufficiently it disperses into a primary wave packet and a tail of smaller dispersive waves. We demonstrated that the nonlinear wave packet interpretation of the wave train of Grimshaw et al. (2012) is appropriate in some parameter regimes, with changes in amplitude reflected in the phase of the nearly solitary wave response. By mapping out the parameter space we have shown that, as expected, the Rossby number is the controlling variable for the dynamics in this problem. For $Ro < 1$, the wave packet propagates with a speed roughly corresponding to the linear group

speed, while for $Ro > 1$, the packet propagates with a speed closer to the linear phase speed. We have further characterized the nonlinear effects present in both the wave packet and the geostrophic state. The effects of nonlinearity were investigated by considering different initial amplitudes and changes in polarity. As in the non-rotating case, the largest nonlinear effects occurred as a result of changes in polarity, both in the geostrophic state and in the wave packet ejected. Surprisingly, the high Rossby number cases yielded nonlinear effects in both the wave train and the geostrophic state. However, as a general rule amplitude effects were smaller than polarity effects.

A different approach to characterizing nonlinear effects is to create different combinations of parameters that yield the same Rossby number. We carried out this process and tracked the time dependent Froude number. While the qualitative features of the evolution were similar in all three cases shown, the variations in the Froude number led to significant differences in the details of the wave train generated. The characterization of these various nonlinear effects in a single simulation is new and significant, providing a guideline for when linear theory can be applied and when nonlinear effects must be considered.

Our results show that the inertial oscillations in the geostrophic state can persist for long times, in agreement with Kuo and Polvani (1997). However, the inertial oscillations never reach large enough amplitudes to induce shear instability. Our results also match the work published by Lelong and Sundermeyer (2005), specifically matching their results for the different Rossby number regions ($Ro < 1$, $Ro = 1$, and $Ro > 1$). By comparing the kinetic energy within the inner geostrophic and outer non-geostrophic regions, we show that the amplitude of the geostrophic oscillations increases quickly as the Rossby number decreases. However, this in turn corresponds to less prominent nonlinear effects within the geostrophic state.

Another significant finding presented is the generation, reflection, and interaction of a wave train propagating in the opposite direction (leftward) during the initial generation. For any physical tank set-up, this reflected wave will impact any measurements of the waves generated and especially any measurements of the geostrophic state.

In addition to the work described in the previous paragraph, future work should consider span-wise variations, especially in the case of the strong geostrophic state for which novel instabilities may be possible (though as noted above, on laboratory scales shear instability is not expected). Systematic studies of the shoaling of rotation-modified solitary waves and undular bores should also be carried out, since it is not known in what manner these may be different from shoaling in the non-rotating case. A more theoretical avenue could quantitatively compare weakly nonlinear and weakly dispersive–strongly nonlinear model equations to the full stratified equations.

Acknowledgements. This research was supported by the Natural Sciences and Engineering Research Council of Canada through Discovery Grant RGPIN 3118442010.

Edited by: R. Grimshaw

References

Boss, E. and Thompson, L.: Energetics of Nonlinear Geostrophic Adjustment, J. Phys. Oceanogr., 25, 1521–1529, doi:10.1175/1520-0485(1995)025<1521:EONGA>2.0.CO;2, 1995.

Cahn, A.: An Investigation of the Free Oscillations of a Simple Current System, J. Meteorol., 2, 113–119, doi:10.1175/1520-0469(1945)002<0113:AIOTFO>2.0.CO;2, 1945.

Carr, M. and Davies, P. A.: The motion of an internal solitary wave of depression over a fixed bottom boundary in a shallow, two-layer fluid, Phys. Fluids, 18, 016601, doi:10.1063/1.2162033, 2006.

Chia, F., Griffiths, R., and Linden, P.: Laboratory experiments on fronts: Part II: The formation of cyclonic eddies at upwelling fronts, Geophys. Astro. Fluid, 19, 189–206, doi:10.1080/03091928208208955, 1982.

Gerkema, T., Zimmerman, J. T. F., Maas, L. R. M., and van Haren, H.: Geophysical and astrophysical fluid dynamics beyond the traditional approximation, Rev. Geophys., 46, RG2004, doi:10.1029/2006RG000220, 2008.

Gill, A.: Atmosphere-ocean dynamics, Academic Press, New York, USA, 1982.

Gill, A. E.: Adjustment under gravity in a rotating channel, J. Fluid Mech., 77, 603–621, doi:10.1017/S0022112076002280, 1976.

Grimshaw, R. and Helfrich, K.: Longtime Solutions of the Ostrovsky Equation, Stud. Appl. Math., 121, 71–88, doi:10.1111/j.1467-9590.2008.00412.x, 2008.

Grimshaw, R. H. J., Ostrovsky, L. A., Shrira, V. I., and Stepanyants, Y. A.: Long Nonlinear Surface and Internal Gravity Waves in a Rotating Ocean, Surv. Geophys., 19, 289–338, doi:10.1023/A:1006587919935, 1998.

Grimshaw, R. H. J., Helfrich, K., and Johnson, E. R.: The Reduced Ostrovsky Equation: Integrability and Breaking, Stud. Appl. Math., 129, 414–436, doi:10.1111/j.1467-9590.2012.00560.x, 2012.

Grimshaw, R. H. J., Helfrich, K. R., and Johnson, E. R.: Experimental study of the effect of rotation on nonlinear internal waves, Phys. Fluids, 25, 056602, doi:10.1063/1.4805092, 2013.

Grue, J., Jensen, A., Rusås, P.-O., and Sveen, J. K.: Breaking and broadening of internal solitary waves, J. Fluid Mech., 413, 181–217, doi:10.1017/S0022112000008648, 2000.

Helfrich, K. R.: Decay and return of internal solitary waves with rotation, Phys. Fluids, 19, 026601, doi:10.1063/1.2472509, 2007.

Helfrich, K. R. and Melville, W. K.: Long Nonlinear Internal Waves, Annu. Rev. Fluid Mech., 38, 395–425, doi:10.1146/annurev.fluid.38.050304.092129, 2006.

Johnson, R.: A modern introduction to the mathematical theory of water waves, vol. 19, Cambridge University Press, Cambridge, UK, 1997.

Killworth, P. D.: The Time-dependent Collapse of a Rotating Fluid Cylinder, J. Phys. Oceanogr., 22, 390–397, doi:10.1175/1520-0485(1992)022<0390:TTDCOA>2.0.CO;2, 1992.

Kuo, A. and Polvani, L.: Time-dependent fully nonlinear geostrophic adjustment, J. Phys. Oceanogr., 27, 1614–1634, doi:10.1175/1520-0485(1997)027<1614:TDFNGA>2.0.CO;2, 1997.

Kuo, A. C. and Polvani, L. M.: Wave–vortex interaction in rotating shallow water. Part 1. One space dimension, J. Fluid Mech., 394, 1–27, 1999.

Lamb, K. G.: Particle transport by nonbreaking, solitary internal waves, J. Geophys. Res.-Oceans, 102, 18641–18660, doi:10.1029/97JC00441, 1997.

Ledwell, J. R., Duda, T. F., Sundermeyer, M. A., and Seim, H. E.: Mixing in a coastal environment: 1. A view from dye dispersion, J. Geophys. Res.-Oceans, 109, C10013, doi:10.1029/2003JC002194, 2004.

Lelong, M. and Dunkerton, T.: Inertia–gravity wave breaking in three dimensions. Part I: Convectively stable waves, J. Atmos. Sci., 55, 2473–2488, doi:10.1175/1520-0469(1998)055<2473:IGWBIT>2.0.CO;2, 1998.

Lelong, M.-P. and Sundermeyer, M. A.: Geostrophic Adjustment of an Isolated Diapycnal Mixing Event and Its Implications for Small-Scale Lateral Dispersion, J. Phys. Oceanogr., 35, 2352–2367, doi:10.1175/JPO2835.1, 2005.

Middleton, J. F.: Energetics of Linear Geostrophic Adjustment, J. Phys. Oceanogr., 17, 735–740, doi:10.1175/1520-0485(1987)017<0735:EOLGA>2.0.CO;2, 1987.

Mihaljan, J.: The exact solution of the Rossby adjustment problem, Tellus A, 15, 150–154. doi:10.1111/j.2153-3490.1963.tb01373.x, 1963.

Oakey, N. and Greenan, B.: Mixing in a coastal environment: 2. A view from microstructure measurements, J. Geophys. Res.-Oceans, 109, C10014, doi:10.1029/2003JC002193, 2004.

Ou, H. W.: Geostrophic Adjustment: A Mechanism for Frontogenesis, J. Phys. Oceanogr., 14, 994–1000, doi:10.1175/1520-0485(1984)014<0994:GAAMFF>2.0.CO;2, 1984.

Rossby, C.: On the mutual adjustment of pressure and velocity distributions in certain simple current systems, J. Mar. Res, 1, 239–263, 1937.

Stastna, M. and Rowe, K.: On weakly nonlinear descriptions of nonlinear internal gravity waves in a rotating reference frame, Atlantic Electronic Journal of Mathematics, 2, 30–54, 2007.

Stastna, M., Poulin, F. J., Rowe, K. L., and Subich, C.: On fully nonlinear, vertically trapped wave packets in a stratified fluid on the f-plane, Phys. Fluids, 21, 106604, doi:10.1063/1.3253400, 2009.

Subich, C. J., Lamb, K. G., and Stastna, M.: Simulation of the Navier–Stokes equations in three dimensions with a spectral collocation method, Int. J. Numer. Meth. Fl., 73, 103–129, doi:10.1002/fld.3788, 2013.

Vallis, G. K.: Atmospheric and Oceanic Fluid Dynamics, Cambridge University Press, Cambridge, UK, 2006.

Washington, W. M.: A note on the adjustment towards geostrophic equilibrium in a simple fluid system, Tellus, 16, 530–534, doi:10.1111/j.2153-3490.1964.tb00189.x, 1964.

Zeitlin, V., Medvedev, S. B., and Plougonven, R.: Frontal geostrophic adjustment, slow manifold and nonlinear wave phenomena in one-dimensional rotating shallow water. Part 1. Theory, J. Fluid Mech., 481, 269–290, doi:10.1017/S0022112003003896, 2003a.

Zeitlin, V., Reznik, G. M., and Jelloul, M. B.: Nonlinear theory of geostrophic adjustment. Part 2. Two-layer and continuously stratified primitive equations, J. Fluid Mech., 491, 207–228, doi:10.1017/S0022112003005457, 2003b.

12

Balanced source terms for wave generation within the Hasselmann equation

Vladimir Zakharov[1,2,3,4], **Donald Resio**[5], **and Andrei Pushkarev**[1,2,3,4]

[1]Department of Mathematics, University of Arizona, Tucson, AZ 85721, USA
[2]Lebedev Physical Institute RAS, Leninsky 53, Moscow 119991, Russia
[3]Novosibirsk State University, Novosibirsk, 630090, Russia
[4]Waves and Solitons LLC, 1719 W. Marlette Ave., Phoenix, AZ 85015, USA
[5]Taylor Engineering Research Institute, University of North Florida, Jacksonville, FL, USA

Correspondence to: Andrei Pushkarev (dr.push@gmail.com)

Abstract. The new Zakharov–Resio–Pushkarev (ZRP) wind input source term (Zakharov et al., 2012) is examined for its theoretical consistency via numerical simulation of the Hasselmann equation. The results are compared to field experimental data, collected at different sites around the world, and theoretical predictions based on self-similarity analysis. Consistent results are obtained for both limited fetch and duration limited statements.

1 Introduction

The scientific description of wind-driven wave seas, inspired by solid state physics statistical ideas (see, for instance, Nordheim, 1928), was proposed by Hasselmann (1962, 1963) in the form of the Hasselmann equation (hereafter HE), also known as the kinetic equation for waves:

$$\frac{\partial \varepsilon}{\partial t} + \frac{\partial \omega_k}{\partial k}\frac{\partial \varepsilon}{\partial r} = S_{\mathrm{nl}} + S_{\mathrm{in}} + S_{\mathrm{diss}}, \qquad (1)$$

where $\varepsilon = \varepsilon(\omega_k, \theta, r, t)$ is the wave energy spectrum as a function of wave dispersion $\omega_k = \omega(k)$, angle θ, two-dimensional real space coordinate $r = (x, y)$ and time t. S_{nl}, S_{in} and S_{diss} are the nonlinear, wind input and wave-breaking dissipation source terms, respectively. Hereafter, only the deep water case $\omega = \sqrt{gk}$ is considered, where g is the gravity acceleration and $k = |k|$ is the absolute value of the vector wave number $k = (k_x, k_y)$.

Since Hasselmann's work, Eq. (1) has become the basis of operational wave forecasting models such as WAM, SWAN and Wavewatch III (Tolman, 2013; SWAN, 2015). While the physical oceanography community consents on the general applicability of Eq. (1), there is no consensus on universal parameterizations of the source terms S_{nl}, S_{in} and S_{diss}.

1.1 The S_{nl} term and weak turbulence theory

The HE is a specific example of the kinetic equation for quasi-particles, widely used in different areas of theoretical physics. There are standard methods for its derivation. In the considered case, two forms of the S_{nl} term were derived by different methods from the Euler equations for free surface incompressible potential flow of a liquid by Hasselmann (1962, 1963) and Zakharov and Filonenko (1966). Resio and Perrie (1991) showed that they are identical on the resonant surface:

$$\omega_{k_1} + \omega_{k_2} = \omega_{k_3} + \omega_{k_4}, \qquad (2)$$
$$k_1 + k_2 = k_3 + k_4. \qquad (3)$$

The S_{nl} term is the complex nonlinear operator acting on ε_k, concealing hidden symmetries (Zakharov and Filonenko, 1967; Zakharov et al., 1992) and cubic with respect to the spectrum ε.

Equation (1) sometimes is called "the Boltzmann equation" by the oceanographic community, although this is a misconception. The Boltzmann equation, derived in the nineteenth century for description of gas kinetics, is quadratic, rather than cubic, with respect to the distribution function.

To understand the relation and difference between the Boltzmann equation and the HE, one should recall the above-mentioned Nordheim (1928) equation. This equation, applicable to quantum quasi-particles, contains both quadratic and cubic terms. Hence, the Boltzmann equation and the HE present opposite limiting cases of a general quantum kinetic equation.

Purely cubical (applicable to classical waves, not to classical particles) systems are relatively new objects in physics. Such equations describe the simplest case of the wave turbulence by the theory called weak turbulence theory (WTT) (Zakharov et al., 1992).

It is clear now that the WTT can be used for the description of a very broad class of physical phenomena, including waves in magneto-hydrodynamics (Galtier et al., 2000), waves in nonlinear optics (Yousefi, 2017), gravitational waves in the universe (Galtier and Nazarenko, 2017; de Oliveira et al., 2013), plasma waves (Balk, 2000; Yoon et al., 2016), capillary waves (Pushkarev and Zakharov, 1996; Yulin, 2017; Tran, 2017), and Kelvin waves in superfluid helium (L'vov and Nazarenko, 2010).

It is unfortunate that the discussion of the HE in the context of WTT has been overlooked by a major part of the oceanographic community for many years now. The community accepts, nevertheless, the HE as the basis for the operational wave forecasting models, thereby believing de facto in WTT without fully appreciating its ramifications.

The WTT essentially differs from the kinetic theory of classical particles and quantum quasi-particles. In the "traditional" gas kinetics (both classical and quantum) the basic solutions are thermodynamic equilibrium spectra, such as Boltzmann and Planck distributions. In the WTT such solutions, though formally existing, play no role – they are non-physical. The physically essential solutions are the non-equilibrium Kolmogorov–Zakharov spectra (or KZ spectra, Zakharov et al., 1992), which are the solutions of the corresponding kinetic equation

$$S_{nl} = 0. \tag{4}$$

The simplest one is the Zakharov–Filonenko (hereafter ZF) solution (Zakharov and Filonenko, 1966), which is the subclass of KZ solutions:

$$\varepsilon \simeq \frac{P^{1/3}}{\omega^4}, \tag{5}$$

where P is the energy flux toward high wave numbers.

The accuracy advantage of knowing the analytical expression for the S_{nl} term, also known in physical oceanography as the XNL, is overshadowed by its computational complexity. Today, none of the operational wave forecasting models can afford to perform XNL computations in real time. Instead, the operational approximation, known as DIA and its derivatives, is used to replace this source term. The implication of such simplification is the inclusion of a tuning coefficient in front of the nonlinear term; however, several publications have shown that DIA does not provide a good approximation of the actual XNL form. The paradigm of replacement of the XNL by DIA and its variations leads to even more grave consequences: other source terms must be adjusted to allow the model Eq. (1) to produce desirable results. In other words, deformations suffered by the XNL model due to the replacement of S_{nl} by its surrogates need to be compensated for by non-physical modification of other source terms to achieve reasonable model behavior in any specific case, leading to a loss of physical universality in the HE model.

1.2 Operational formulations for the wind energy input S_{in} and wave energy dissipation S_{diss} terms

In contrast to S_{nl}, the knowledge of S_{in} and S_{diss} source terms is poor; furthermore, both include many heuristic factors and coefficients. The creation of a reliable, well-justified theory of S_{in} has been hindered by strong turbulent fluctuations, uncorrelated with the wave motions, in the boundary layer over the sea surface. Even one of the most crucial elements of this theory, the vertical distribution of horizontal wind velocity in the region closest to the ocean surface, where wave motions strongly interact with atmospheric motions, is still the subject of debate. The history of the development of different wind input forms is full of heuristic assumptions, which fundamentally restrict the magnitude and directional distribution of this term. As a result, the values of different wind input terms scatter by a factor of 300–500 % (Badulin et al., 2005; Pushkarev and Zakharov, 2016). For example, experimental determination of S_{in}, as provided by direct measurements of the momentum flux from the air to the water, cannot be rigorously performed in a laboratory due to gravity wave dispersion dependence on the water depth, as well as problems with scale effects in laboratory winds. Additional information on the detailed analysis of the current state of the art of wind input terms can be found in Pushkarev and Zakharov (2016).

Similar to the wind input term, there is little consensus on the parameterization of the source dissipation term S_{diss}. The physical dissipation mechanism, which most physical oceanographers agree on, is the effect of wave energy loss due to wave breaking, while there are also other dubious ad hoc "long wave" dissipation source terms with heuristically justified physical explanations. Currently, there is not even an agreement on the location of wave breaking events in Fourier space. The approach currently utilized in operational wave forecasting models mostly relies on the dissipation localized in the vicinity of the spectral energy peak. Recent numerical experiments show (Pushkarev and Zakharov, 2016; Dyachenko et al., 2015; Zakharov et al., 2009), however, that such an approach does not pass most of the tests associated with the essentially nonlinear nature of the HE Eq. (1).

1.3 Road map for the construction and verification of balanced source terms

The next chapters present a balanced set of wind energy input and wave energy dissipation source terms, based on WTT and experimental data analysis. Further, they are numerically checked to comply with WTT predictions and experimental observations. As mentioned above, contrary to previous attempts to build the detailed-balance source terms, the current approach is based neither on the development of a rigorous analytic theory of turbulent atmospheric boundary layers nor on reliable and repeatable air to ocean momentum measurements. The new S_{in} is constructed in the artificial way realizing, in a sense, "the poor man approach", based on the finding of a two-parameter family of HE self-similar solutions and their restriction to the single-parameter one with the help of comparison with the data of experimental observations, accumulated for several decades.

Section 2 presents experimental evidence of wave energy spectrum characteristics in the form of a specific regression line, found by Resio et al. (2004). The analytic form of this regression line will play a crucial role in narrowing the circle of possible outcomes, obtained with WTT analysis.

Section 3 studies self-similar solutions of the HE – a kinetic equation for surface ocean waves, starting with the analysis of the behavior of the dissipationless HE in infinite space, containing the wind source term in power function form. This approach is similar in spirit to one realized by Zakharov and Filonenko (1967) for finding the solution of the equation $S_{nl} = 0$ in the infinite Fourier domain, which derived the ZF spectrum $\varepsilon \sim \frac{P^{1/3}}{\omega^4}$, where P is the energy flux toward high wave numbers. The Fourier domain in both situations does not contain any dissipation function: its role is played by infinite phase volume as the effective energy sink at infinitely high wave numbers.

Such a situation is similar to one realized in incompressible liquid turbulence for large Reynolds numbers, where the energy distribution is given by the famous Kolmogorov spectrum, transferring the energy from large to small scales, where the energy dissipation is realized due to viscosity, but the viscosity coefficients, i.e., the dissipation details, are not included in the final Kolmogorov spectrum expression. The ZF spectrum and its further KZ generalizations are in this sense the ideological Kolmogorov spectrum counterparts, having the significant difference that the Kolmogorov spectrum is a plausible conjecture, while KZ spectra are the exact solutions of the wave kinetic equation.

Since the current research is application oriented, it is important to understand why this formally academic approach is connected with reality. In this context, there is no such thing as the dissipation at infinitely small waves in nature: however, it is clear that the existence of an absorption at sufficiently high finite frequencies provides a wave scale in real applications that still preserves the KZ solutions, found from the HE equation in infinite space.

As was mentioned before, this statement was confirmed in a different physical context with radically different inertial ranges (the wave-number band between characteristic energy input and characteristic wave energy dissipation), showing KZ solutions with different corresponding indices. As for the considered case of gravity waves on the surface of a deep fluid, KZ spectra have been routinely observed in multiple experiments. The results, published before 1985, are summarized by Phillips (1985). Thereafter, they were observed and discussed by Long and Resio (2007). A complete survey of all measurements requires a separate comprehensive paper, which is in our plans for the future.

The assumed close relation of the HE in the infinite space and finite domain, bounded by high-frequency dissipation, also has a much deeper meaning, consisting in the fact that S_{nl} is the leading term of the HE (Zakharov, 2010; Zakharov and Badulin, 2011). This allows further use of the solutions found from the "zero-dissipation" HE Eq. (13) in infinite space for "practical" Fourier domains with the dissipation localized at finite high enough wave numbers. They take the form of a two-parameter family of self-similar solutions, which can be further restricted to the single-parameter one using experimental regression dependence, presented in Sect. 2. These self-similar solutions present realistic HE solutions and describe a broad class of wave-energy spectra observed in ocean and wave-tank experiments.

The indices, corresponding to self-similar solutions, allow one to wrap up Sect. 3 with the specific form of the wind input term in infinite phase space, called the ZRP wind input term (Zakharov et al., 2012; Pushkarev and Zakharov, 2016), with an arbitrary coefficient in front of it. Now, the theoretical part of the wind source term S_{in} construction is finished, but the obtained model is not suitable yet for numerical simulation, since to perform in finite phase space, it has to be augmented with the wave-breaking dissipation term.

Section 4 explains the dissipation function used in the presented model. The wave-breaking dissipation, also known as "white-capping dissipation", is an important physical phenomenon not properly studied yet for reasons of mathematical and technological complexity. Longuet-Higgins (1980a, b) achieved important results, but did not accomplish the theory completely. Irisov and Voronovich (2011) studied the wave-breaking of short waves, "squeezed" by surface currents, caused by longer waves, and showed that they become steep and unstable. Our explanation is simpler but has the same consequences: the "wedge" formation, preceding the wave breaking, causes the "fat tail" appearance in Fourier space. Subsequent smoothing of the tip of the wedge is equivalent to a "chopping off" of the developed high-frequency tail in Fourier space – a sort of natural low-pass filtering – leading to the loss of the wave energy. Both scenarios lead to smoothing of the wave surface, and are indirectly confirmed by the numerical experiments presented in the current study.

There is considerable freedom in choosing a specific analytic form of such a high-frequency dissipation term, given the lack of a generally accepted rigorous derivation for this mechanism. Consequently, one can choose a preferred one and possibly justify it, but any particular choice will be questioned since it will remain somewhat artificial. Because of that, our motivation was that at the current stage of development, we considered simplicity as a primary motivating factor. Instead of following the previous path of time-consuming numerical and empirical formulations based on field experiments, the authors decided to continue the spectrum from some specific frequency point, well above the spectral peak, with the Phillips law $\sim \omega^{-5}$, which decays faster than the equilibrium spectrum ω^{-4} and therefore corresponds to a net wave energy absorption. Although a version of this concept was incorporated by Janssen (2009), detailed forms of this source term have not been developed to date, other than that the spectrum at high frequencies appears to consistently tend toward an $\sim \omega^{-5}$ form as noted by Phillips (1985). Additional evidence for a transition from $\sim \omega^{-4}$ to $\sim \omega^{-5}$ at frequencies above the equilibrium range comes from analysis of multiple data sets by Resio et al. (2004). In that paper the transition from $\sim \omega^{-4}$ to $\sim \omega^{-5}$ occurs approximately at $f_{\rm d} = 1.1$ Hz; i.e., the physical spectrum has to be continued from this point by $\sim \omega^{-5}$.

The spectrum amplitude at the junction frequency $f_{\rm d}$ is dynamically changing in time. It is important that this analytic continuation contributes to a differential in inverse action, which also affects frequencies lower than $f_{\rm d}$, since the nonlinear interaction term $S_{\rm nl}$ is calculated over both "dynamic" and fixed Phillips areas. Therefore, the Phillips part of the spectrum "sends" the information about the presence of the dissipation above $f_{\rm d}$ to the rest of the spectrum.

At this point, all that remains for source-term closure in the HE model is the coefficient in front of the wind input term, since it is not well defined experimentally. If we carry out the numerical simulation with some arbitrary chosen coefficient, we could obtain a range of spectral energies but would retain the qualitative properties of the HE, like the $\sim \omega^{-4}$ spectrum, spectral peak down-shift and peak frequency behavior in accordance with self-similar laws.

To solve this, we choose the wind source coefficient to reproduce the same wave energy growth as was observed in field experiments. The value of this coefficient, found from the comparison with field observations of wave energy growth, is equal to 0.05. This step completes the construction of the HE model.

In the next sections we proceed with numerical simulations based on the HE model described above. Section 5 discusses the details of the numerical model setup. Section 6 describes the duration limited numerical simulation, which is the subject of more academic than applied interest, targeted at self-similarity concept support, while the limited fetch numerical simulation results, described in Sect. 7, besides academic interest, are the subject of comparison with the field

experiments. A check of the compliance of numerical results with field experimental measurements is presented in Sect. 8.

2 Experimental evidence

Here we examine the empirical evidence from around the world, which has been utilized to quantify energy levels within the equilibrium spectral range by Resio et al. (2004). For convenience, we shall also use the same notation used by Resio et al. (2004) in their study, for the angular averaged spectral energy densities in frequency and wavenumber spaces:

$$E_4(f) = \frac{2\pi \alpha_4 V g}{(2\pi f)^4}, \tag{6}$$

$$F_4(k) = \beta k^{-5/2}, \tag{7}$$

where $f = \frac{\omega}{2\pi}$, α_4 is the constant, V is some characteristic velocity and $\beta = \frac{1}{2}\alpha_4 V g^{-1/2}$. These notations are based on relation of spectral densities $E(f)$ and $F(k)$ in frequency $f = \frac{\omega}{2\pi}$ and wave-number k bases:

$$F(k) = \frac{c_{\rm g}}{2\pi} E(f), \tag{8}$$

where $c_{\rm g} = \frac{{\rm d}\omega}{{\rm d}k} = \frac{1}{2 \cdot 2\pi} \frac{g}{f}$ is the group velocity.

The notations in Eqs. (6) and (7) are connected with the spectral energy density $\epsilon(\omega, \theta)$ through

$$E(f) = 2\pi \int\limits_0^{2\pi} \epsilon(\omega, \theta) {\rm d}\theta. \tag{9}$$

The Resio et al. (2004) analysis showed that experimental energy spectra $F(k)$, estimated through averaging $\langle k^{5/2} F(k) \rangle$, can be approximated by a linear regression line as the function of $(u_\lambda^2 c_{\rm p})^{1/3} g^{-1/2}$. Figure 1 shows that the regression line

$$\beta = \frac{1}{2} \alpha_4 \left[(u_\lambda^2 c_{\rm p})^{1/3} - u_0 \right] g^{-1/2}, \tag{10}$$

indeed, seems to be a reasonable approximation of these observations.

Here $\alpha_4 = 0.00553$, $u_0 = 1.93\,{\rm m\,s^{-1}}$, $c_{\rm p}$ is the spectral peak phase speed and u_λ is the wind speed at the elevation equal to a fixed fraction $\lambda = 0.065$ of the spectral peak wavelength $2\pi/k_{\rm p}$, where $k_{\rm p}$ is the spectral peak wave number. It is important to emphasize that the Resio et al. (2004) experiments show that parameter β increases with development of the wind-driven sea, when $f_{\rm p}$ decreases and $C_{\rm p}$ increases. This observation is consistent with the weak turbulent theory, where $\beta \sim P^{1/3}$ (Zakharov et al., 1992); here P is the wave energy flux toward small scales.

Resio et al. (2004) assumed that the near-surface boundary layer can be treated as neutral and thus follows a conventional logarithmic profile

$$u_\lambda = \frac{u_*}{\kappa} \ln \frac{z}{z_0} \tag{11}$$

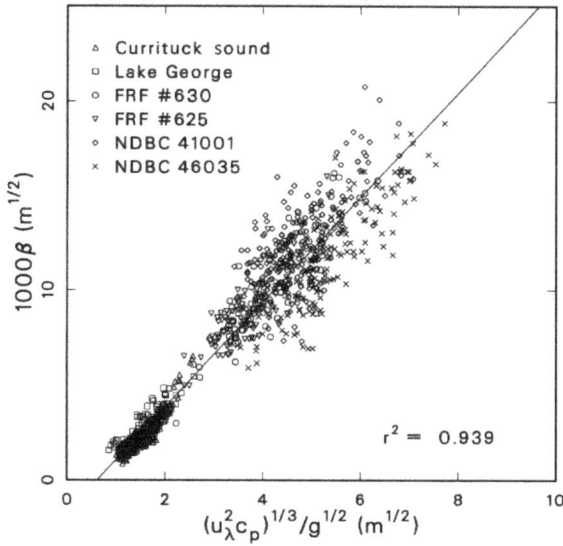

Figure 1. Correlation of the equilibrium range coefficient β with $(u_\lambda^2 c_p)^{1/3}/g^{1/2}$ based on data from six disparate sources. Adapted from Resio et al. (2004).

with a Von Karman coefficient $\kappa = 0.41$, where $z = \lambda \cdot 2\pi/k_p$ is the elevation equal to a fixed fraction $\lambda = 0.065$ of the spectral peak wavelength $2\pi/k_p$, where k_p is the spectral peak wave number, and $z_0 = \alpha_C u_*^2/g$ is subject to Charnock (1955) surface roughness with $\alpha_C = 0.015$.

3 Theoretical considerations

Self-similar solutions consistent with the conservative kinetic equation

$$\frac{\partial \epsilon(\omega, \theta)}{\partial t} = S_{nl} \tag{12}$$

were studied in Zakharov (2005) and Badulin et al. (2005). In this section we study self-similar solutions of the forced kinetic equation

$$\frac{\partial \epsilon(\omega, \theta)}{\partial t} = S_{nl} + \gamma(\omega, \theta)\epsilon(\omega, \theta) \tag{13}$$

where $\epsilon(\omega, \theta) = \frac{2\omega^4}{g} N(\boldsymbol{k}, \theta)$ is the energy spectrum.

One should note that this equation does not contain any explicit wave dissipation term; the role of dissipation is played by the existence of the energy sink at infinitely high wave numbers, in the spirit of the WTT; see Zakharov and Filonenko (1967) and Zakharov et al. (1992).

For our purposes, it is sufficient to simply use the dimensional estimate for S_{nl},

$$S_{nl} \simeq \omega \left(\frac{\omega^5 \epsilon}{g^2}\right)^2 \epsilon. \tag{14}$$

Eq. (13) has a self-similar solution if

$$\gamma(\omega, \theta) = \alpha \omega^{1+s} f(\theta) \tag{15}$$

where s is a constant. Looking for a self-similar solution in the form

$$\epsilon(\omega, t) = t^{p+q} F(\omega t^q), \tag{16}$$

we find

$$q = \frac{1}{s+1}, \tag{17}$$

$$p = \frac{9q-1}{2} = \frac{8-s}{2(s+1)}. \tag{18}$$

The function $F(\xi)$ has a maximum at $\xi \sim \xi_p$; thus, the frequency of the spectral peak is

$$\omega_p \simeq \xi_p t^{-q}. \tag{19}$$

The phase velocity at the spectral peak is

$$c_p = \frac{g}{\omega_p} = \frac{g}{\xi_p} t^q = \frac{g}{\xi_p} t^{\frac{1}{s+1}}. \tag{20}$$

According to experimental data, the main energy input into the spectrum occurs in the vicinity of the spectral peak, i.e., at $\omega \simeq \omega_p$. For $\omega \gg \omega_p$, the spectrum is described by the Zakharov–Filonenko tail

$$\epsilon(\omega) \sim P^{1/3} \omega^{-4}. \tag{21}$$

Here

$$P = \int_0^\infty \int_0^{2\pi} \gamma(\omega, \theta)\epsilon(\omega, \theta)\,d\omega\,d\theta. \tag{22}$$

This integral converges if $s < 2$. For large ω,

$$\epsilon(\omega, t) \simeq \frac{t^{p-3q}}{\omega^4} \simeq \frac{t^{\frac{2-s}{2(s+1)}}}{\omega^4}. \tag{23}$$

More accurately,

$$\epsilon(\omega, t) \simeq \frac{\mu g}{\omega^4} u^{1-\eta} c_p^\eta g(\theta), \tag{24}$$

$$\eta = \frac{2-s}{2}. \tag{25}$$

Now, supposing $s = 4/3$ and $\gamma \simeq \omega^{7/3}$, we get $\eta = 1/3$, which is exactly the experimental regression line prediction. Because it is known from the regression line in Fig. 1 that $\xi = 1/3$, we immediately get $s = 4/3$ and the wind input term

$$S_{wind} \simeq \omega^{7/3} \epsilon. \tag{26}$$

For many years, the assumption has been that there could be a net input or dissipation within the equilibrium range;

however, Thomson et al. (2013) recently used extensive data from Ocean Station Papa to show that there was minimal wind input into the wave spectrum in the equilibrium range. Resio et al. (2004) suggest that the existence of significant net energy input or dissipation within the frequency range would tend to force the spectrum away from an f^{-4} form, contrary to the pattern found in field measurements. If we assume that the wind source is primarily centered on the spectral peak, the only missing component in our numerical solution is an unknown coefficient in front of it, which will be defined later from the comparison with total energy growth in experimental observations.

Another important theoretical relationship that can be derived from joint consideration of Eqs. (6), (8) and (24) is

$$1000\beta = \lambda \frac{(u^2 c_\mathrm{p})^{1/3}}{g^{1/2}}, \tag{27}$$

which shows a theoretical equivalence to the experimental regression, where λ is an unknown constant, defined experimentally.

At the end of the section, we present the summary of important relationships.

Wave action N, energy E and momentum M in frequency-angle presentation are

$$N = \frac{2}{g^2} \int\limits_0^\infty \int\limits_0^{2\pi} \omega^3 n \, d\omega \, d\phi, \tag{28}$$

$$E = \frac{2}{g^2} \int\limits_0^\infty \int\limits_0^{2\pi} \omega^4 n \, d\omega \, d\phi, \tag{29}$$

$$M = \frac{2}{g^3} \int\limits_0^\infty \int\limits_0^{2\pi} \omega^5 n \cos\phi \, d\omega \, d\phi. \tag{30}$$

The self-similar relations for the duration limited case are given by

$$\epsilon = t^{p+q} F(\omega t^q), \tag{31}$$

$$9q - 2p = 1, \quad p = 10/7, \quad q = 3/7, \quad s = 4/3, \tag{32}$$

$$N \sim t^{p+q}, \tag{33}$$

$$E \sim t^p, \tag{34}$$

$$M \sim t^{p-q}, \tag{35}$$

$$\langle\omega\rangle \sim t^{-q}. \tag{36}$$

The same sort of self-similar analysis gives self-similar relations for the fetch limited case:

$$\epsilon = \chi^{p+q} F(\omega \chi^q), \tag{37}$$

$$10q - 2p = 1, \quad p = 1, \quad q = 3/10, \quad s = 4/3, \tag{38}$$

$$N \sim \chi^{p+q}, \tag{39}$$

$$E \sim \chi^p, \tag{40}$$

$$M \sim \chi^{p-q}, \tag{41}$$

$$\langle\omega\rangle \sim \chi^{-q}. \tag{42}$$

4 The details of "implicit" dissipation

Now that the construction of the ZRP wind input term with the unknown coefficient has been accomplished in the spirit of WTT in the previous chapter, the HE model, suitable for numerical simulation, still misses the dissipation term localized at finite wave numbers – there is no such thing as the infinite phase volume in reality: the real ocean Fourier space is confined by a characteristic wave number corresponding to the start of the dissipation effects caused by the wave-breaking events.

There is a lot of freedom in choosing the dissipation term. Since there is no current interpretation of the wave-breaking dissipation mechanism, one can choose it in whatever shape it is preferred, but any particular choice will be questioned since it is an artificial one.

Because of that, the motivation consisted in the fact that at the current "proof of concept" stage one needs to know the effective sink with the simplest structure. Continuation of the spectrum from ω_d with the Phillips law $A(\omega_\mathrm{d}) \cdot \omega^{-5}$ (see Phillips, 1966), decaying faster than the equilibrium spectrum ω^{-4}, will get high-frequency dissipation. The corresponding analytic parameterization of this dissipation term will be unknown, while not in principle impossible to figure out in some way. One should note that this method of dissipation is not our invention: it is described in Janssen (2009).

Specifically, the coefficient $A(\omega_\mathrm{d})$ in front of ω^{-5} is unknown but is not required to be defined in an explicit form. Instead, it is dynamically determined from the continuity condition of the spectrum, at frequency ω_d, on every time step. In other words, the starting point of the Phillips spectrum coincides with the last point of the dynamically changing spectrum, at the frequency point $\omega_\mathrm{d} = 2\pi f_\mathrm{d}$, where $f_\mathrm{d} \simeq 1.1\,\mathrm{Hz}$, as per Long and Resio (2007). This is the way the high-frequency "implicit" damping is incorporated into the alternative computational framework of the HE. The question of the finer details of the high-frequency "implicit" damping structure is of secondary importance, at the current "proof of concept" stage.

The whole set of the input and dissipation terms is accomplished now with one uncertainty: the explained approach leaves one parameter arbitrary – the constant in front of the wind input term. We choose it to be equal to 0.05 from the condition of the reproduction of the field observations of wave energy growth along the fetches, analyzed in Badulin et al. (2007).

5 Numerical validation of the relationship

To check the self-similar hypothesis posed in Eq. (26), we performed a series of numerical simulations of Eq. (1) in the spatially homogeneous duration limited $\frac{\partial N}{\partial r} = 0$ and spatially inhomogeneous fetch limited $\frac{\partial N}{\partial t} = 0$ situations.

All simulations used the WRT (Webb–Resio–Tracy) method (see Tracy and Resio, 1982), which calculates the nonlinear interaction term in the exact form. The presented numerical simulation utilized the version of the WRT method, previously used in Webb (1978), Resio and Perrie (1989), Perrie and Zakharov (1999), Pushkarev et al. (2003), Long and Resio (2007), Korotkevich et al. (2008), Zakharov and Badulin (2015), and Pushkarev and Zakharov (2016), and used the grid of 71 logarithmically spaced points in the frequency range from 0.1 to 2.0 Hz and 36 equidistant points in the angle domain. The constant time step in the range between 1 and 2 s has been used for explicit first-order accuracy integration in time.

There is a balance between the number of nodes of the grid and the volume of the calculation to be performed. The particular version of the WRT model has been tuned to the minimum grid number of nodes to solve realistic physical problems, but is still fast enough to simulate them over a reasonable time span. The correctness of this statement is confirmed by the multiple numerical experiments cited above, reproducing mathematical properties of the Hasselmann equation.

For convenience, we present the pseudo-code used for the main cycle of the described model.

1. Calculate $S_{\mathrm{nl}}(\varepsilon(f, \theta))$.

2. Overwrite $\varepsilon(f, \theta)$ to f^{-5} for $f > 1.1\,\mathrm{Hz}$.

3. Update $\varepsilon(f, \theta) = \varepsilon(f, \theta) + \mathrm{d}t \cdot S_{\mathrm{nl}}(f, \theta)$.

4. Solve analytically $\frac{\partial \varepsilon(f, \theta)}{\partial t} = S_{\mathrm{wind}}(f, \theta)\varepsilon(f, \theta)$ for time $\mathrm{d}t$.

5. Return to step 1.

All numerical simulations discussed in the current paper have been started from a uniform noise energy distribution in Fourier space $\varepsilon(\omega, \theta) = 10^{-6}\,\mathrm{m}^4$, corresponding to a small initial wave height with an effectively negligible nonlinearity level. The constant wind of speed $10\,\mathrm{m\,s^{-1}}$ was assumed to blow away from the shoreline, along the fetch. The assumption of constant wind speed is a necessary simplification due to the fact that the numerical simulation is being compared to various data from field experiments, and the considered setup is the simplest physical situation which can be modeled.

The same ZRP wind input term Eq. (26) has been used in both cases as

$$S_{\mathrm{in}}(\omega, \theta) = \gamma(\omega, \theta) \cdot \varepsilon(\omega, \theta), \tag{43}$$

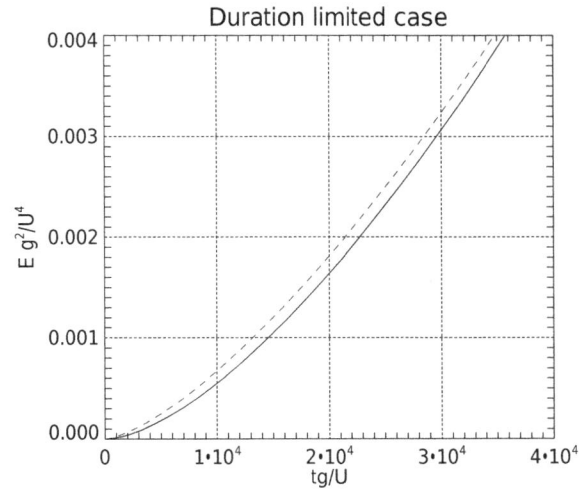

Figure 2. Dimensionless energy Eg^2/U^4 vs. dimensionless time tg/U for the wind speed $U = 10\,\mathrm{m\,s^{-1}}$ duration limited case – solid line. Self-similar solution with the empirical coefficient in front of it: $1.3 \times 10^{-9}(tg/U)^{10/7}$ – dashed line.

$$\gamma(\omega, \theta) = \begin{cases} 0.05\dfrac{\rho_{\mathrm{air}}}{\rho_{\mathrm{water}}}\omega\left(\dfrac{\omega}{\omega_0}\right)^{4/3} q(\theta) \\ \qquad \text{for} \quad f_{\min} \le f \le f_{\mathrm{d}}, \quad \omega = 2\pi f \\ 0 \text{ otherwise}, \end{cases} \tag{44}$$

$$q(\theta) = \begin{cases} \cos 2\theta \text{ for } -\pi/4 \le \theta \le \pi/4 \\ 0 \text{ otherwise}, \end{cases} \tag{45}$$

$$\omega_0 = \frac{g}{U}, \quad \frac{\rho_{\mathrm{air}}}{\rho_{\mathrm{water}}} = 1.3 \times 10^{-3}, \tag{46}$$

where U is the wind speed at the reference level of 10 m, and ρ_{air} and ρ_{water} are the air and water density, respectively. It is conceivable to use a more sophisticated expression for $q(\theta)$, for instance $q(\theta) = q(\theta) - q(0)$. To make direct comparison with experimental results of Resio et al. (2004), we used the relation $u_* \simeq U/28$ (see Golitsyn, 2010) in Eq. (11). Frequencies $f_{\min}$ and f_{d} depend on the wind speed and should be found empirically. In current numerical experiments for $U = 10$ and $U = 5\,\mathrm{m\,s^{-1}}$, $f_{\min} = 0.1\,\mathrm{Hz}$ and $f_{\mathrm{d}} = 1.1\,\mathrm{Hz}$. This choice is justified by the obtained numerical results.

The above-described "implicit dissipation" term S_{diss} has played the dual role of a direct energy cascade flux sink due to wave breaking as well as a numerical scheme stabilization factor at high wave numbers.

6 Duration limited numerical simulation

The duration limited simulation has been performed for a wind speed of $U = 10\,\mathrm{m\,s^{-1}}$.

Figure 2 shows the total energy growth as the function of time, consistent with the self-similar prediction Eq. (34) for index $p = 10/7$, supplied with the empirical coefficient in front of it; see Fig. 3.

Duration limited case

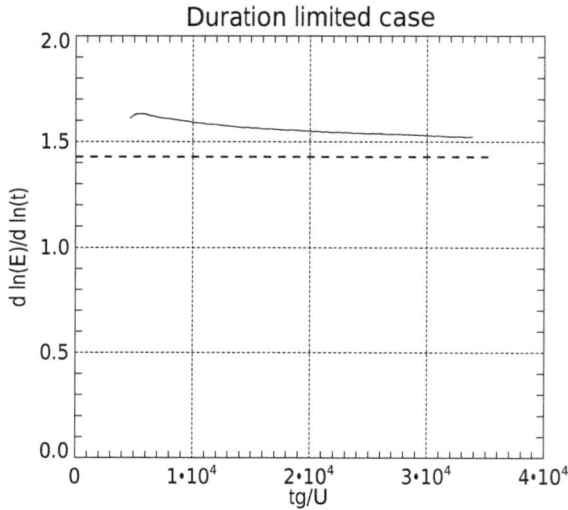

Figure 3. Energy local power function index $p = \frac{d \ln E}{d \ln t}$ as a function of dimensionless time tg/U for the wind speed $U = 10\,\mathrm{m\,s^{-1}}$ duration limited case – solid line. Theoretical value of the self-similar index $p = 10/7$ – thick horizontal dashed line.

Duration limited case

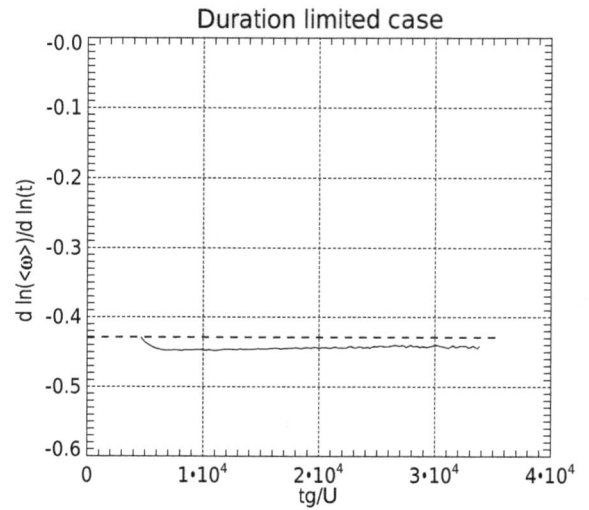

Figure 5. Mean frequency local power function index $-q = \frac{d \ln \langle \omega \rangle}{d \ln t}$ as the function of dimensionless time tg/U for the wind speed $U = 10\,\mathrm{m\,s^{-1}}$ duration limited case (solid line). Theoretical value of self-similar exponent $q = -3/7$ – thick horizontal dashed line.

Duration limited case

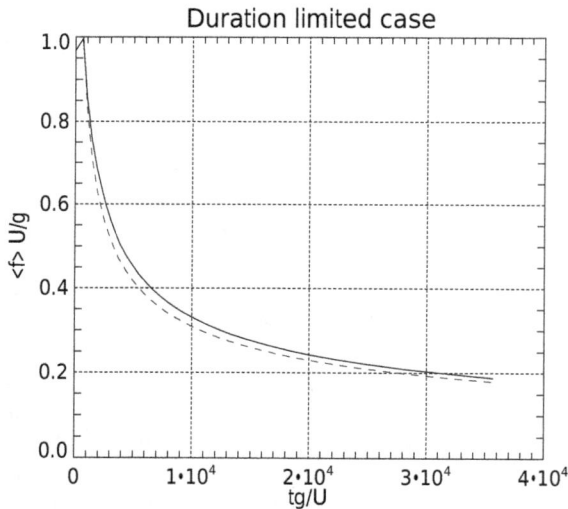

Figure 4. Dimensionless mean frequency $\langle f \rangle \cdot U/g = E/N \cdot U/g$ (solid line) vs. dimensionless time tg/U for the wind speed $U = 10\,\mathrm{m\,s^{-1}}$ duration limited case – solid line, self-similar solution with the empirical coefficient in front of it: $16.0 \cdot (tg/U)^{-3/7}$ – dashed line.

Duration limited case

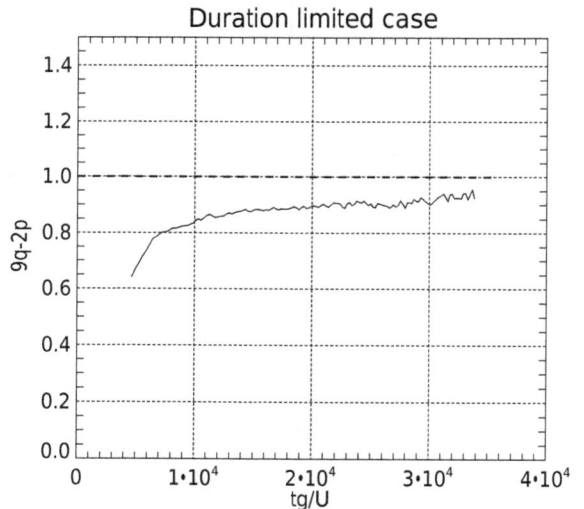

Figure 6. "Magic number" $9q - 2p$ as the function of dimensionless time tg/U for the wind speed $U = 10\,\mathrm{m\,s^{-1}}$ duration limited case – solid line. The target value 1 for the self-similar relation Eq. (32) is represented by the horizontal dashed line.

One should specifically elaborate on the local index p numerical calculation procedure for Fig. 3. First, the total energy function was smoothed via a moving average, then the corresponding derivative is estimated numerically via finite differences, and finally a moving average is used to obtain the time-varying index value.

The relatively small systematic deviation from self-similar behavior, visible in Fig. 2, is connected with the following two facts.

First, the transition process in the beginning of the simulation, when the wave system behavior is far from a self-similar one. The self-similar solution is a pure power function, which does not take into account the initial transition process, and which causes the systematic difference. This systematic difference could be diminished via a parallel shift, which would

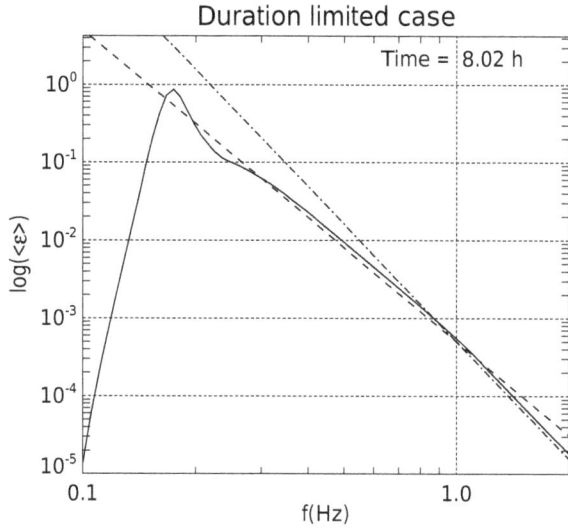

Figure 7. Decimal logarithm of the angle averaged spectrum as the function of the decimal logarithm of the frequency for the wind speed $U = 10\,\mathrm{m\,s^{-1}}$ duration limited case – solid line. Spectrum $\sim f^{-4}$ – dashed line; spectrum $\sim f^{-5}$ – dash-dotted line.

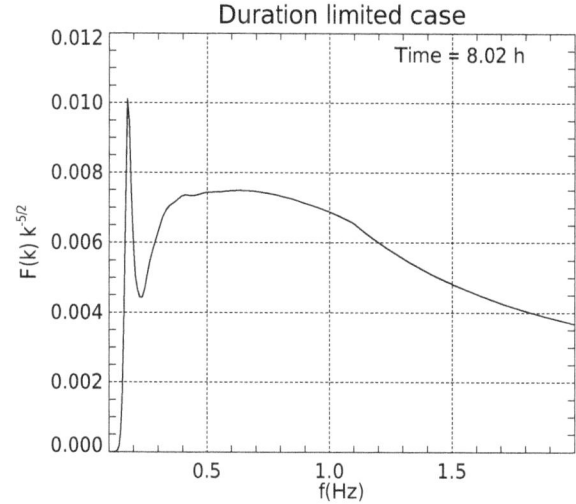

Figure 8. Compensated spectrum as the function of linear frequency f for the wind speed $U = 10\,\mathrm{m\,s^{-1}}$ duration limited case.

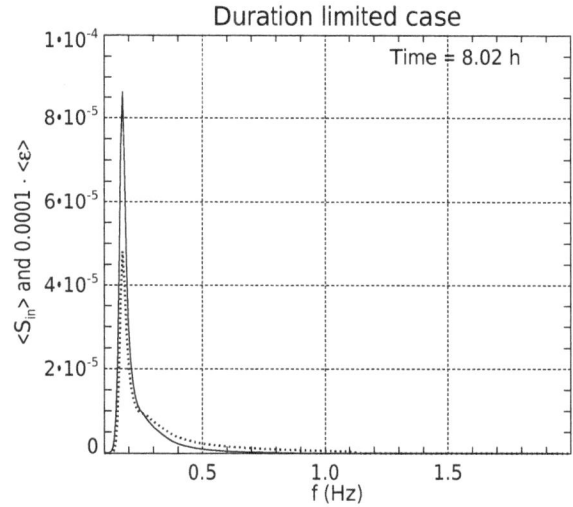

Figure 9. Typical, angle averaged, wind input function density $\langle S_{\mathrm{in}} \rangle = \frac{1}{2\pi} \int \gamma(\omega, \theta) \varepsilon(\omega, \theta) \mathrm{d}\theta$ (dotted line) and angle averaged spectrum $\langle \varepsilon \rangle = \frac{1}{2\pi} \int \varepsilon(\omega, \theta) \mathrm{d}\theta$ (solid line) as the functions of the frequency $f = \frac{\omega}{2\pi}$ for the wind speed $U = 10\,\mathrm{m\,s^{-1}}$ duration limited case.

take into account the initial transition process. Such a parallel shift is equivalent to starting the simulation from different initial conditions.

The second fact is the asymptotic nature of the self-similar solution, producing an evolution of the simulated wave system toward self-similar behavior with increasing time. As seen in Fig. 3, the numerical value of the local exponent converges to the theoretical value $p = 10/7$, reaching approximately 6 % accuracy for a sufficient dimensionless time 3×10^4.

The dependence of the mean frequency on time, shown in Fig. 4, is consistent with the self-similar dependence found in Eq. (36) for $q = 3/7$, supplied with the empirical coefficient in front of it: see Fig. 5.

The systematic deviation of two lines in Fig. 4 remains within 3 % of the target value $q = 3/7$ for the same reasons as for wave-energy behavior – the transition process in the beginning of the simulation and the asymptotic nature of the self-similar solution.

A check of the consistency with the "magic number" $9q - 2p = 1$ (see Eq. 32) is presented in Fig. 6. The reason for systematic deviation from the target value 1 is obviously connected with the reasons for the systematic deviations of p and q, as the "magic number" is calculated as their linear combination, reaching the accuracy of approximately 10 % for a long enough dimensionless time of 3×10^4.

One should note that indices p and q and the "magic relation" $9q - 2p$ exhibit asymptotic convergence to the corresponding target values.

Figure 7 presents an angle-integrated energy spectrum as the function of frequency, in logarithmic coordinates. One can see that it consists of the segments of

– the spectral peak region,

– the inertial (equilibrium) range ω^{-4} spanning from the spectral peak to the beginning of the "implicit dissipation" $f_{\mathrm{d}} = 1.1\,\mathrm{Hz}$, and

– a Phillips high-frequency tail ω^{-5} starting approximately from $f_{\mathrm{d}} = 1.1\,\mathrm{Hz}$.

Duration limited case

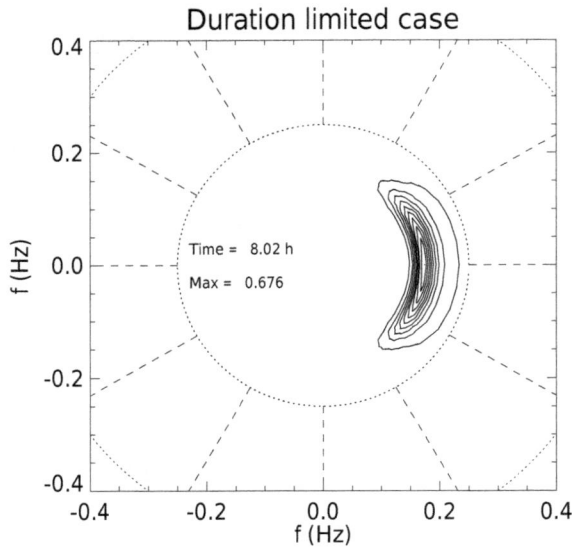

Figure 10. Angular spectrum for the wind speed $U = 10 \, \text{m s}^{-1}$ duration limited case.

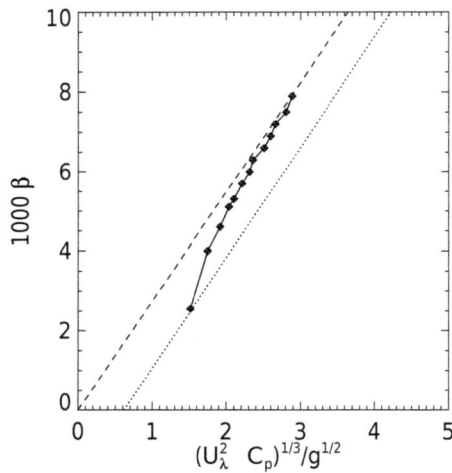

Figure 11. Experimental, theoretical and numerical evidence of the dependence of 1000β on $(u_\lambda^2 c_p)^{1/3}/g^{1/2}$. Dashed line – theoretical prediction Eq. (10) for $\lambda = 2.74$; dotted line – experimental regression line from Resio et al. (2004) and Long and Resio (2007). Line connected diamonds – results of numerical calculations for the wind speed $U = 10 \, \text{m s}^{-1}$ duration limited case. Being parameterized by dimensionless time tg/U, the numerical simulation trajectory evolves from the left to the right on the graph, covering a time span from $tg/U = 0$ to $tg/U \simeq 3.5 \times 10^5$.

The compensated spectrum $F(k) \cdot k^{5/2}$ is presented in Fig. 8.

One can see a plateau-like region responsible for $k^{-5/2}$ behavior, equivalent to the $\sim f^{-4}$ tail in Fig. 7. This shape of the spectrum is similar to that observed by Long and Resio (2007). This exact solution of Eq. (12), known as the KZ spectrum, was found by Zakharov and Filonenko (1967). The

universality of f^{-4} asymptotic for the "inertial" (also known as "equilibrium" in oceanography) range between spectral peak energy input and high-frequency energy dissipation areas has been observed in multiple experimental field observations and is accepted by the oceanographic community after the seminal work of Phillips (1985). One should note that most of the energy flux into the system comes in the vicinity of the spectral peak, as shown in Fig. 9, providing a significant inertial interval for the KZ spectrum.

The angular spectral distribution of energy, presented in Fig. 10, is consistent with the results of experimental observations (Resio et al., 2011) that show a broadening of the angular spreading in both directions away from the spectral peak frequency.

To compare the duration limited numerical simulation results with the experimental analysis by Resio et al. (2004), presented in Fig. 1, Fig. 11 shows the function $\beta = F(k) \cdot k^{5/2}$ as the function of $(u_\lambda^2 C_p)^{1/3}/g^{1/2}$ for wind speed $U = 10 \, \text{m s}^{-1}$, along with the regression line from Resio et al. (2004) and the theoretical prediction Eq. (27) for $\lambda = 2.74$. The numerical results and theoretical prediction line fall within a very small rms deviation ($r^2 = 0.939$; see Fig. 1) from the regression line. One should note asymptotic convergence of the numerical simulation results to the theoretical line.

7   Limited fetch numerical simulation

The limited fetch simulation was performed in the framework of the stationary version of Eq. (1):

$$\frac{1}{2} \frac{g \cos \theta}{\omega} \frac{\partial \epsilon}{\partial x} = S_{\text{nl}}(\epsilon) + S_{\text{wind}} + S_{\text{diss}}, \qquad (47)$$

where x is chosen as the coordinate axis orthogonal to the shore and θ is the angle between the individual wave number k and the axis x. To find the dependence on the wind speed, directed off the shore, two numerical simulations for wind speeds of $U = 5 \, \text{m s}^{-1}$ and $U = 10 \, \text{m s}^{-1}$ have been performed.

The stationarity in Eq. (47) is somewhat difficult for numerical simulation, since it contains a singularity in the form of $\cos \theta$ in front of $\frac{\partial \epsilon}{\partial x}$. This problem was overcome by zeroing one-half of the Fourier space of the system for the waves propagating toward the shore. Since the energy in such waves is small with respect to waves propagating in the offshore direction, such an approximation is quite reasonable for our purposes.

Since the wind forcing index s in the fetch limited case is similar to that in the duration limited case, the numerical simulation of Eq. (47) has been performed for the same input functions as in the duration limited case with the same low-level energy noise initial conditions in Fourier space.

Figure 12 presents total energy growth as a function of fetch, consistent with the self-similar solution Eq. (40) for

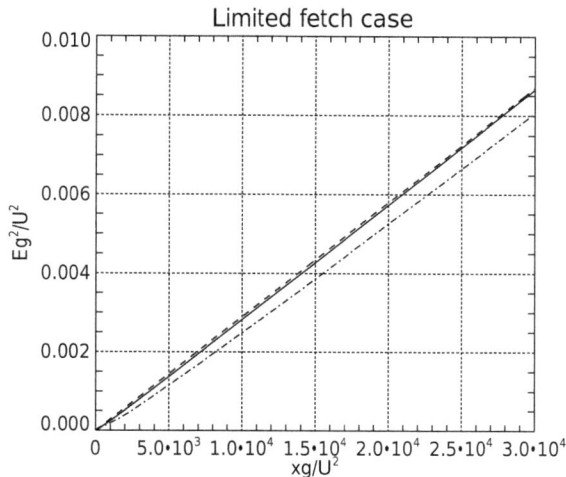

Figure 12. Dimensionless energy Eg^2/U^4 vs. dimensionless fetch xg/U^2 for the fetch limited case: wind speed $U = 10\,\mathrm{m\,s^{-1}}$ – solid line; wind speed $U = 5\,\mathrm{m\,s^{-1}}$ – dash-dotted line. Self-similar solution with the empirical coefficient in front of it: $2.9 \times 10^{-7} xg/U^2$ – dashed line.

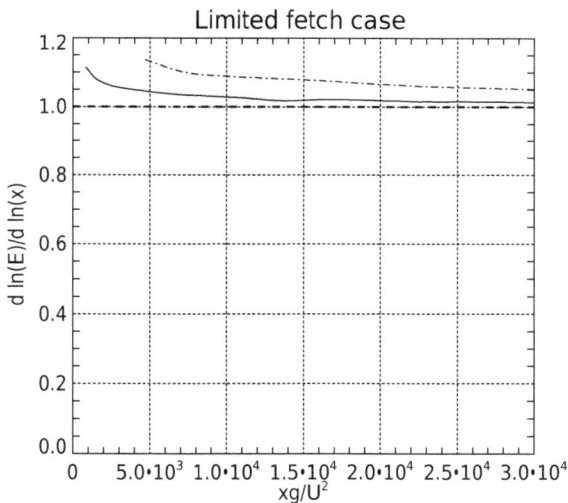

Figure 13. Energy local power function index $p = \frac{\mathrm{d}\ln E}{\mathrm{d}\ln x}$ as the function of dimensionless fetch xg/U^2 for the fetch limited case: wind speed $U = 10\,\mathrm{m\,s^{-1}}$ – solid line; wind speed $U = 5\,\mathrm{m\,s^{-1}}$ – dash-dotted line. Theoretical value of self-similar index $p = 1$ – thick horizontal dashed line.

index $p = 1$, and its appropriate empirical coefficient. The corresponding values of indices p along the fetch are presented in Fig. 13. The small amplitude oscillations observed in the index behavior can be attributed to the finite grid resolution used in the simulation.

The wave evolution for the wind speed $U = 5\,\mathrm{m\,s^{-1}}$ case is expected to be slower than for the $U = 10\,\mathrm{m\,s^{-1}}$ case due to the weaker nonlinear interaction term. One can see, indeed, slower asymptotic convergence of the calculated total

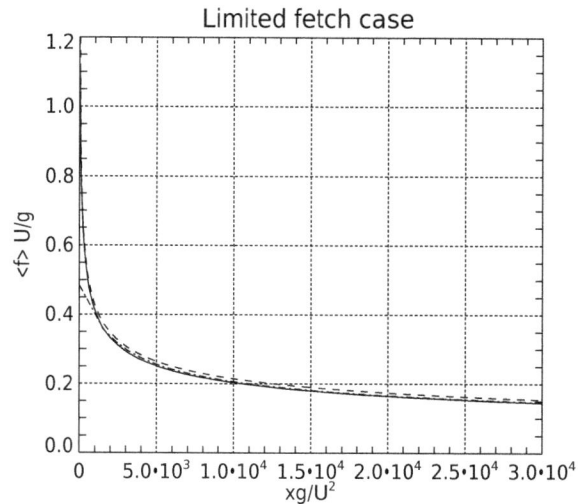

Figure 14. Dimensionless mean frequency as the function of the dimensionless fetch, calculated as $\langle f \rangle = \frac{1}{2\pi} \frac{\int \omega n \mathrm{d}\omega \mathrm{d}\theta}{\int n \mathrm{d}\omega \mathrm{d}\theta}$, where $n(\omega, \theta) = \frac{\varepsilon(\omega,\theta)}{\omega}$ is the wave action spectrum, for wind speed $10\,\mathrm{m\,s^{-1}}$ (solid line) and $5\,\mathrm{m\,s^{-1}}$ (dashed line). The dash-dotted line is the self-similar dependence $3.4 \cdot (\frac{xg}{U^2})^{-0.3}$ with the empirical coefficient in front of it.

energy local power index to the target value $p = 1$ for the $U = 5\,\mathrm{m\,s^{-1}}$ case compared to the $U = 10\,\mathrm{m\,s^{-1}}$ case. The deviation of results from the $U = 10\,\mathrm{m\,s^{-1}}$ case relative to the target value does not exceed an error of about 5 %, while for the $U = 5\,\mathrm{m\,s^{-1}}$ case the error does not exceed 20 %. The role of the relatively short (in time) non-self-similar development of the wave system at the very beginning of the fetch should be noted as well as the factor contributing to the deviation from the target value of index $p = 1$: the wave system obviously needs some time to evolve into a fully self-similar mode.

The dependence of the mean frequency on the fetch, shown in Fig. 14, is consistent with the self-similar dependence Eq. (42) for index $q = 0.3$, supplied with the empirical coefficient in front of it. The small amplitude oscillations observed in index behavior can be attributed to the finite grid resolution used in the simulation, since the spectral peak moves continuously between discrete frequencies in a manner that cannot be matched in these discretized simulations.

The local values of indices q for two different wind speed amplitudes are presented in Fig. 15 along with the target value of the self-similar index $q = 0.3$. After sufficient fetch one can see only about 14 % deviation from the target value for the $U = 10\,\mathrm{m\,s^{-1}}$ case and about 2.5 % for the $U = 5\,\mathrm{m\,s^{-1}}$ case.

The reasons for the 10 % systematic deviation from the self-similar solutions observed in the lines in Fig. 14, corresponding to the wind speeds of $U = 5\,\mathrm{m\,s^{-1}}$ and $U = 10\,\mathrm{m\,s^{-1}}$, are the same as noted previously for wave energy

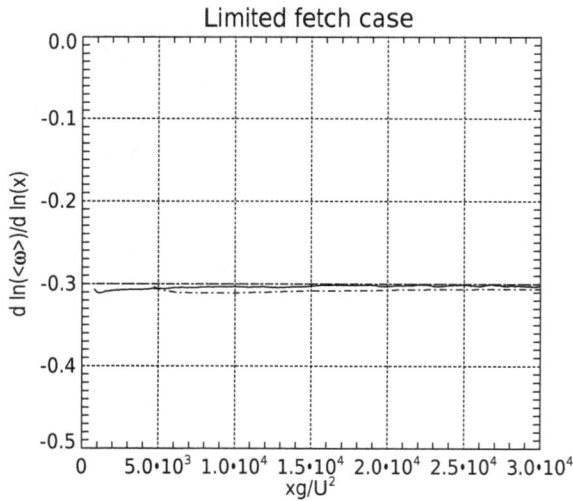

Figure 15. Local mean frequency exponent $-q = \frac{\mathrm{d}\ln\langle\omega\rangle}{\mathrm{d}\ln x}$ as the function of dimensionless fetch xg/U^2 for the limited fetch case. Wind speed $U = 10\,\mathrm{m\,s}^{-1}$ – solid line; wind speed $U = 5\,\mathrm{m\,s}^{-1}$ – dashed line. Horizontal dashed line – target value of the self-similar exponent $q = 0.3$.

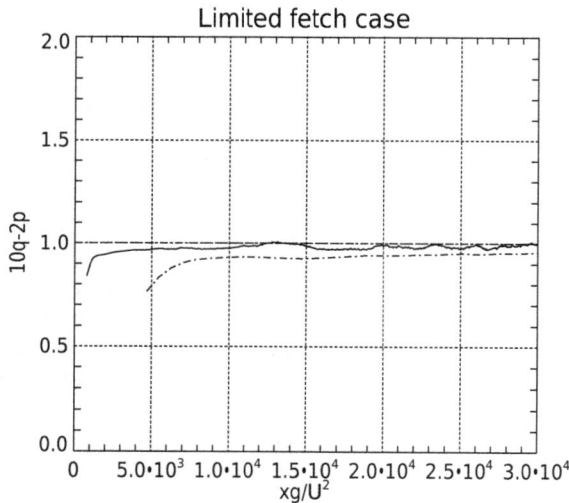

Figure 17. Decimal logarithm of the angle averaged spectrum as the function of the decimal logarithm of the frequency for the wind speed $U = 10\,\mathrm{m\,s}^{-1}$ limited fetch case – solid line. Spectrum $\sim f^{-4}$ – dashed line; spectrum $\sim f^{-5}$ – dash-dotted line.

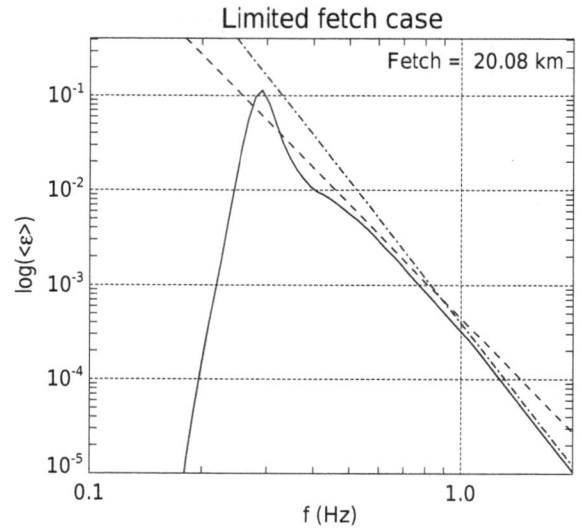

Figure 16. "Magic number" $10q - 2p$ as a function of dimensionless fetch xg/U^2 for the limited fetch case. Wind speed $U = 10\,\mathrm{m\,s}^{-1}$ – solid line; wind speed $U = 5\,\mathrm{m\,s}^{-1}$ – dashed line. Horizontal dashed line – self-similar target value $10q - 2p = 1$.

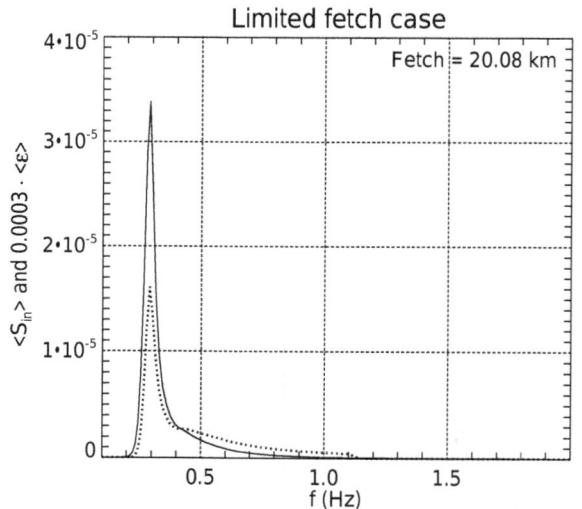

Figure 18. Typical, angle averaged, wind input function density $\langle S_{\mathrm{in}}\rangle = \frac{1}{2\pi}\int \gamma(\omega,\theta)\varepsilon(\omega,\theta)\mathrm{d}\theta$ (dotted line) and angle averaged spectrum $\langle\varepsilon\rangle = \frac{1}{2\pi}\int \varepsilon(\omega,\theta)\mathrm{d}\theta$ (solid line) as the functions of the frequency $f = \frac{\omega}{2\pi}$ for the wind speed $U = 10\,\mathrm{m\,s}^{-1}$ limited fetch case.

behavior – the transition process in the beginning of the simulation and the asymptotic nature of the self-similar solution.

The check on the consistency of the calculated "magic number" $(10q - 2p)$ (see Eq. 38) is presented in Fig. 16. The reason for systematic deviation from the target value 1 is obviously connected with the systematic deviations of p and q, as the "magic number" is calculated as their linear combination, reaching the accuracy of approximately 10 % for fetches 3×10^4. As noted previously, the small amplitude oscillations

observed in the indices' behavior can be attributed to the finite grid resolution used in the simulation.

One should note that indices p and q and the "magic relation" $10q - 2p$ exhibit asymptotic convergence to the corresponding target values.

Figure 17 presents an angle-integrated energy spectrum, as the function of frequency, in logarithmic coordinates. As

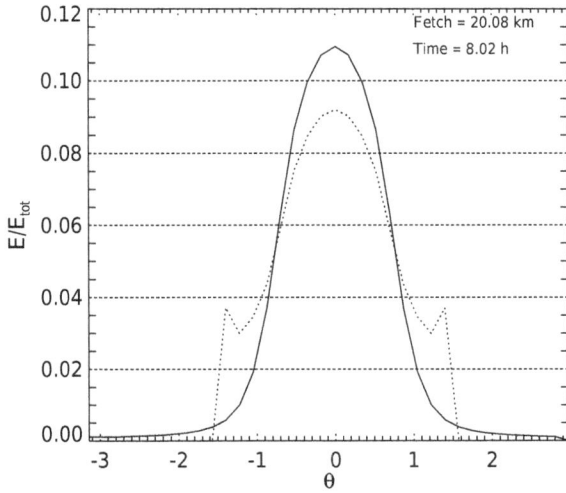

Figure 19. Relative wave energy distribution $E(\theta)/E_{\text{tot}} = \int_{f_{\text{min}}}^{f_d} \varepsilon(\omega, \theta) d\omega / \int_{f_{\text{min}}}^{f_d} \int_0^{2\pi} \varepsilon(\omega, \theta) d\omega d\theta$ as the function of angle θ for the duration limited (solid line) and limited fetch (dotted line) cases.

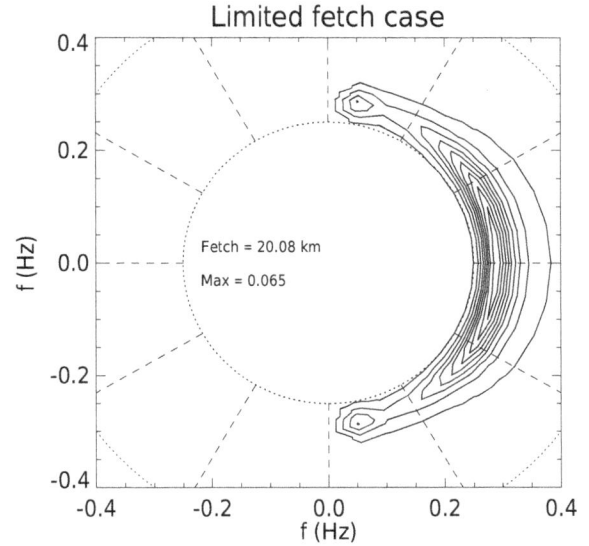

Figure 21. Angular spectrum for the wind speed $U = 10 \, \text{m s}^{-1}$ limited fetch case.

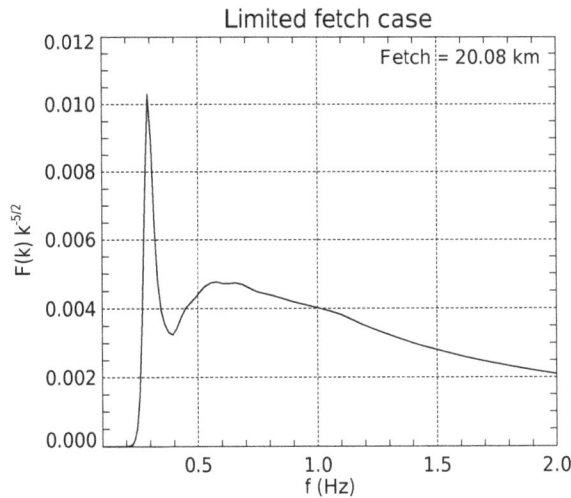

Figure 20. Compensated spectrum as the function of linear frequency f for the wind speed $10 \, \text{m s}^{-1}$ limited fetch case.

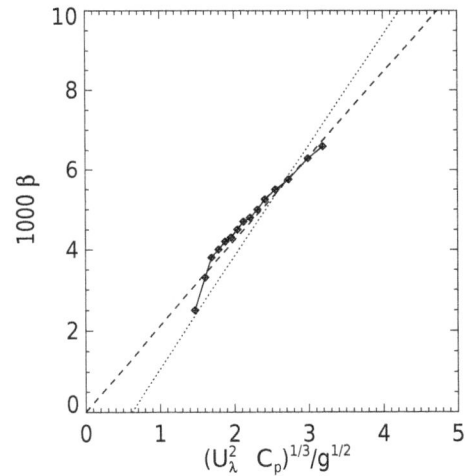

Figure 22. Experimental, theoretical and numerical evidence of the dependence of 1000β on $(u_\lambda^2 c_p)^{1/3}/g^{1/2}$. Dashed line – theoretical prediction Eq. (10) for $\lambda = 2.11$; dotted line – experimental regression line from Resio et al. (2004) and Long and Resio (2007). Line connected diamonds – the results of numerical calculations for the wind speed $U = 10 \, \text{m s}^{-1}$ limited fetch case. Being parameterized by the dimensionless fetch coordinate $\chi = \frac{xg}{U^2}$, the numerical simulation trajectory evolves from the left to the right on the graph, covering a fetch span from $\chi = 0.$ to $\chi \simeq 3.0 \times 10^4$.

could be seen in the duration limited case, one can see that it consists of three process-related segments:

- the spectral peak region,

- the inertial (equilibrium) range ω^{-4} spanning from the spectral peak to the beginning of the "implicit dissipation" $f_d = 1.1 \, \text{Hz}$, and

- a Phillips high-frequency tail ω^{-5} starting approximately at $f_d = 1.1 \, \text{Hz}$.

The compensated spectrum $F(k) \cdot k^{5/2}$ is presented in Fig. 20. One can see a plateau-like region responsible for

$k^{-5/2}$ behavior, equivalent to the ω^{-4} tail in Fig. 17 and similar to that observed by Long and Resio (2007). As in the duration limited case, the KZ solution (Zakharov and Filonenko, 1967) also holds for the fetch limited case, and most of the energy flux into the system comes in the vicinity of the spectral peak as well, as shown in Fig. 18, providing a significant inertial (equilibrium) range for the KZ spectrum

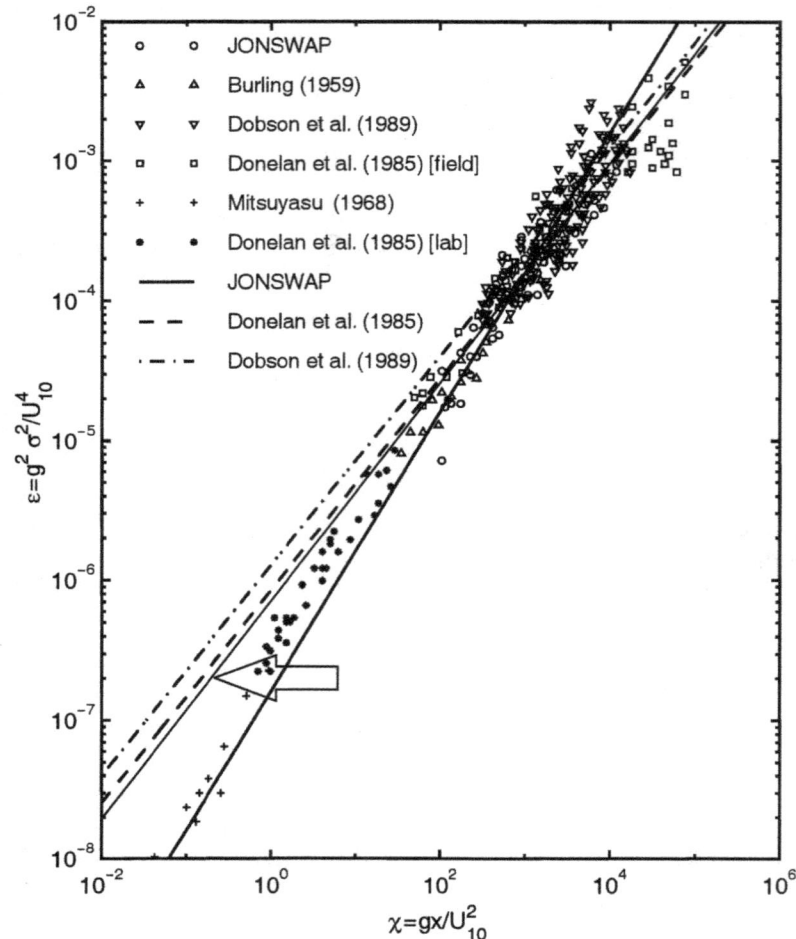

Figure 23. The solid line (pointed to by arrows) presents non-dimensional total energy from the limited fetch numerical experiments, superimposed onto Fig. 5.4, which is adapted from Young (1999). The original caption is "A composite of data from variety of studies showing the development of the non-dimensional energy, ε as a function of non-dimensional fetch, χ. The original JONSWAP study (Hasselmann et al., 1973) used the data marked, JONSWAP, together with that of Burling (1959) and Mitsuyasu (1968). Also shown are a number of growth curves obtained from the various data sets. These include JONSWAP Eq. (5.27), Donelan et al. (1985) Eq. (5.33) and Dobson et al. (1989) Eq. (5.38)."

between spectral peak energy input and high-frequency energy dissipation areas.

The angular spectral distribution of energy, presented in Fig. 21, as in the duration limited case, is consistent with the results of experimental observations by Resio et al. (2011) that show a broadening of the angular spreading in both directions away from the spectral peak frequency.

The excess spectral energy at very oblique angles is a numerical artifact connected with the specifics of how the limited fetch statement is simulated here, i.e., the above-mentioned singularity presence on the left-hand side of Eq. (47) at $\theta = \pm\pi/2$.

The detailed structure of angular spreading for both the duration and limited fetch cases is given in Fig. 19. The time that would be required to produce such a pattern is far in excess of the time for this excess energy to be removed from

the equilibrium range by the nonlinear flux and can be shown to vanish when a time–space simulation is used instead of the stationary solution assumed here.

It is clearly seen that the "blobs" in the limited fetch case contain no more than 5 % of the total energy of the corresponding spectrum and could be neglected for the purposes of the presented research.

To compare the limited fetch numerical simulation results with the experimental analysis by Resio et al. (2004), presented in Fig. 1, Fig. 22 shows the function $\beta = F(k) \cdot k^{5/2}$ as a function of $(u_\lambda^2 C_p)^{1/3}/g^{1/2}$ for wind speed $U = 10.0 \, \text{m s}^{-1}$, along with the regression line from Resio et al. (2004) and its theoretical prediction Eq. (27) for $\lambda = 2.11$. The numerical results and theoretical prediction line fall within the rms deviation ($r^2 = 0.939$; see Fig. 1) from the regression line. One should note asymptotic convergence of the numerical simu-

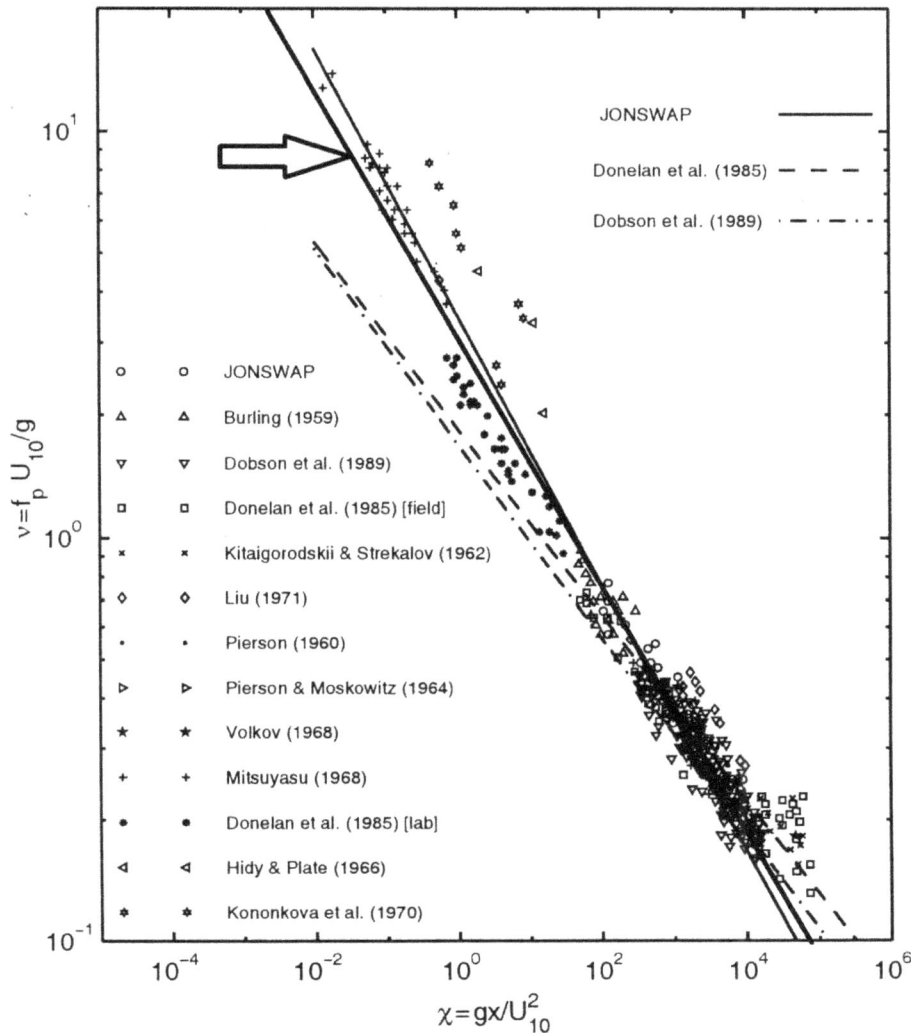

Figure 24. The solid line (pointed to by arrows) presents non-dimensional average frequency as the function of the fetch for limited fetch numerical experiments, superimposed on Fig. 5.5 adapted from Young (1999). The original caption is "A composite of data from a variety of studies showing the development of the non-dimensional peak frequency, v as a function of non-dimensional fetch, χ. The original JONSWAP study (Hasselmann et al., 1973) used all the data shown with the exception of that marked Donelan et al. (1985) and Dobson et al. (1989). Also shown are a number of growth curves obtained from the various data sets. These include JONSWAP Eq. (5.28), Donelan et al. (1985) Eq. (5.34) and Dobson et al. (1989) Eq. (5.39)."

lation results to the theoretical line. Being parameterized by the fetch coordinate, the numerical simulation results evolve from the left to the right on the graph, from the dimensionless fetch equal to 0, to 3.0×10^4.

8   Comparison with the experiments

A comparison of limited fetch and duration limited simulations with the experimental results by Long and Resio (2007) and the theoretical prediction based on Eq. (10) is presented in Figs. 11 and 22. One should note that the numerical results and theoretical prediction line with corresponding values of

λ fall into the rms deviation ($r^2 = 0.939$; see Fig. 1) relative to the experimental regression line Eq. (10).

The dependencies of the dimensionless energy and the frequency on the dimensionless fetch for the limited fetch simulation, superimposed on the experimental observations collected by Young (1999), are presented in Figs. 23 and 24, showing good consistency of the presented numerical results and the experimental observations.

9   Conclusions

We have analyzed the new ZRP form for wind input, proposed in Zakharov et al. (2012) in terms of both fetch limited and duration limited wave growth. The approach proposed here for the development of a set of balanced source terms uses only two empirical coefficients: one in the magnitude of the wind source term and the second in the location of the transition from the $\sim \omega^{-4}$ to $\sim \omega^{-5}$ spectrum. This approach focuses on the combination of the theoretical finding of the self-similar solutions and the extraction of the relevant one through the comparison with the field experimental data.

The numerical simulations for both the duration limited and fetch limited cases, using the ZRP wind input term, XNL nonlinear term S_{nl} and "implicit" high-frequency dissipation, show remarkable consistency with predicted self-similar properties of the HE and with the regression line from field studies, relating energy levels in the equilibrium range to wind speed by Resio et al. (2004) and Long and Resio (2007).

The proposed model is the proof of the concept, providing strong support for simplified assumptions, such as discontinuity and the fixed frequency transition point of the source terms. The influence of these effects will be studied, in particular, using a more sophisticated approach by Zakharov and Badulin (2012) in the future.

Although the integral parameters of the model have been verified against the experimental observations, the verification of the spectral details, such as angular spreading, requires additional studies.

Observed oscillations of self-similar indices are interpreted as the effects of the discreteness of the model, which suggests that a study of the influence of the grid resolution on such oscillations is desirable in future research.

A test of the model invariance with respect to wind speed change from 5 to $10\,\mathrm{m\,s}^{-1}$ has already been performed, but further study of the effects of a wider range of wind speed variation on self-similar properties of the model is desired in the future.

At the moment of submission of the manuscript, the main technical obstacle to effective development of a new generation of physically based HE models was insufficiently fast calculation of the exact nonlinear interaction. The transition to the 2-D case requires a radical increase in the calculation speed. We hope that such improvements will be made in the near future.

The authors hope that this new framework will offer additional guidance for the source terms in operational models.

Acknowledgements. The research presented in Sect. 6 has been accomplished due to the support of the grant "Wave turbulence: the theory, mathematical modeling and experiment" of the Russian Scientific Foundation no. 14-22-00174. The research set forth in Sect. 1 was funded by the program of the presidium of RAS: "Nonlinear dynamics in mathematical and physical sciences". The research presented in other chapters was supported by ONR grant N00014-10-1-0991.

The authors gratefully acknowledge the support of these foundations.

Edited by: Victor Shrira

References

Badulin, S., Babanin, A. V., Resio, D. T., and Zakharov, V.: Weakly turbulent laws of wind-wave growth, J. Fluid Mech., 591, 339–378, 2007.

Zakharov, V. E. and Badulin, S. I: The generalized Phillips' spectra and new dissipation function for wind-driven seas, arXiv:1212.0963v2 [physics.ao-ph], 1–16, https://arxiv.org/abs/1212.0963v2, 2015.

Badulin, S. I., Pushkarev, A. N., Resio, D., and Zakharov, V. E.: Self-similarity of wind-driven seas, Nonlin. Proc. Geoph., 12, 891–945, https://doi.org/10.5194/npg-12-891-2005, 2005.

Badulin, S. I., Pushkarev, A. N., Resio, D., and Zakharov, V. E.: Self-similarity of wind-driven sea, Nonlinear Proc. Geoph., 12, 891–945, 2005.

Balk, A. M.: On the Kolmogorov–Zakharov spectra of weak turbulence, Physica D, 139, 137–157, 2000.

Charnock, H.: Wind stress on a water surface, Q. J. Roy. Meteor. Soc., 81, 639–640, 1955.

de Oliveira, H. P., Zayas, L. A. P., and Rodrigues, E. L.: Kolmogorov–Zakharov spectrum in AdS gravitational collapse, Phys. Rev. Lett., 111, 051101, https://doi.org/10.1103/PhysRevLett.111.051101, 2013.

Dyachenko, A. I., Kachulin, D. I., and Zakharov, V. E.: Evolution of one-dimensional wind-driven sea spectra, JETP Lett., 102, 577–581, 2015.

Galtier, S. and Nazarenko, S.: Turbulence of weak gravitational waves in the early universe, 6 pp., available at: https://arxiv.org/abs/1703.09069v2, last acces: 22 September 2017.

Galtier, S., Nazarenko, S., Newell, A., and Pouquet, A.: A weak turbulence theory for incompressible magnetohydrodynamics, J. Plasma Phys., 63, 447–488, 2000.

Golitsyn, G.: The energy cycle of wind waves on the sea surface, Izv. Atmos. Ocean. Phy., 46, 6–13, 2010.

Hasselmann, K.: On the non-linear energy transfer in a gravity-wave spectrum. Part 1. General theory, J. Fluid Mech., 12, 481–500, 1962.

Hasselmann, K.: On the non-linear energy transfer in a gravity wave spectrum. Part 2. Conservation theorems; wave-particle analogy; irrevesibility, J. Fluid Mech., 15, 273–281, 1963.

Irisov, V. and Voronovich, A.: Numerical Simulation of Wave Breaking, J. Phys. Oceanogr., 41, 346–364, 2011.

Janssen, P.: The Interaction of Ocean Waves and Wind, Cambridge monographs on mechanics and applied mathematics, Cambridge U.P., 2009.

Korotkevich, A. O., Pushkarev, A. N., Resio, D., and Zakharov, V. E.: Numerical verification of the weak turbulent model for swell evolution, Eur. J. Mech. B-Fluid., 27, 361–387, 2008.

Long, C. and Resio, D.: Wind wave spectral observations in Currituck Sound, North Carolina, J. Geophys. Res., 112, C05001, https://doi.org/10.1029/2006JC003835, 2007.

Longuet-Higgins, M. S.: A technique for time-dependent, free-surface flow, Proc. R. Soc. Lon. Ser. A, 371, 441–451, 1980a.

Longuet-Higgins, M. S.: On the forming of sharp corners at a free surface, Proc. R. Soc. Lon. Ser. A, 371, 453–478, 1980b.

L'vov, V. S. and Nazarenko, S.: Spectrum of Kelvin-wave turbulence in superfluids, JETP Lett., 91, 428–434, 2010.

Nordheim, L. W.: On the kinetic method in the new statistics and its application in the electron theory of conductivity, Proc. R. Soc. Lon. Ser. A, 119, 689–698, 1928.

Perrie, W. and Zakharov, V. E.: The equilibrium range cascades of wind-generated waves, Eur. J. Mech. B-Fluid., 18, 365–371, 1999.

Phillips, O. M.: The dynamics of the upper ocean, Cambridge monographs on mechanics and applied mathematics, Cambridge U. P., 1966.

Phillips, O. M.: Spectral and statistical properties of the equilibrium range in wind-generated gravity waves, J. Fluid Mech., 156, 505–531, https://doi.org/10.1017/S0022112085002221, 1985.

Pushkarev, A. and Zakharov, V.: Limited fetch revisited: comparison of wind input terms, in surface wave modeling, Ocean Model., 103, 18–37, https://doi.org/10.1016/j.ocemod.2016.03.005, 2016.

Pushkarev, A., Resio, D., and Zakharov, V.: Weak turbulent approach to the wind-generated gravity sea waves, Physica D, 184, 29–63, 2003.

Pushkarev, A. N. and Zakharov, V. E.: Turbulence of capillary waves, Phys. Rev. Lett., 76, 3320–3323, https://doi.org/10.1103/PhysRevLett.76.3320, 1996.

Resio, D. and Perrie, W.: Implications of an f^{-4} equilibrium range for wind-generated waves, J. Phys. Oceanogr., 19, 193–204, 1989.

Resio, D., Long, C., and Perrie, W.: The role of nonlinear momentum fluxes on the evolution of directional wind-wave spectra, J. Phys. Oceanogr., 41, 781–801, 2011.

Resio, D. T. and Perrie, W.: A numerical study of nonlinear energy fluxes due to wave-wave interactions in a wave spectrum. Part I: Methodology and basic results, J. Fluid Mech., 223, 603–629, 1991.

Resio, D. T., Long, C. E., and Vincent, C. L.: Equilibrium-range constant in wind-generated wave spectra, J. Geophys. Res., 109, C01018, https://doi.org/10.1029/2003JC001788, 2004.

SWAN: available at: http://swanmodel.sourceforge.net/, (last access: 22 September 2017), 2015.

Thomson, J., D'Asaro, E. A., Cronin, M. F., Rogers, W. E., Harcourt, R. R., and Shcherbina, A.: Waves and the equilibrium range at Ocean Weather Station P, J. Geophys. Res., 118, 1–12, 2013.

Tolman, H. L.: User Manual and System Documentation of WAVEWATCH III, Environmental Modeling Center, Marine Modeling and Analysis Branch, 2013.

Tracy, B. and Resio, D.: Theory and calculation of the nonlinear energy transfer between sea waves in deep water, WES report 11, US Army Engineer Waterways Experiment Station, Vicksburg, MS, 1982.

Tran, M. B.: On a quantum Boltzmann type equation in Zakharov's wave turbulence theory, available at: https://nttoan81.wordpress.com/, last access: 22 September 2017.

Webb, D. J.: Non-linear transfers between sea waves, Deep-Sea Res., 25, 279–298, 1978.

Yoon, P. H., Ziebell, L. F., Kontar, E. P., and Schlickeiser, R.: Weak turbulence theory for collisional plasmas, Phys. Rev. E, 93, 033203, https://doi.org/10.1103/PhysRevE.93.033203, 2016.

Young, I. R.: Wind Generated Ocean Waves, Elsevier, Elsevier Science Ltd., The Boulevard, Langford Lane Kidlington, Oxford OX5 1GB, UK, 1999.

Yousefi, M. I.: The Kolmogorov–Zakharov model for optical fiber communication, IEEE T. Inform. Theory, 63, 377–391, 2017.

Yulin, P.: Understanding of weak turbulence of capillary waves, available at: http://hdl.handle.net/1721.1/108837, last access: 22 September 2017.

Zakharov, V. E.: Theoretical interpretation of fetch limited wind-drivensea observations, Nonlin. Processes Geophys., 12, 1011–1020, https://doi.org/10.5194/npg-12-1011-2005, 2005.

Zakharov, V. E.: Energy balances in a wind-driven sea, Phys. Scripta, 2010, T142, http://stacks.iop.org/1402-4896/2010/i=T142/a=014052, 2010.

Zakharov, V. E. and Badulin, S. I.: On energy balance in wind-driven sea, Dokl. Akad. Nauk+, 440, 691–695, 2011.

Zakharov, V. E. and Badulin, S. I.: The generalized Phillips' spectra and new dissipation function for wind-driven seas, available at: http://arxiv.org/abs/arXiv:1212.0963v2, last access: 22 September 2017, 2012.

Zakharov, V. E. and Filonenko, N. N.: The energy spectrum for stochastic oscillation of a fluid's surface, Dokl. Akad. Nauk, 170, 1992–1995, 1966.

Zakharov, V. E. and Filonenko, N. N.: The energy spectrum for stochastic oscillations of a fluid surface, Sov. Phys. Docl., 11, 881–884, 1967.

Zakharov, V. E., L'vov, V. S., and Falkovich, G.: Kolmogorov Spectra of Turbulence I: Wave Turbulence, Springer-Verlag, 1992.

Zakharov, V. E., Korotkevich, A. O., and Prokofiev, A. O.: On dissipation function of ocean waves due to whitecapping, in: American Institute of Physics Conference Series, edited by: Simos, T. E., G.Psihoyios, and Tsitouras, C., vol. 1168, 1229–1237, 2009.

Zakharov, V. E., Resio, D., and Pushkarev, A.: New wind input term consistent with experimental, theoretical and numerical considerations, arXiv:1212.1069v1 [physics.ao-ph], 1–21, http://arxiv.org/abs/1212.1069/, 2012.

13

A simple kinematic model for the Lagrangian description of relevant nonlinear processes in the stratospheric polar vortex

Víctor José García-Garrido[1], **Jezabel Curbelo**[1,2], **Carlos Roberto Mechoso**[3], **Ana María Mancho**[1], and **Stephen Wiggins**[4]

[1]Instituto de Ciencias Matemáticas, CSIC-UAM-UC3M-UCM, C/ Nicolás Cabrera 15, Campus de Cantoblanco UAM, 28049 Madrid, Spain
[2]Departamento de Matemáticas, Facultad de Ciencias, Universidad Autonóma de Madrid, Madrid, Spain
[3]Department of Atmospheric and Oceanic Sciences, University of California at Los Angeles, Los Angeles, CA, USA
[4]School of Mathematics, University of Bristol, Bristol, UK

Correspondence to: Ana María Mancho (a.m.mancho@icmat.es)

Abstract. In this work, we study the Lagrangian footprint of the planetary waves present in the Southern Hemisphere stratosphere during the exceptional sudden Stratospheric warming event that took place during September 2002. Our focus is on constructing a simple kinematic model that retains the fundamental mechanisms responsible for complex fluid parcel evolution, during the polar vortex breakdown and its previous stages. The construction of the kinematic model is guided by the Fourier decomposition of the geopotential field. The study of Lagrangian transport phenomena in the ERA-Interim reanalysis data highlights hyperbolic trajectories, and these trajectories are Lagrangian objects that are the kinematic mechanism for the observed filamentation phenomena. Our analysis shows that the breaking and splitting of the polar vortex is justified in our model by the sudden growth of a planetary wave and the decay of the axisymmetric flow.

1 Introduction

The availability of high-resolution and high-quality reanalysis data sets provides us with a powerful tool for obtaining a detailed view of the space–time evolution of the stratospheric polar night vortex (SPV), which has implications for the geophysical fluid dynamics of the entire Earth. The complexity of such a detailed view, however, makes it difficult to extract the physical mechanisms underlying notable transport features in the observed behaviour. The goal of this work is to gain new insights into the fundamental mechanisms responsible for complex fluid parcel evolution, since these lie at the heart of our understanding of the dynamics and chemistry of the stratosphere. To this end, we extract, directly from the data, a simple model with stripped-down dynamics in order to directly probe, in a controlled and systematic manner, the physical mechanisms responsible for the key observed transport features of the SPV. Models of this kind, termed "kinematic models", have provided a simple approach for studying Lagrangian transport and exchange associated with flow structures such as meandering jets and travelling waves (Bower, 1991; Samelson, 1992; Malhotra and Wiggins, 1998; Samelson and Wiggins, 2006). Other works have used analytical kinematic models to illustrate phenomena in planetary atmospheres (e.g. Rypina et al., 2007; Morales-Juberías et al., 2015). In the present paper, we focus on SPV transport processes associated with filamentation and vortex breaking, of which the dynamical structure is not fully understood.

The importance of an increased understanding of the SPV was dramatically demonstrated by the intense research effort that followed the discovery of the "Antarctic ozone hole" phenomenon in the 1970s (Chubachi, 1984; Farman et al., 1985; Solomon, 1988). In the following decades, during which monitoring of ozone in atmospheric columns above Antarctica showed little interannual variability, in situ measurements corroborated by satellite data revealed that ozone was systematically decreasing in the Antarctic lower strato-sphere during the southern spring season. Whilst this was immediately associated with the simultaneous increase in at-mospheric pollution by anthropogenic activities, several key questions arose (Solomon, 1999):

1. Why does this occur over Antarctica and not over the Arctic (since pollution sources are stronger in the North-ern than in the Southern Hemisphere)?

2. Why does this occur in the spring season?

3. Will ozone depletion extend worldwide?

The research demonstrated that, indeed, increased atmo-spheric pollution was to be blamed for the ozone depletion and identified the participating substances and special mech-anisms. The research also demonstrated that the unique at-mospheric conditions above Antarctica were responsible for the geographic preference for ozone destruction. In particu-lar, it was shown that the strong circumpolar and westerly SPV characteristic of the southern winter and spring strato-sphere contributes to the isolation of the cold polar region, setting up a favourable environment for the special chem-istry to act. The new knowledge led to the formulation of in-ternational agreements that resulted in a negative answer to question (3) above. The analysis of transport of fluid parcels outside the region isolated by the SPV also showed strong stirring and mixing of the flow. In this "surf zone" (McIn-tyre and Palmer, 1984), air parcels can travel long distances away from the SPV in an environment where contours of long-lived tracers, such as potential vorticity, can stretch and form complex patterns. In this region, Rossby wave breaking (RWB) is associated with irreversible deformation that pulls material filaments of the outer edge of the SPV and enhances mixing with the exterior flow (McIntyre and Palmer, 1983, 1984, 1985). Such a process makes the SPV edge a barrier to horizontal transport of air parcels (Juckes and McIntyre, 1987) while continuously eroding and regenerating the SPV edge by filamentation (Bowman, 1993). Polvani and Plumb (1992) and Nakamura and Plumb (1994) examined, in an ide-alized setting, the way in which Rossby waves break and eject SPV material outward. The latter conceived a similar setting in which Rossby waves also break inward.

Dynamical systems theory provides valuable insights into the transport processes described in the previous paragraph. Tools of the theory include the geometrical structures, re-ferred to as hyperbolic trajectories (HTs), their stable and unstable manifolds, and their intersection in homoclinic and heteroclinic trajectories that provide the theoretical and com-putational basis for describing the filamentation process. A challenge in the application of these concepts to realistic geo-physical flows is that while the structures mentioned are de-fined for infinite-time autonomous or periodic systems, geo-physical flows are typically defined as finite-time data sets and are not periodic. Mancho et al. (2006b) addressed this

challenge for realistic ocean flows by identifying special hy-perbolic trajectories in the finite data set, called distinguished hyperbolic trajectories (DHTs), and by computing stable and unstable manifolds as curves advected by the velocity field. A pioneering effort for identifying HTs for the stratosphere was due to Bowman (1993). McIntyre and Palmer (1983), Bowman (1996) and de la Cámara et al. (2013) suggested that HTs are responsible for the cat-eye structures associated with planetary wave breaking at the critical levels, i.e. where the wave phase speed matches the background velocity (Stewart-son, 1977; Warn and Warn, 1978). HTs are at the locations where the cat-eye structures meet. Perturbation of the cat eyes results in irreversible deformation of material contours, signifying Rossby wave breaking. de la Cámara et al. (2013) and Guha et al. (2016) identified HTs both within and out-side the SPV, thus suggesting that Rossby wave breaking can occur in either of those regions. The former authors worked with reanalysis data, while Guha et al. (2016) used a dynam-ical model based on the shallow-water equations in which the perturbing waves are produced in a controlled manner. Therefore, HTs are essential features for tracer mixing both outside and inside the vortex, and for occasional air crossings of the vortex edge.

We focus on the SPV behaviour during the major strato-spheric sudden warming that occurred in the southern strato-sphere during September 2002. In this unusual event, the SPV broke down in the middle stratosphere (Mechoso et al., 1988; Varotsos, 2002, 2003, 2004; Allen et al., 2003; Konopka et al., 2005; Esler and Scott, 2005; Manney et al., 2006; Charlton et al., 2006; Taguchi, 2014). We begin by identifying key Lagrangian features of the flow in reanaly-sis data fields. Next, we build a kinematic model of the event that emulates the behaviour of planetary waves observed in the data. We show that our model produces strikingly sim-ilar transport features to those found in the reanalysis data, confirming the key role played by the HTs during vortex fil-amentation and breakdown.

The structure of the paper is as follows. Section 2 de-scribes the data and methods we used. Section 3 describes the planetary waves in the reanalysis data in the year 2002 in the stratosphere at selected pressure levels (10 hPa). We relate these to filamentation phenomena and the polar vor-tex breakdown that occurred in that year. Section 4 repro-duces the findings obtained with our analytical kinematical model, confirming the role played by the HTs in the 2002 vortex filamentation and breakdown. Section 5 discusses the consistency of the kinematic model as representative of at-mospheric flows that conserve potential vorticity. Finally, in Sect. 6, we present the conclusions.

2 Data and methods

2.1 ERA-Interim reanalysis data

To achieve a realistic representation of the atmospheric transport processes, it is crucial to use a reliable and high-quality data set. We use in this work the ERA-Interim reanalysis data set produced by a weather forecast assimilation system developed by the European Centre for Medium-Range Weather Forecasts (ECMWF; Simmons et al., 2007). de la Cámara et al. (2013) obtained encouraging results on the suitability of the ERA-Interim data set for Lagrangian studies of stratospheric motions in their comparison of parcel trajectories on the 475 K isentropic surface (around 20 km) using this data set and the trajectories of super-pressure balloons released from Antarctica by the VORCORE project during the spring of 2005 (Rabier et al., 2010).

The ERA-Interim data set that we selected for this study is available four times daily (00:00, 06:00, 12:00 and 18:00 UTC), with a horizontal resolution of $1° \times 1°$ in longitude and latitude and 60σ levels in the vertical from 1000 to 0.1 hPa. The data cover the period from 1979 to the present day (Dee et al., 2011) and they can be downloaded from http://apps.ecmwf.int/datasets/data/interim-full-daily/levtype=sfc/. In particular, we will use the data for the geopotential height on surfaces of constant pressure and wind fields on isentropic surfaces for the period August–September 2002.

The geopotential height Z on constant pressure surfaces p is defined as the normalization to $g_0 = 9.80665 \, \mathrm{ms^{-2}}$ (standard gravity at mean sea level) of the gravitational potential energy per unit mass at an elevation s (over the Earth's surface) and has the form

$$Z(\lambda, \phi, p, t) = \frac{1}{g_0} \int_0^{s(p,t)} g(\lambda, \phi, z) \, \mathrm{d}z \,, \tag{1}$$

where g is the acceleration due to gravity, λ is longitude, ϕ is latitude and z is the geometric height (Holton, 2004). In the quasi-geostrophic approximation, the geopotential height is proportional to the streamfunction of the geostrophic flow (Holton, 2004).

For the analysis of planetary waves, we apply a zonal Fourier decomposition to the geopotential height field on the 10 hPa pressure level (approximately 850 K potential temperature). The zonal wave decomposition yields

$$Z = \mathcal{Z}_0(\phi, p, t) + \sum_{k=1}^{\infty} \mathcal{Z}_k(\lambda, \phi, p, t) \,. \tag{2}$$

The mean flow is defined as

$$\mathcal{Z}_0(\phi, p, t) = \frac{1}{2\pi} \int_0^{2\pi} Z(\lambda, \phi, p, t) \, d\lambda \,, \tag{3}$$

and the different modes $\mathcal{Z}_k$ with wave number $k \geq 1$ have the sinusoidal description:

$$\mathcal{Z}_k(\lambda, \phi, p, t) = \mathcal{B}_k(\phi, t) \cos(k\lambda + \varphi_k(\phi, p, t)), \tag{4}$$

where $\lambda \in [0, 2\pi)$ is longitude, $\phi \in [-\pi/2, \pi/2]$ is latitude, $\mathcal{B}_k$ is the amplitude of the wave and φ_k its phase. During the warming event that occurred in the southern stratosphere during September 2002, the flow was dominated by the contributions of the mean flow and the two longest planetary waves ($\mathcal{Z}_1$ and $\mathcal{Z}_2$; Krüger et al., 2005)

2.2 Lagrangian descriptors

Dynamical systems theory provides a qualitative description of the evolution of particle trajectories by means of geometrical objects that partition the phase space (the atmosphere in our case) into regions in which the system shows distinct dynamical behaviours. These geometrical structures act as material barriers to fluid parcels and are closely related to flow regions known as hyperbolic, where rapid contraction and expansion takes place. Several Lagrangian techniques have been developed in order to detect such structures in geophysical fluids. This is challenging because, while classical dynamical systems theory is defined for infinite-time autonomous or periodic systems, in geophysical contexts, the velocity fields are generally time dependent, aperiodic in time and defined over a finite discrete space–time domain. Among others, techniques developed are finite-size Lyapunov exponents (FSLEs) (Aurell et al., 1997) and finite-time Lyapunov exponents (FTLEs) (Haller, 2000; Haller and Yuan, 2000; Haller, 2001; Shadden et al., 2005). Other techniques include DHTs (Ide et al., 2002; Ju et al., 2003) and the direct calculation of manifolds as material surfaces (Mancho et al., 2003, 2004, 2006b), the geodesic theory of Lagrangian coherent structures (LCS) (Haller and Beron-Vera, 2012) and the variational theory of LCS (Farazmand and Haller, 2012), etc. Our choice in this work will be the use of the Lagrangian descriptor (LD) function M introduced by Madrid and Mancho (2009) and Mendoza and Mancho (2010). The function M has been applied in a variety of geophysical contexts. For example, in the ocean, it has been used to analyse the structure of the Kuroshio current (Mendoza and Mancho, 2012), to discuss the performance of different oceanic data sets (Mendoza et al., 2014), to analyse and develop search and rescue strategies at sea (García-Garrido et al., 2015) and to efficiently manage in real time the environmental impact of marine oil spills (García-Garrido et al., 2016). In the field of atmospheric sciences, M has been used to study transport processes across the southern SPV and RWB de la Cámara et al., 2012, 2013; Smith and McDonald, 2014; Guha et al., 2016 and to investigate the Northern Hemisphere major stratospheric final warming in 2016 (Manney and Lawrence, 2016).

The dynamical system that governs the atmospheric flow is given by

$$\dot{x} = v(x(t), t) \ , \ x(t_0) = x_0 \ , \tag{5}$$

where $x(t; x_0)$ represents the trajectory of a parcel that, at time t_0, is at position x_0, and v is the wind velocity field. Since our interest is in the timescale of stratospheric sudden warmings (~ 10 days), we can assume to a good approximation that the fluid parcels evolve adiabatically. Therefore, trajectories are constrained to surfaces of constant specific potential temperature (isentropic surfaces). We will concentrate on the 850 K surface, which is in the middle stratosphere and approximately corresponds to the 10 hPa level. In Sect. 3, we expand on the reasons for this choice.

To compute fluid parcels trajectories, it is necessary to integrate Eq. (5). As the velocity field is provided on a discrete spatiotemporal grid, the first issue to deal with is that of interpolation. We apply bicubic interpolation in space and third-order Lagrangian polynomials in time (see Mancho et al., 2006a for details). Moreover, for the time evolution, we have used an adaptive Cash–Karp Runge–Kutta method. It is important to remark that, as done in de la Cámara et al. (2012) for the computation of particle trajectories, we use Cartesian coordinates in order to avoid the singularity problem arising at the poles from the description of the Earth's system in spherical coordinates. For our Lagrangian diagnostic, we use the M function defined as follows:

$$M(x_0, t_0, \tau) = \int_{t_0-\tau}^{t_0+\tau} \|v(x(t; x_0), t)\| \, dt, \tag{6}$$

where $\| \cdot \|$ stands for the modulus of the velocity vector. At a given time t_0, the function $M(x_0, t_0, \tau)$ measures the arc length traced by the trajectory starting at $x_0 = x(t_0)$ as it evolves forward and backward in time for a time interval τ. Sharp changes of M values (what we call singular features of M) occur for sufficiently large τ, for very close initial conditions and highlight stable and unstable manifolds.

Mendoza and Mancho (2010, 2012) have performed systematic numerical computations of invariant manifolds and found that they are aligned with singular features of M. They also provide examples in geophysical flows where manifolds are defined in a constructive way. Invariant manifolds are mathematical objects classically defined for infinite time intervals. The unstable (stable) manifold of a hyperbolic fixed point or periodic trajectory is formed by the set of trajectories that in negative (positive) infinity time approach these special trajectories. In geophysical contexts, this definition is not realizable, because only finite-time aperiodic data sets are possible. Nevertheless, manifolds can still be defined constructively with the following procedure. At the beginning time, these curves are approximated by segments with short length, aligned with the stable and unstable subspaces of the DHT identified with algorithms described in Ide et al. (2002) and

Madrid and Mancho (2009). This starting step aims to build a finite-time version of the asymptotic property of manifolds. Next, segments are advected forward and backward in time by the velocity field. Due to the large expansion and contraction rates in the neighbourhood of the DHT, the curves grow rapidly in forward and backward time and specific issues are addressed by the procedure described in Mancho et al. (2003) and Mancho et al. (2004). The procedure provides curves, manifolds, that by construction are barriers to transport in geophysical flows. In this way, since manifolds are aligned with singular features of M, the latter belong to invariant curves of system Eq. (5), and therefore their crossing points are indeed trajectories of system Eq. (5). The capability of LDs, in general, and M in particular, to reveal invariant manifolds was analysed in detail in Mancho et al. (2013). Lopesino et al. (2015) and Lopesino et al. (2017) have discussed, in discrete and continuous-time dynamical systems, respectively, a theoretical framework for some particular versions of LDs in specific examples.

The consistency between the output field of Eq. (6) and FTLE ridges has been discussed in some references (see Mendoza and Mancho, 2010; de la Cámara et al., 2012; Mancho et al., 2013). The integral expression in Eq. (6) can be split in two terms: one for forward time and the other for backward time integration. Explicit calculations discussed in Mancho et al. (2013) for a linear saddle, show that singular features of the first term are aligned with the stable manifolds while those for the backward time integration are aligned with the unstable manifolds. This is similar to what is obtained with FTLEs that highlight stable and unstable manifolds, respectively, for forward and backward time integration intervals. The fact that we choose to add both fields is advantageous for highlighting hyperbolic trajectories at the crossing points of the singular features.

As an example relevant to the case that motivates the present study, we show in Fig. 1 the evaluation of M over the Southern Hemisphere using $\tau = 15$ on the 850 K isentropic level for 5 August 2002. The representation shows a stereographic projection (see Snyder, 1987) in which the SPV is clearly visible by the bright yellow colour as well as the filamentation phenomena ejecting material both from the outer and inner parts of the jet. These filaments are related to the presence of hyperbolic trajectories highlighted in the figure. The fact that these saddle points of the LD field are hyperbolic trajectories of system Eq. (5) is numerically supported. To this end, de la Cámara et al. (2013) show that, for similar ERA-Interim fields, these points belong to the intersection of stable and unstable manifolds highlighted by the singular features of the field (see their Fig. 2). In what follows, all figures showing M were computed with $\tau = 15$. This choice of τ is made based on the fact that diabatic heating/cooling processes in the extratropical stratosphere generally have longer timescales than those of horizontal advection. Hence, air parcels move on two-dimensional isentropic surfaces to a good approximation (they stay within

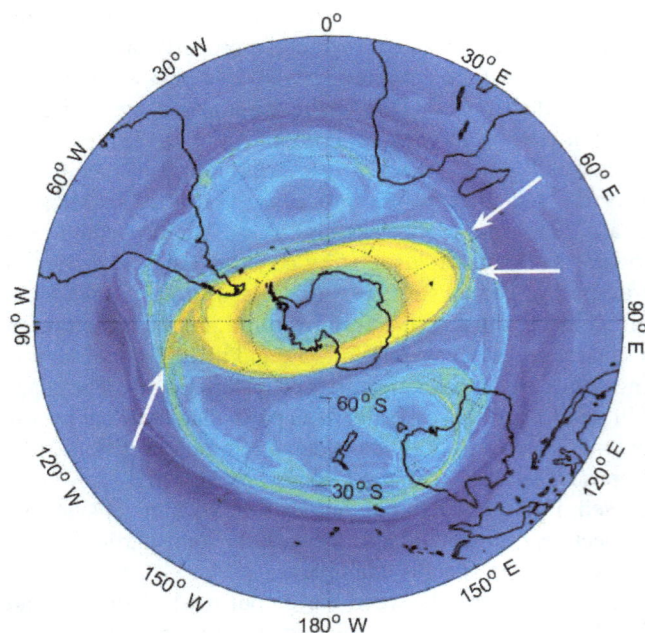

Figure 1. Stereographic projection of Lagrangian descriptors evaluated using $\tau = 15$ on the 850 K isentropic level for 5 August 2002 at 00:00:00 UTC. The SPV is clearly visible, as well as three hyperbolic trajectories (HTs) outside the vortex (marked with white arrows), two northeast and one southwest of it. Filamentation phenomena, which occur in the neighbourhood of HTs, are visible both inside and outside the vortex, where the outer filamentous structures play the role of eroding the jet barrier. Notice also the presence of two eddy-coherent structures over the South Atlantic and south of Australia.

850 K for 30 days) (Plumb, 2007). Moreover, diabatic heating rates in the Antarctic mid-stratosphere are on the order of $0.5\,\mathrm{K\,day}^{-1}$, although uncertainties in this magnitude remain large (Fueglistaler et al., 2009). During the time interval of our calculations of isentropic trajectories ($\tau = 15$ days, i.e. time period of 30 days), the material surface would experience an increase of potential temperature of around 15 K. Nevertheless, calculations of M using wind fields at 850 K and 700 K (not shown) produce qualitatively similar results. This suggests that horizontal motions of the parcels will be affected by similar geometric structures at those isentropic levels and that the isentropic approach is justified in our problem.

3 Data analysis

As we indicated in the previous section, in order to characterize the planetary waves that propagate in the stratosphere, we carry out a Fourier decomposition of the geopotential height. In Fig. 2, we show the axisymmetric mean flow together with waves 1 and 2 in the geopotential field for 22 September 2002 on the 10 hPa pressure surface. The time evolution of these

waves is described in the movies S1–S4 in the Supplement. Movies S1–S3 show components 0, 1 and 2 separately for the time period of interest, while S4 shows the superposition of these three waves. It is important to reiterate that, since the geopotential provides a good approximation of the streamfunction of the large-scale flow in the extratropical regions, its analysis will provide us with guidance on the building of the simple kinematic model presented in the next section.

On the 10 hPa pressure level, the winter SPV in the Southern Hemisphere can be broadly defined as a circumpolar westerly jet. Figure 3a and b illustrate the evolution of the circulation during August–September 2002. We can clearly see the gradual deceleration of the SPV and the abrupt change in direction from westerly to easterly velocities at high latitudes that occurred on 22 September. This was a unique major sudden stratospheric warming (SSW) in the southern stratosphere. Planetary waves in the southern stratosphere were very active during the period where the 2002 SSW developed. Figure 3c presents a time series of the ratio between the amplitudes of waves 1 and 2. Increased wave 1 amplitude results in a displacement of the SPV vortex from a circumpolar configuration, while increased wave 2 amplitude results in a stretching the SPV in one direction and contraction (or "pinching") in the orthogonal direction. According to Fig. 3c, the amplitude of wave 1 was generally larger than that of wave 2 during the entire period, confirming the major role of this wave. Finally, Fig. 3d displays the variations in time of the ridges of wave 1 and wave 2. Note that wave 1 is quasi-stationary, while wave 2 propagates eastward, as is typical in the southern stratosphere during early spring (Manney et al., 1991; Quintanar and Mechoso, 1995).

The contribution of these different waves to the evolution of the SPV and their transport implications are clearly observed in movie S5. A regime giving rise to the stretching of material lines and the appearance of hyperbolic regions and the associated filamentation processes is observed. These filamentous structures and HTs are clearly highlighted by the application of LDs to the wind fields, as shown in Figs. 1 and 4. Filamentation phenomena occur both inside and outside the vortex, where the outer filamentous structures play the role of eroding the jet material barrier. Also, the presence of HTs in the flow (see captions of Figs. 1 and 4) indicates regions subjected to intense deformation and mixing (see Ottino, 1989). We emphasize that HTs appear both inside and outside the SPV. Finally, the breakup of the SPV on 24 September 2002 depicted in Fig. 4b (see also movie S5) occurs when manifolds associated with an HT that forms within the SPV connect the interior and the exterior of the jet, allowing for the interchange of parcels through the barrier. The pinching of the SPV takes place off the pole because $\mathcal{Z}_1$ has large amplitudes in the days preceding the breakup. As we approach 24 September, $\mathcal{Z}_2$ becomes the same order as $\mathcal{Z}_0$, and the jet elongates and flattens. At this point, the mean flow reversal is crucial for completing the pinching process

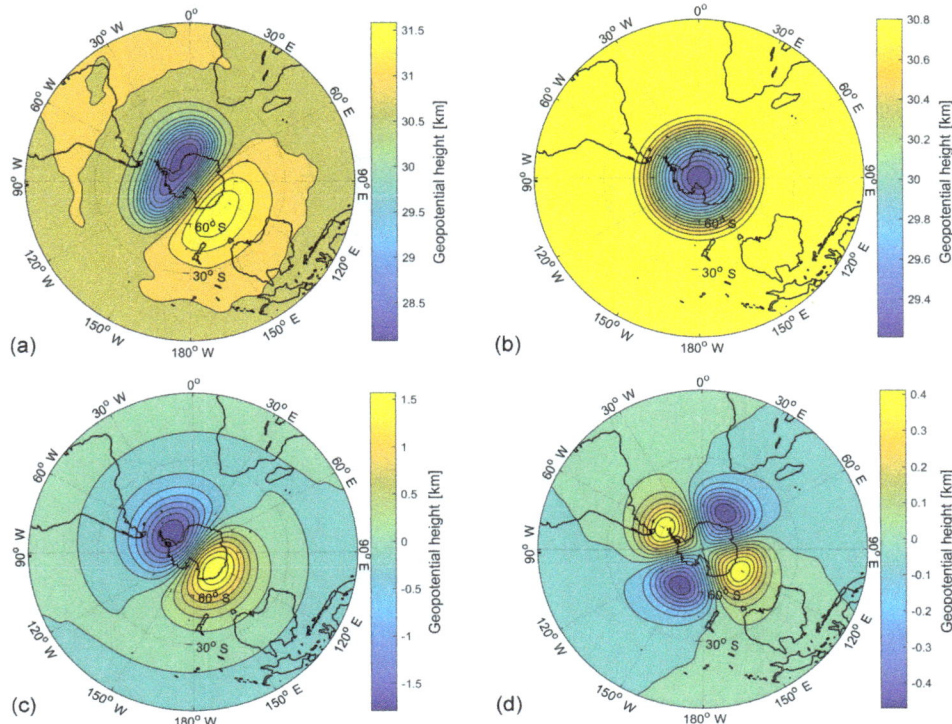

Figure 2. Stereographic projection of the geopotential height field and its Fourier decomposition for the 10 hPa pressure level on 22 September 2002 at 00:00:00 UTC: **(a)** geopotential height; **(b)** mean flow; **(c)** Fourier component $\mathcal{Z}_1$; **(d)** Fourier component $\mathcal{Z}_2$. Observe how the amplitude of the planetary wave with wave number 1 can be at least 3 times larger than that of wave number 2.

and the appearance of a HT in the interior of the vortex as this splits apart.

4 The kinematic model

Kinematic models have a long history in the geophysical fluid dynamics community. They allow for a detailed parametric study of the influence of identified flow structures on transport and exchange of fluid parcels. All early studies utilizing the dynamical systems approach for understanding Lagrangian transport and exchange associated with flow structures, such as meandering jets and travelling waves, have employed kinematic models (see Samelson and Wiggins, 2006).

Continuing in this spirit, we propose a kinematic model that allows us to identify, in a controlled fashion, the characteristics of the distinct propagating waves that are responsible for the different Lagrangian features observed in the SPV. Our kinematic model is inspired by the Fourier component decomposition of the geopotential extracted from the ERA-Interim data as discussed in the previous section. The analysis of data from August and September 2002 shows a mean axisymmetric flow, disturbed mainly by waves with planetary wave numbers 1 and 2 whose amplitudes and phase speeds vary in a time-dependent fashion. Therefore, we propose a kinematic model in the form of a streamfunction given by

$$\Psi = \varepsilon_0 \Psi_0 + \varepsilon_1 \Psi_1 + \varepsilon_2 \Psi_2 \,, \tag{7}$$

where $\varepsilon_0, \varepsilon_1, \varepsilon_1$ are the perturbation parameters, which we will refer to as amplitudes, and Ψ_i are the Fourier components, along the azimuthal direction with wave numbers $i = 0, 1, 2$ respectively, which we describe next.

We will work in a plane (x, y) that is the orthographic projection of the Southern Hemisphere onto the equatorial plane (Snyder, 1987). For simplicity, and in order to highlight the periodicity along the azimuthal direction, the components of the streamfunction are given in terms of polar coordinates satisfying $x = r \cos(\lambda)$ and $y = r \sin(\lambda)$, where the azimuthal direction λ is related to the geographical longitude and r is related to the geographical latitude.

The particular forms of Ψ_0, Ψ_1 and Ψ_2 are inspired by the Fourier decomposition of the geopotential field shown in Fig. 2 for the 10 hPa pressure level on 22 September 2002. Starting with the mean zonal velocity, we will assume a jet with the following expression:

$$v_\lambda = r(r - a)e^{-r}. \tag{8}$$

Therefore, $v_\lambda = 0$ only at $r = 0$ and $r = a$, and the velocity decreases exponentially away from the pole. Changing the values of a will allow us to consider variations in the position

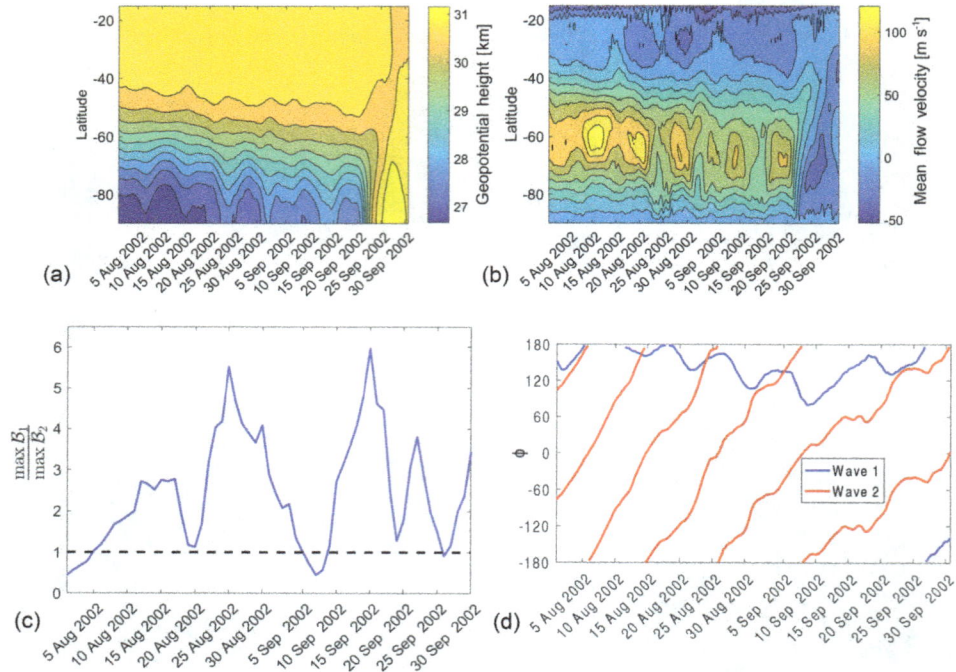

(a)   (b)   (c)   (d)

Figure 3. On the 10 hPa pressure level: **(a)** time evolution of the geopotential height corresponding to the mean flow. **(b)** Time evolution of the mean flow velocity. Notice the change in wind direction from westerly to easterly that takes place from 22 to 24 September 2002. **(c)** Time series of the ratio of the amplitudes of waves 1 and 2. **(d)** Hovmöller (time–latitude) showing the position of the ridges of waves 1 and 2 at latitude 60° S.

of the jet maxima. Integration with respect to r gives

$$\Psi_0 = e^{-r}(ar + a - r(r+2) - 2). \qquad (9)$$

The other streamfunction components are

$$\Psi_1 = -r^2 e^{-r^2} \sin(\lambda) \qquad (10)$$

and

$$\Psi_2 = (r/d)^2 e^{-r^2/d} \sin(2\lambda + \omega_2 t + \pi/4), \qquad (11)$$

where d and w_2 are also tunable constants, and the phase $\pi/4$ was added so that the relative positions of waves 1 and 2 at $t = 0$ resemble those in Fig. 2. Positive values of ω_2 correspond to clockwise rotation. Note that Eq. (11) can represent a wave that propagates in the azimuthal direction λ if w_2 is different than zero. Figure 5 shows the streamfunctions Eqs. (9), (10) and (11) in the horizontal plane for the particular set of parameters indicated in the corresponding caption. In the panels of Fig. 5 and the following, the centre represents the South Pole and the circular dashed line indicates the Equator. The similarity between Figs. 2 and 5 for the selected set of parameters is evident, taking into consideration that they correspond to stereographic and orthographic projections, respectively.

The velocity of fluid parcels in the Cartesian coordinates (x, y) is given by Hamilton's equations:

$$\frac{dx}{dt} = -\frac{\partial \Psi}{\partial y}, \quad \frac{dy}{dt} = \frac{\partial \Psi}{\partial x}. \qquad (12)$$

We take the amplitudes to be time dependent in order to emulate changes in magnitudes. Let us start with ε_0 constant and

$$\varepsilon_1 = \eta_1(1 + \sin(\mu t + \pi)), \quad \varepsilon_2 = \eta_2(1 + \sin(\mu t)). \qquad (13)$$

Here, η_1 and η_2 are constants. The time dependence of ε_1 and ε_2 allows us to analyse each wave either separately or together and their transient effect on the observed Lagrangian structures and therefore their transport implications. The time dependence in Eq. (13) is such that one amplitude decreases when the other increases, roughly allowing conservation of the total energy when both waves are present. In the simulations presented below, $\mu = 2\pi/10$.

We begin by considering the case of a mean flow with $a = 2$ and just wave 2 rotating at different speeds. Furthermore, $d = 1$ and $\eta_2 = 1$. Let us start with $\omega_2 = 0$, i.e. the stationary case. For this case, the dotted line in Fig. 6a shows the azimuthal velocity of the mean flow for $\varepsilon_0 = -2.5$, the dashed line is the azimuthal velocity of wave 2 at $\lambda = 0$, where the radial velocity cancels, the solid line is the total azimuthal velocity and the blue line is the wave phase speed. According to Fig. 6a, there are two points where the total velocity cancels, one being the origin. We can also easily see that there are additional fixed points at the r coordinate where the dotted and dashed curves intersect, but placed along the lines $\lambda = \pi/2, 3\pi/4$. This gives a total of five points in the

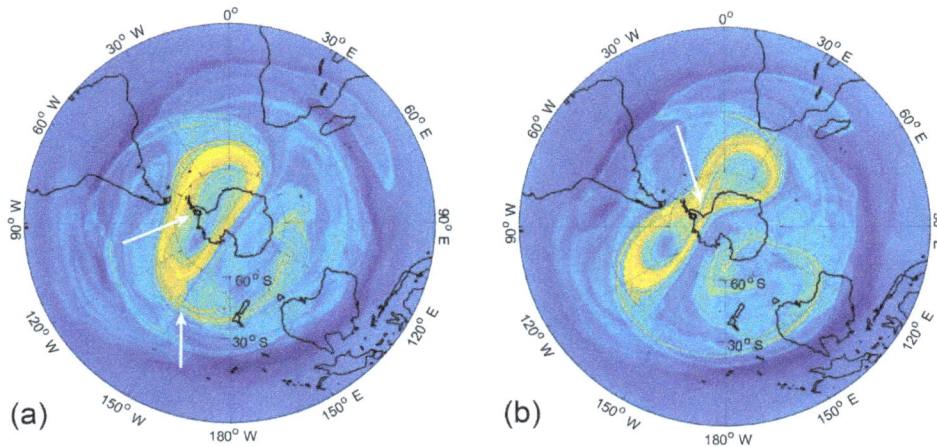

Figure 4. Stereographic projection of the M function calculated using $\tau = 15$ on the 850 K isentropic level for **(a)** 22 September 2002 at 00:00:00 UTC and **(b)** 24 September 2002 at 00:00:00 UTC. Filamentation phenomena and hyperbolic trajectories (marked with white arrows) are nicely captured in these simulations both in the interior and the exterior of the SPV. Observe how the vortex breakdown on 24 September occurs when, in the interior of the vortex, a HT allows the transport and mixing of parcels across the barrier.

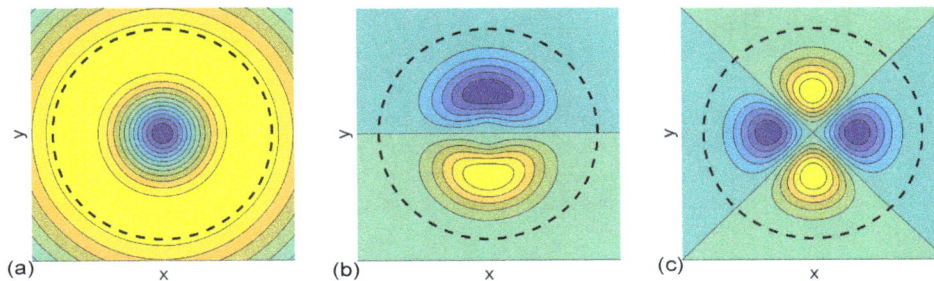

Figure 5. Representation of the three components of the streamfunction. Panel **(a)** indicates $\varepsilon_0 \Psi_0$ for $a = 2$ and $\varepsilon_0 < 0$; **(b)** Ψ_1; and **(c)** Ψ_2 for $d = 1$, $w_2 = 0$.

hemisphere. Figure 6b shows the M function for $\tau = 15$ evaluated on this stationary field at $t = 0$. The minima of M highlighting the five fixed points are evident. Moreover, we can see two hyperbolic points in the outer part of the vortex.

Next, we consider the case with the same parameters except for ω_2. Figure 6c shows how this picture changes when $\omega_2 = 0.1$, i.e. for the slow rotation rate of wave 2. The total azimuthal velocity of the wave, in this case, is given by the dashed line in Fig. 6a plus the phase velocity represented by the green line in the figure. If this total azimuthal velocity of the wave is added up to the mean flow, two points are found in which the total azimuthal velocity cancels. Additionally, for a slow rotating wave, similarly to the previous case, the total azimuthal velocity of the wave can still be equal to the zonal mean velocity at some points in the domain. Therefore, Fig. 6c is similar to Fig. 6b except for a rotation. However, for a fast rotation of wave 2 ($\omega_2 = 4\pi$; red line), the total azimuthal velocity of the wave will be larger than the zonal mean velocity at all points in the domain. In this case, the

pattern of M (Fig. 6d) is very different from the pattern in Fig. 6b showing no HTs.

Figure 7 displays the function M obtained from the kinematic model for the same mean flow of Fig. 6a and different parameters for waves 1 and 2. All the representations are for $t = 0$ and $\tau = 15$. Figure 7a is for the same case as Fig. 6b, except that the amplitude of wave 2 changes in time ($\eta_1 = 0, \eta_2 = 1$). Again, two HTs are visible in the external jet boundary along which filamentation occurs. Figure 7b corresponds to just wave number 1, changing amplitude in time ($\eta_1 = 1, \eta_2 = 0$). We can see one HT at the outer boundary of the jet where material of the vortex is being ejected. In these figures, transport processes producing filamentation-ejecting material have close connections to those present in Figs. 1 and 4a), which have been linked to Rossby wave breaking at midlatitudes (Guha et al., 2016). In Fig. 7c, the mean flow is perturbed by the not-rotating wave 2 of Fig. 7a and wave 1 of Fig. 7b ($\eta_1 = 1, \eta_2 = 1$). In Fig. 7d, the parameters are the same as Fig. 7c, except that wave 2 rotates ($\omega_2 = 2\pi/15$). The jet shape and filamentary struc-

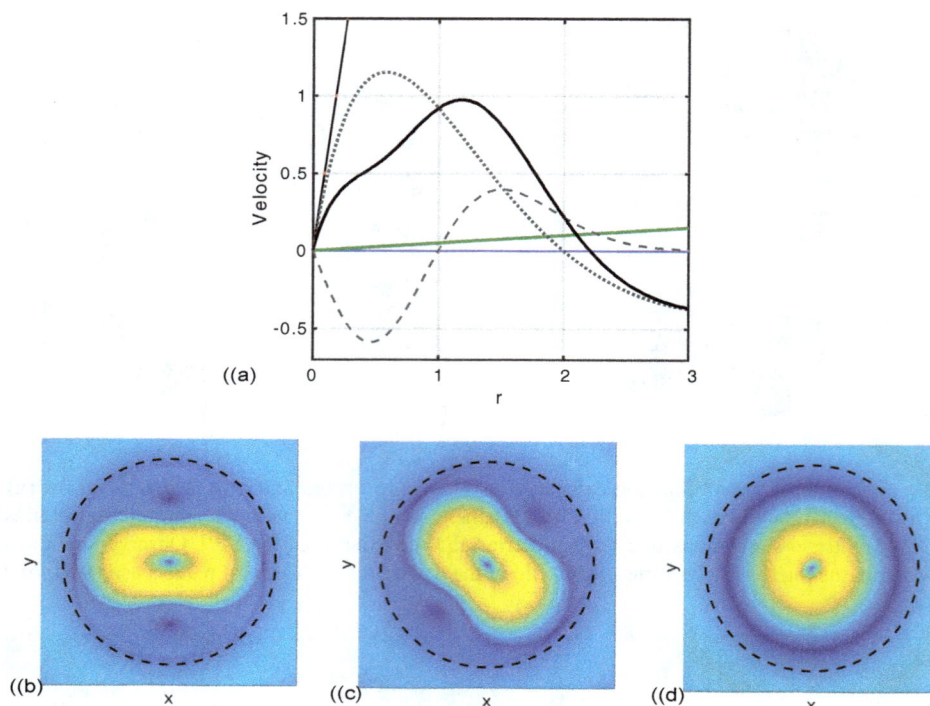

Figure 6. Some illustrative parameter choices for the kinematic model. **(a)** A representation of the mean flow azimuthal velocity (dotted line), the azimuthal velocity of wave 2 for the stationary case along $\lambda = 0$ (dashed line), the total azimuthal velocity along $\lambda = 0$ (solid line), the phase velocity for $\omega_2 = 0.1$ (green line) and the phase velocity for $\omega_2 = 4\pi$ (red line); **(b)** representation of the M function for a kinematic model considering a mean flow ($a = 2$) plus a stationary wave 2 ($d = \eta_2 = 1$); **(c)** the same as **(b)** for a rotating wave 2 with $\omega_2 = 0.1$; **(d)** the same as **(b)** for a rotating wave 2 with $\omega_2 = 4\pi$.

tures greatly resemble those present in the reanalysis data, as shown in Figs. 1 and 4a.

Figure 8 present a jet which in the interior is eroded by waves 2 and 1, respectively. To achieve such a configuration, free parameters are specifically tuned, including a zonal mean flow with negative velocities near the pole. In Fig. 8a, the mean flow obtained with parameters $\varepsilon_0 = 2.6$ and $a = 0.75$ is perturbed by just a travelling wave 2 ($\eta_1 = 0, \eta_2 = 1, \omega_2 = -4\pi/25$) with $d = 2$. Two filaments projecting material from the interior of the vortex are observed, and they are related to the presence of interior HTs. In Fig. 8b, the mean flow is obtained with the parameters $\varepsilon_0 = 2.5$ and $a = 0.5$. This mean flow is perturbed by just wave 1 with amplitude that varies in time ($\eta_1 = 1, \eta_2 = 0$). A protruding material filament from the interior of the vortex is observed, which is related to the presence of an interior HT. The interior filaments in these figures recover features that are identified as interior Rossby wave breaking phenomena in de la Cámara et al. (2013) and Guha et al. (2016), and are also visible from the reanalysis data, as shown in Figs. 1 and 4a.

Figure 4b shows the pinching of the SPV in the observations on 24 September 2002, which is before its breakup. In the kinematic model, this structure can be obtained with a strong Ψ_2 and a substantial contribution from Ψ_1 to have a

displacement from the pole. Movies S1, S2, S3 and S4 in the Supplement illustrate such structures. In order to reproduce the splitting, we do not need to consider the displacement, and thus we neglect mode 1 in what follows. Figure 9 shows a sequence of M patterns obtained with the amplitude of mean flow is given by

$$\varepsilon_0 = \eta_0(1 + \sin(\mu t + \pi)), \tag{14}$$

where $\eta_0 = -2.5$ and $\mu = 2\pi/10$, and a stationary wave 2 ($\omega_2 = 0$) with amplitude given by Eq. (13). Note that, in this way, the mean flow weakens as wave 2 strengthens, and vice versa. The parameters fit a streamfunction which at $t = 0$ coincides with that used in Fig. 7a. The development of an hyperbolic point at the pole in the observations (Fig. 4b) can be clearly seen in Fig. 9a. The two vortices have completely split at $t = 6$.

5 Kinematic models and conservation of potential vorticity

In this section, we discuss the connection between the kinematic model introduced in the previous section and a fundamental dynamical principle of geophysical fluids. Geophysical flows that are adiabatic and frictionless conserve the po-

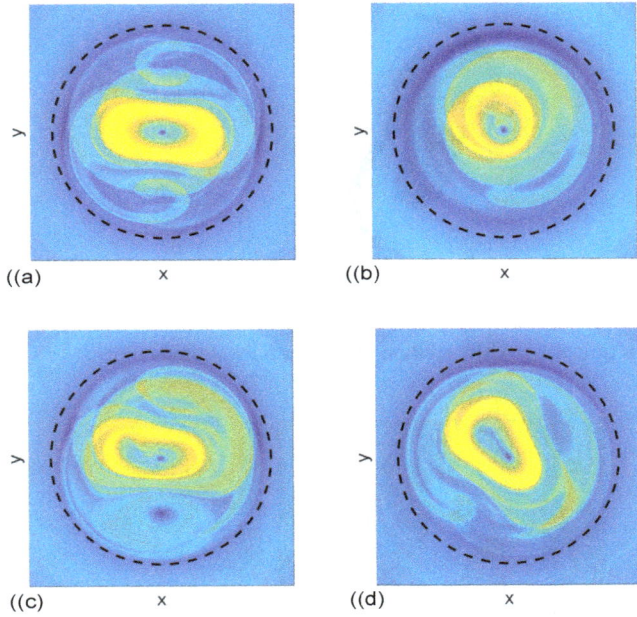

Figure 7. Lagrangian patterns obtained at $t = 0$ for $\tau = 15$ and different parameter settings in the kinematic model. **(a)** Fourier components Ψ_0 and Ψ_2 with the latter adjusted to perturb the vortex in its outer part; **(b)** Fourier components Ψ_0 and Ψ_1 with the latter adjusted to perturb the vortex in its outer part; **(c)** the model keeps Ψ_0, Ψ_1 and Ψ_2; **(d)** the model keeps Ψ_0, Ψ_1 and Ψ_2 with parameters adjusted differently to **(c)**.

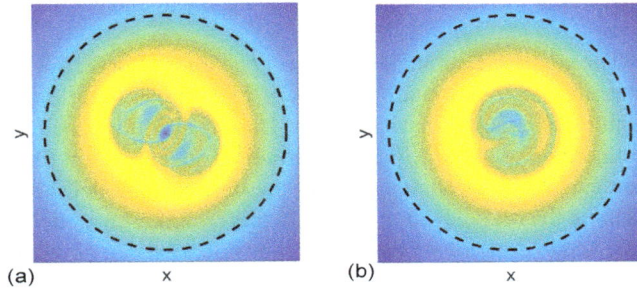

Figure 8. Lagrangian patterns obtained at $t = 0$ for $\tau = 15$ and different parameter settings in the kinematic model. In panel **(a)**, the model keeps Ψ_0 and Ψ_2 adjusted to perturb the vortex in its interior part; in panel **(b)**, the model keeps Ψ_0 and Ψ_1 adjusted to perturb the vortex in its interior part.

tential vorticity Q along trajectories. Conservation of Q is expressed as follows:

$$\frac{dQ}{dt} = 0. \tag{15}$$

Here, d/dt stands for the material derivative. A natural question here is to discuss whether the proposed kinematic model conserves Q. Let us assume that our setting is described by the quasi-geostrophic motion of simple vortices in a shallow-water system (see Polvani and Plumb, 1992; Nakamura and

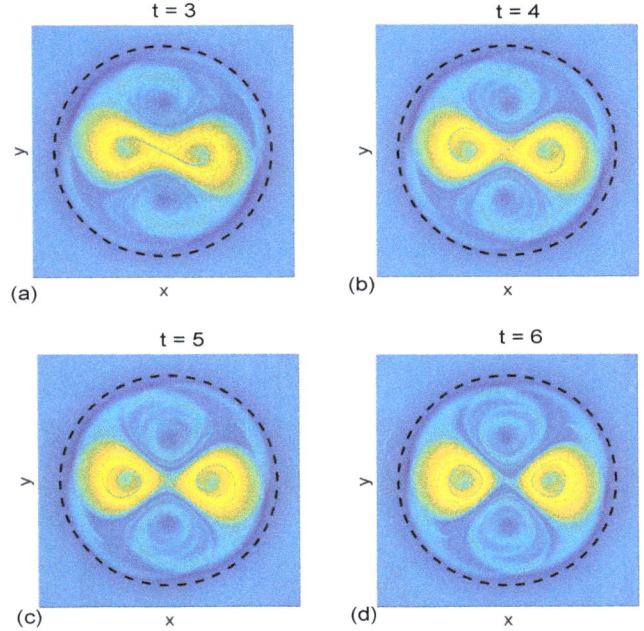

Figure 9. Evolution of the Lagrangian template for the case in which the mean flow decreases and wave 2 increases. The sequence reproduces many of the Lagrangian features observed in the splitting event that occurred at the end of September 2002 (see movie S5). **(a)** $t = 3$; **(b)** $t = 4$; **(c)** $t = 5$; **(d)** $t = 6$.

Plumb, 1994) in which Q is given by

$$Q = f_0 + \nabla^2 \Psi - \gamma^2 \Psi + f_0 \frac{h}{D}. \tag{16}$$

Here, f_0 is a constant related to the rotation rate, D is the mean depth of the shallow-water system, $D - h$ is the total depth, h is the bottom boundary of the fluid layer, which is small when compared to D, and $\gamma = f_0/\sqrt{g_0 D}$, where g_0 is the gravity constant. Ψ is the geostrophic streamfunction for the horizontal velocity field, in our case given by Eq. (7), with parameters corresponding to those of Fig. 7d, i.e. $\varepsilon_0 = -2.5$, $\eta_1 = 1$, $\eta_2 = 1$, $a = 2$, $d = 1$ and $\omega_2 = 2\pi/15$.

We assume that at the initial time, $t = 0$, the vorticity Q consists of a circular patch with constant vorticity Q_0 inside and vorticity Q_1 outside. At a later time, $t = 2$, the vorticity distribution that preserves Eq. (15) is obtained by advecting the circular contour at $t = 0$, according to motion Eq. (12), with algorithms described in Mancho et al. (2004). Figure 10 summarizes the evolution of the vorticity.

In order to preserve Eq. (16) from time $t = 0$ to time $t = 2$, and assuming the barotropic approach in which $\gamma = 0$, h is solved from Eq. (16) as

$$\frac{h}{D} = \frac{Q}{f_0} - \frac{\nabla^2 \Psi}{f_0} - 1. \tag{17}$$

Figure 11 shows the evolution of the function h/D between $t = 0$ and $t = 2$. In particular, the figure shows results for

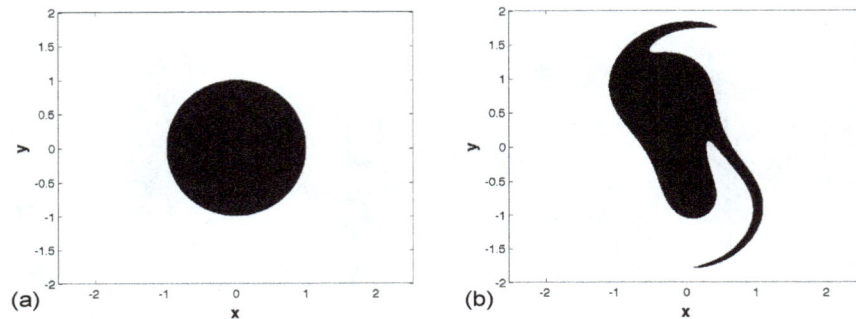

Figure 10. Evolution of a vorticity patch. **(a)** Initial vorticity distribution at time $t = 0$; **(b)** evolution of the vorticity at time $t = 2$.

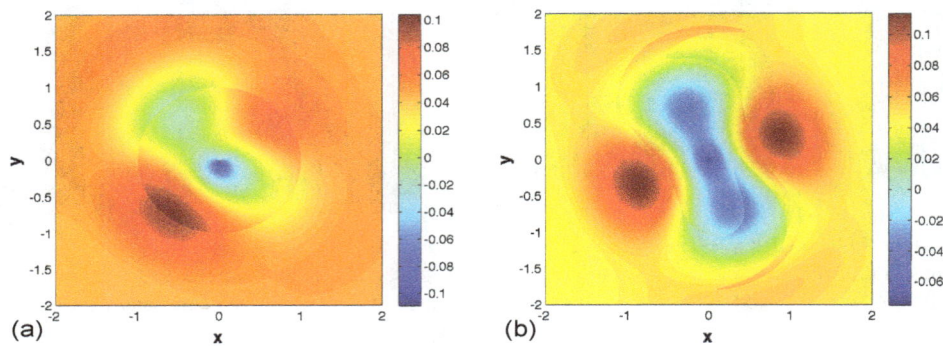

Figure 11. Evolution of the scaled lower boundary h. **(a)** The function h/D at time $t = 0$; **(b)** evolution of h/D at time $t = 2$.

$Q_0 = 2$, $Q_1 = 1.8$ and $f_0 = 20$. We note that this calculation could have been repeated for any initial distribution of Q defined as a piecewise constant function. The lower boundary h is thus a time-dependent function adjusted to preserve the conservation of the potential vorticity. Without this forcing, kinematic models would not preserve potential vorticity.

6   Conclusions

In this work, we propose a simple kinematic model for studying transport phenomena in the Antarctic polar vortex. We are interested in gaining insights into the processes which carry material outward from the vortex structure and inward to the vortex structure.

The construction of the kinematic model is realized by analysing geopotential height data produced by the ECMWF. In particular, our focus is on the stratospheric sudden warming event that took place in 2002, producing the pinching and then breaking of the stratospheric polar vortex. We identify the prevalent Fourier components during this period, which consist of a mean axisymmetric flow and waves with wave numbers 1 and 2. The kinematic model is based on analytical expressions which recover the spatial structures of these representative Fourier components. The model can be controlled so that waves with wave numbers 1 and 2 can be switched on and off independently. We are also able to adjust the relative position of the waves with respect to the mean axisymmetric flow.

The study of Lagrangian transport phenomena in the ERA-Interim reanalysis data by means of Lagrangian descriptors highlights hyperbolic trajectories. These trajectories are Lagrangian objects "seeding" the observed filamentation phenomena. The Lagrangian study of the kinematic model sheds light on the role played by waves in this regard. The model is adjusted to a stationary case which considers a mean flow and a stationary wave 2 that perturbs the mean flow in its outer part, producing hyperbolic trajectories. For the stationary case, hyperbolic trajectories are easily identified. This framework is modified by transforming it to a time-dependent problem by making the wave phase speed different from zero or by introducing time-dependent amplitudes. This allows to relate the time-dependent structures with those easily identified in the stationary case. The setting is repeated with wave 1 and both wave 1 and wave 2 together. The joint presence of these waves produces complex Lagrangian patterns remarkably similar to those observed from the analysis of the complex reanalysis data and confirms the findings discussed by Guha et al. (2016). Further adjustment of some model parameters is able to produce erosion by means of filaments just in the interior part of the flow. Finally, we point out that our analysis shows that the breaking and splitting of the polar vortex is justified in our model by the sudden growth of wave 2 and the decay of the axisymmetric flow.

Competing interests. The authors declare that they have no conflict of interest.

Acknowledgements. V. J. García-Garrido, J. Curbelo and A. M. Mancho are supported by MINECO grant MTM2014-56392-R. The research of C. R. Mechoso is supported by the U.S. NSF grant AGS-1245069. The research of S. Wiggins is supported by ONR grant no. N00014- 01-1-0769. We also acknowledge support from ONR grant no. N00014-16-1-2492.

Edited by: C. López

References

Allen, D. R., Bevilacqua, R. M., Nedoluha, G. E., Randall, C. E., and Manney, G. L.: Unusual stratospheric transport and mixing during the 2002 Antarctic winter, Geophys. Res. Lett., 30, 1599, https://doi.org/10.1029/2003GL017117, 2003.

Aurell, E., Boffeta, G., Crisanti, A., Paladin, G., and Vulpiani, A.: Predictability in the large: An extension of the concept of Lyapunov exponent, J. Phys. A, 30, 1–26, 1997.

Bower, A. S.: A simple kinematic mechanism for mixing fluid parcels across a menadering jet, J. Phys. Oceanogr., 21, 173–180, 1991.

Bowman, K. P.: Large-scale isentropic mixing properties of the Antarctic polar vortex from analyzed winds, J. Geophys. Res., 98, 23013–23027, https://doi.org/10.1029/93JD02599, 1993.

Bowman, K. P.: Rossby Wave Phase Speeds and Mixing Barriers in the Stratosphere. Part I: Observations, J. Atmos. Sci., 53, 905–916, 1996.

Charlton, A. J., O'Neill, A., Lahoz, W. A., and Berrisford, P.: The Splitting of the Stratospheric Polar Vortex in the Southern Hemisphere, September 2002: Dynamical Evolution, J. Atmos. Sci., 66, 590–602, 2006.

Chubachi, S.: Preliminary result of ozone observations at Syowa station from February 1982 to January 1983, Mem. Nat. Inst. Polar Res., 34, 13–19, 1984.

Dee, D. P., Uppala, S. M., Simmons, A. J., Berrisford, P., Poli, P., Kobayashi, S., Andrae, U., Balmaseda, M. A., Balsamo, G., Bauer, P., Bechtold, P., Beljaars, A. C. M., van de Berg, L., Bidlot, J., Bormann, N., Delsol, C., Dragani, R., Fuentes, M., Geer, A. J., Haimberger, L., Healy, S. B., Hersbach, H., Hólm, E. V., Isaksen, L., Kållberg, P., Köhler, M., Matricardi, M., McNally, A. P., Monge-Sanz, B. M., Morcrette, J.-J., Park, B.-K., Peubey, C., de Rosnay, P., Tavolato, C., Thépaut, J.-N., and Vitart, F.: The ERA-Interim reanalysis: configuration and performance of the data assimilation system, Q. J. Roy. Meteor. Soc. 137, 553–597, https://doi.org/10.1002/qj.828, 2011.

de la Cámara, A., Mancho, A. M., Ide, K., Serrano, E., and Mechoso, C.: Routes of transport across the Antarctic polar vortex in the southern spring., J. Atmos. Sci., 69, 753–767, 2012.

de la Cámara, A., Mechoso, R., Mancho, A. M., Serrano, E., and Ide., K.: Isentropic transport within the Antarctic polar night vortex: Rossby wave breaking evidence and Lagrangian structures., J. Atmos. Sci., 70, 2982–3001, 2013.

Esler, J. G. and Scott, R. K.: Excitation of Transient Rossby Waves on the Stratospheric Polar Vortex and the Barotropic Sudden Warming, Phys. Rev. E, 62, 3661–3681, 2005.

Farazmand, M. and Haller, G.: Computing Lagrangian Coherent Structures from variational LCS theory, Chaos, 22, 013128, https://doi.org/10.1063/1.3690153, 2012.

Farman, J. D., Gardiner, B. G., and Shanklin, J. D.: Large losses of total ozone in Antarctica reveal seasonal ClOx/NOx interaction, Nature, 315, 207–210, 1985.

Fueglistaler, S., Legras, B., Beljaars, A., Morcrette, J.-J., Simmons, A., Tompkins, A. M., and Uppala, S.: The diabatic heat budget of the upper troposphere and lower/mid stratosphere in ECMWF reanalyses, Q. J. Roy. Meteor. Soci., 135, 21–37, https://doi.org/10.1002/qj.361, 2009.

García-Garrido, V. J., Mancho, A. M., Wiggins, S., and Mendoza, C.: A dynamical systems approach to the surface search for debris associated with the disappearance of flight MH370, Nonlin. Processes Geophys., 22, 701–712, https://doi.org/10.5194/npg-22-701-2015, 2015.

García-Garrido, V. J., Ramos, A., Mancho, A. M., Coca, J., and Wiggins, S.: A dynamical systems perspective for a real-time response to a marine oil spill, Mar. Pollut. Bull., 112, 201–210, https://doi.org/10.1016/j.marpolbul.2016.08.018, 2016.

Guha, A., Mechoso, C. R., Konor, C. S., and Heikes, R. P.: Modeling Rossby Wave Breaking in the Southern Spring Stratosphere, J. Atmos. Sci., 73, 393–406, 2016.

Haller, G.: Finding finite-time invariant manifolds in two-dimensional velocity fields, Chaos, 10, 99–108, 2000.

Haller, G.: Distinguished material surfaces and coherent structures in three-dimensional fluid flows, Physica D, 149, 248–277, 2001.

Haller, G. and Beron-Vera, F. J.: Geodesic theory of transport barriers in two-dimensional flows, Physica D, 241, 1680–1702, 2012.

Haller, G. and Yuan, G.: Lagrangian coherent structures and mixing in two-dimensional turbulence, Physica D, 147, 352–370, 2000.

Holton, J. R.: An Introduction to Dynamic Meteorology, Elsevier Academic Press, 2004.

Ide, K., Small, D., and Wiggins, S.: Distinguished hyperbolic trajectories in time-dependent fluid flows: analytical and computational approach for velocity fields defined as data sets, Nonlin. Processes Geophys., 9, 237–263, https://doi.org/10.5194/npg-9-237-2002, 2002.

Ju, N., Small, D., and Wiggins, S.: Existence and computation of hyperbolic trajectories of aperiodically time-dependent vector fields and their approximations, Int. J. Bif. Chaos, 13, 1449–1457, 2003.

Juckes, M. N. and McIntyre, M. E.: A high-resolution one-layer model of breaking planetary waves in the stratosphere, Nature, 328, 590–596, 1987.

Konopka, O., Groo, J.-U., Hoppel, K. W., Steinhorst, H.-M., and Muller, R.: Mixing and Chemical Ozone Loss during and after the Antarctic Polar Vortex Major Warming in September 2002, J. Atmos. Sci., 62, 848–859, 2005.

Krüger, K., Naujokat, B., and Labitzke, K.: The Unusual Midwinter Warming in the Southern Hemisphere Stratosphere 2002: A comparison to Northern Hemisphere Phenomena., J. Atmos. Sci., 62, 603–613, 2005.

Lopesino, C., Balibrea-Iniesta, F., Wiggins, S., and Mancho, A. M.: Lagrangian Descriptors for Two Dimensional, Area Preserving Autonomous and Nonautonomous Maps, Commun. Nonlinear Sci., 27, 40–51, 2015.

Lopesino, C., Balibrea-Iniesta, F., García-Garrido, V. J., Wiggins, S., and Mancho, A. M.: A theoretical framework for lagrangian descriptors, Int. J. Bifurcat. Chaos, 27, 1730001, https://doi.org/10.1142/S0218127417300014, 2017.

Madrid, J. A. J. and Mancho, A. M.: Distinguished trajectories in time dependent vector fields, Chaos, 19, 013111, https://doi.org/10.1063/1.3056050, 2009.

Malhotra, N. and Wiggins, S.: Geometric Structures, Lobe Dynamics, and Lagrangian Transport in Flows with Aperiodic Time-Dependence, with Applications to Rossby Wave Flow, J. Nonlinear Sci., 8, 401–456, 1998.

Mancho, A. M., Small, D., Wiggins, S., and Ide, K.: Computation of Stable and Unstable Manifolds of Hyperbolic Trajectories in Two-Dimensional, Aperiodically Time-Dependent Vectors Fields, Physica D, 182, 188–222, 2003.

Mancho, A. M., Small, D., and Wiggins, S.: Computation of hyperbolic trajectories and their stable and unstable manifolds for oceanographic flows represented as data sets, Nonlin. Processes Geophys., 11, 17–33, https://doi.org/10.5194/npg-11-17-2004, 2004.

Mancho, A. M., Small, D., and Wiggins, S.: A comparison of methods for interpolating chaotic flows from discrete velocity data, Comput. Fluids, 35, 416–428, 2006a.

Mancho, A. M., Small, D., and Wiggins, S.: A tutorial on dynamical systems concepts applied to Lagrangian transport in oceanic flows defined as finite time data sets: Theoretical and computational issues, Phys. Rep., 237, 55–124, 2006b.

Mancho, A. M., Wiggins, S., Curbelo, J., and Mendoza, C.: Lagrangian descriptors: A Method for Revealing Phase Space Structures of General Time Dependent Dynamical Systems, Commun. Nonlinear Sci., 18, 3530–3557, 2013.

Manney, G. L. and Lawrence, Z. D.: The major stratospheric final warming in 2016: dispersal of vortex air and termination of Arctic chemical ozone loss, Atmos. Chem. Phys., 16, 15371–15396, https://doi.org/10.5194/acp-16-15371-2016, 2016.

Manney, G. L., Farrara, J. D., and Mechoso, C. R.: The behavior of wave 2 in the southern hemisphere stratosphere during late winter and early spring, J. Atmos. Sci., 48, 976–998, 1991.

Manney, G. L., Sabutis, J. L., Alley, D. R., Lahoz, W. A., Scaife, A. A., Randall, C. E., Pawson, S., Naujokat, B., and Swinbank, R.: Simulations of Dynamics and Transport during the September 2002 Antarctic Major Warming, J. Atmos. Sci., 66, 2006.

McIntyre, M. E. and Palmer, T. N.: Breaking planetary waves in the stratosphere, Nature, 305, 593–600, https://doi.org/10.1038/305593a0, 1983.

McIntyre, M. E. and Palmer, T. N.: The surf zone in the stratosphere, J. Atmos. Terr. Phys., 46, 825–849, 1984.

McIntyre, M. E. and Palmer, T. N.: A note on the general concept of wave breaking for Rossby and gravity waves, Pure Appl. Geophys., 123, 964–975, https://doi.org/10.1007/BF00876984, 1985.

Mechoso, C. R., O'Neill, A., Pope, V. D., and Farrara, J. D.: A study of the stratospheric final warming of 1982 in the southern hemisphere, Q. J. Roy. Meteor. Soc., 114, 1365–1384, https://doi.org/10.1002/qj.49711448402, 1988.

Mendoza, C. and Mancho, A. M.: The hidden geometry of ocean flows, Phys. Rev. Lett., 105, 038501, https://doi.org/10.1103/PhysRevLett.105.038501, 2010.

Mendoza, C. and Mancho, A. M.: Review Article: "The Lagrangian description of aperiodic flows: a case study of the Kuroshio Current", Nonlin. Processes Geophys., 19, 449–472, https://doi.org/10.5194/npg-19-449-2012, 2012.

Mendoza, C., Mancho, A. M., and Wiggins, S.: Lagrangian descriptors and the assessment of the predictive capacity of oceanic data sets, Nonlin. Processes Geophys., 21, 677–689, https://doi.org/10.5194/npg-21-677-2014, 2014.

Morales-Juberías, R., Sayanagi, K. M., Simon, A. A., Fletcher, L. N., and Cosentino, R. G.: Meandering shallow atmospheric jet as a model of Saturn's North-Polar hexagon, Astrophys. J. Lett., 806:L18, https://doi.org/10.1088/2041-8205/806/1/L18, 2015.

Nakamura, M. and Plumb, R. A.: The Effects of Flow Asymmetry on the Direction of Rossby Wave Breaking, J. Atmos. Sci., 51, 2031–2044, 1994.

Ottino, J. M.: The Kinematics of Mixing: Stretching, Chaos, and Transport, Cambridge University Press, Cambridge, England, reprinted 2004, 1989.

Plumb, R. A.: Tracer interrelationships in the stratosphere, Rev. Geophys., 45, RG4005, https://doi.org/10.1029/2005RG000179, 2007.

Polvani, L. M. and Plumb, R. A.: Rossby Wave Breaking, Microbreaking, Filamentation, and Secondary Vortex Formation: The Dynamics of a Perturbed Vortex, J. Atmos. Sci., 49, 462–476, 1992.

Quintanar, A. I. and Mechoso, C. R.: Quasi-stationary waves in the Southern Hemisphere. Part I: Observational data, J. Climate, 8, 2659–2672, 1995.

Rabier, F., Bouchard, A., Brun, E., Doerenbecher, A., Guedj, S., Guidard, V., Karbou, F., Peuch, V., El Amraoui, L., Puech, D., Genthon, C., Picard, G., Town, M., Hertzog, A., Vial, F., Cocquerez, P., Cohn, S. A., Hock, T., Fox, J., Cole, H., Parsons, D., Powers, J., Romberg, K., VanAndel, J., Deshler, T., Mercer, J., Haase, J. S., Avallone, L., Kalnajs, L., Mechoso, C. R., Tangborn, A., Pellegrini, A., Frenot, Y., Thépaut, J., McNally, A., Balsamo, G., and Steinle, P.: The Concordiasi Project in Antarctica, B. Am. Meteorol. Soc., 91, 69–86, https://doi.org/10.1175/2009BAMS2764.1, 2010.

Rypina, I. I., Brown, M. G., Beron-Vera, F. J., Kocak, H., Olascoaga, M. J., and Udovydchenkov, I. A.: On the Lagrangian dynamics of atmospheric zonal jets and the permeability of the stratospheric polar vortex, J. Atmos. Sci., 64, 3595–3610, 2007.

Samelson, R. and Wiggins, S.: Lagrangian Transport in Geophysical Jets and Waves: The Dynamical Systems Approach, Springer-Verlag, New York, 2006.

Samelson, R. M.: Fluid exchange across a meandering jet, J. Phys. Oceanogr., 22, 431–440, 1992.

Shadden, S. C., Lekien, F., and Marsden, J. E.: Definition and properties of Lagrangian Coherent Structures from finite-time Lyapunov exponents in two-dimensional aperiodic flows, Physica D, 212, 271–304, 2005.

Simmons, A., Uppala, S., Dee, D., and Kobayashi, S.: ERA-Interim: New ECMWF reanalysis products from 1989 onwards, ECMWF Newsletter, 110, 25–35, 2007.

Smith, M. L. and McDonald, A. J.: A quantitative measure of polar vortex strength using the function M, J. Geophys. Res.-Atmos., 119, 5966–5985, 2014.

Snyder, J. P.: Map Projections–A Working Manual, U.S. Geological Survey professional paper, U.S. Government Printing Office, available at: https://books.google.es/books?id= nPdOAAAAMAAJ (last access: 14 March 2017), 1987.

Solomon, S.: The mystery of the Antarctic Ozone "Hole", Rev. Geophys., 26, 131–148, 1988.

Solomon, S.: Stratospheric ozone depletion: review of concepts and history, Rev. Geophys., 37, 375–316, 1999.

Stewartson, K.: The evolution of the critical layer of a Rossby wave, Geophys. Astrophys. Fluid, 9, 185–200, https://doi.org/10.1080/03091927708242326, 1977.

Taguchi, M.: Predictability of Major Stratospheric Sudden Warmings of the Vortex Split Type: Case Study of the 2002 Southern Event and the 2009 and 1989 Northern Events, J. Atmos. Sci., 71, 2886–2904, 2014.

Varotsos, C.: The southern hemisphere ozone hole split in 2002, Environ. Sci. Pollut. R. Int., 9, 375–376, 2002.

Varotsos, C.: What is the lesson from the unprecedented event over antarctica in 2002, Environ. Sci. Pollut. R. Int., 10, 80–81, 2003.

Varotsos, C.: The extraordinary events of the major, sudden stratospheric warming, the diminutive antarctic ozone hole, and its split in 2002, Environ. Sci. Pollut. R. Int., 11, 405–411, 2004.

Warn, T. and Warn, H.: The Evolution of a Nonlinear Critical Level, Stud. Appl. Math., 59, 37–71, https://doi.org/10.1002/sapm197859137, 1978.

Conditional nonlinear optimal perturbations based on the particle swarm optimization and their applications to the predictability problems

Qin Zheng[1], **Zubin Yang**[1], **Jianxin Sha**[2], and **Jun Yan**[1]

[1]Institute of Science, PLA University of Science and Technology, Nanjing, 211101, China
[2]Troop 94906, People's Liberation Army, Suzhou, 215157, China

Correspondence to: Zubin Yang (qinzheng@mail.iap.ac.cn)

Abstract. In predictability problem research, the conditional nonlinear optimal perturbation (CNOP) describes the initial perturbation that satisfies a certain constraint condition and causes the largest prediction error at the prediction time. The CNOP has been successfully applied in estimation of the lower bound of maximum predictable time (LBMPT). Generally, CNOPs are calculated by a gradient descent algorithm based on the adjoint model, which is called ADJ-CNOP. This study, through the two-dimensional Ikeda model, investigates the impacts of the nonlinearity on ADJ-CNOP and the corresponding precision problems when using ADJ-CNOP to estimate the LBMPT. Our conclusions are that (1) when the initial perturbation is large or the prediction time is long, the strong nonlinearity of the dynamical model in the prediction variable will lead to failure of the ADJ-CNOP method, and (2) when the objective function has multiple extreme values, ADJ-CNOP has a large probability of producing local CNOPs, hence making a false estimation of the LBMPT. Furthermore, the particle swarm optimization (PSO) algorithm, one kind of intelligent algorithm, is introduced to solve this problem. The method using PSO to compute CNOP is called PSO-CNOP. The results of numerical experiments show that even with a large initial perturbation and long prediction time, or when the objective function has multiple extreme values, PSO-CNOP can always obtain the global CNOP. Since the PSO algorithm is a heuristic search algorithm based on the population, it can overcome the impact of nonlinearity and the disturbance from multiple extremes of the objective function. In addition, to check the estimation accuracy of the LBMPT presented by PSO-CNOP and ADJ-CNOP, we partition the constraint domain of initial perturbations into sufficiently fine grid meshes and take the LBMPT obtained by the filtering method as a benchmark. The result shows that the estimation presented by PSO-CNOP is closer to the true value than the one by ADJ-CNOP with the forecast time increasing.

1 Introduction

Weather and climate predictability problems are attractive and significant in atmospheric and oceanic sciences. The goal of studying predictability problems is to investigate and understand the reasons for and mechanisms of the prediction uncertainty. Early in 1975, Lorenz divided the predictability problem into two types based on the consideration that prediction uncertainties were mainly caused by the initial and model errors. The first type of predictability problem is concerned with the uncertainty in the forecast results induced by initial errors, and the second type deals with the uncertainty caused by model errors. He further introduced the singular vector (SV) into the predictability problem study. The leading singular vector is the initial perturbation with the greatest linear growth, and, accordingly, can be used to estimate the evolution of initial errors in the tangent linear regime during the course of a forecast (Buizza and Palmer, 1995; Lacarra and Talagrand, 1988; Farrell, 1990; Borges and Hartmann, 1992; Kalnay, 2003). Since the atmospheric and oceanic movement is a nonlinear physical process, an SV based on linear theory cannot effectively express the procedure. This limits its applications to the predictability problems (Mu and Duan, 2003). To this end, Mu et al. (2003) presented the conditional nonlinear optimal perturbation (CNOP) method. CNOP refers to the initial perturbation that satisfies a certain constraint condition and has the maximum nonlinear evolution at the prediction time. Therefore, it stands for the initial uncertainty which leads to the largest prediction error. CNOP has been widely applied to weather and climate predictability studies (Duan et al., 2004, 2009; Mu et al., 2007a, b, 2009; Terwisscha and Dijkstra, 2008; Yu et al., 2012; Wang et al., 2012, 2013). Riviere et al. (2010) used an extension of the CNOP approach, i.e., the nonlinear singular vectors, to estimate the predictability of atmospheric moisture processes so as to reveal the effects of nonlinear processes.

Based on actual demands, Mu et al. (2002) separated the predictability problem into three sub-problems, i.e., the problem of the LBMPT, the problem of the upper bound of maximum prediction error, and the problem of the lower bound of maximum allowable initial error and parameter error. Duan and Luo (2010) formulated these three problems into three constrained nonlinear optimization problems. Meanwhile, they used the CNOP method to find the solutions of the three sub-problems.

Capturing CNOPs is a kind of constraint optimization problem, and optimization algorithms commonly used in solving CNOPs are based on the gradient descent method, including the spectral projected gradient 2 (SPG2; Brigin et al., 2000), sequential quadratic programming (SQP; Powell, 1982), and limited memory BFGS (L-BFGS; Liu and Nocedal, 1989). Among these algorithms, the gradient information is always provided by the backward integral of the corresponding adjoint model of the prediction model (Duan et al., 2004, 2008; Mu and Zhang, 2006; Mu et al., 2009; Jiang and Wang, 2010; Yu et al., 2012; Wang et al., 2012, 2013). But the optimal algorithms based on gradient information involve the forward integral of the tangent model and backward integration of the adjoint model. It might cause the following two problems: (1) for the numerical prediction model with complex physical processes, the validity of the tangent linear approximation cannot be guaranteed when the forecast period is long; (2) for the actual prediction model, it is quite difficult and time consuming to develop the adjoint model. Recently, Zheng et al. (2012, 2014) attempted to apply genetic algorithms (GAs) to capture CNOPs of the dynamical model containing discontinuous "on–off" switches. They concluded that GAs, with proper genetic operator configuration, can overcome the non-smooth influences and obtain the global CNOP with high probability. Thus, in non-smooth cases, using GAs to solve predictability problems is more effective than using the conventional optimization algorithm.

The particle swarm optimization (PSO) algorithm is an intelligent algorithm proposed by Kennedy and Eberhart (1995) which imitated the process of bird foraging. In the PSO algorithm, each particle, as a vector in solution space, represents a potential solution of the optimal prob-

lem. Compared to the gradient descent algorithm based on the adjoint model, the PSO algorithm has better effectiveness in searching the global optimal solution for nonlinear and non-smooth optimal problems; PSO does not require a gradient of the objective function. PSO is similar to the GA in the sense that they are both population-based search approaches and that they both depend on information sharing among their population members to enhance their search processes using a combination of deterministic and probabilistic rules (Hassan et al., 2005). PSO has the same effectiveness as the GA but with significantly better computational efficiency; it has memory and constructive cooperation between particles (Fang and Zheng, 2009). This paper tries to use PSO to calculate the CNOP of the forecast model, and obtains a highly precise estimation of the lower bound of maximum predictable time.

The Ikeda model was originally proposed by Ikeda (1979) as a model describing light going across a nonlinear optical resonator. The Ikeda model has strong nonlinearity, and the two-dimensional difference scheme is its most common form. Li et al. (2016) investigated the stability of solutions of the Ikeda model and tested the dependence of the solutions on the model parameter. In addition, they provided the solution description of various shapes corresponding to parameter values of different regions. This paper takes the two-dimensional Ikeda model as the prediction model to reveal how the nonlinearity impacts the precision when estimating the LBMPT based on the ADJ-CNOP method. A new method, PSO-CNOP, is presented to solve this problem.

This paper is organized as follows: Sect. 2 is devoted to describing the three predictability sub-problems, the definition of CNOP and the two-dimensional Ikeda model. Section 3 provides the knowledge about the particle swarm optimization (PSO) algorithm. In Sect. 4, the performances of ADJ-CNOP and PSO-CNOP are compared when solving CNOPs through numerical experiments. The impacts on the estimation precision of the lower bound of maximum predictable time are also demonstrated in this section. The conclusion and discussion are presented in Sect. 5.

2　Related conceptions and the forecast model

The three predictability sub-problems, the definition of CNOP and the two-dimensional Ikeda model are briefly described as follows, and more detailed introductions can be found in Mu et al. (2002, 2003) and Li and Zheng (2016).

2.1　Three sub-problems of the predictability problem

The three predictability sub-problems associated with the first kind of predictability problem are introduced, and the forecast model is supposed to be perfect in the following.

Problem 1: the lower bound of maximum predictable time (LBMPT)

Assuming that there is an error in the initial condition (IC) of the forecast model, it will lead to a prediction error when integrating forward the model from the IC to predict the atmospheric or oceanic states in the future.

Let $\mathbf{u}_0$ be the IC, $\mathbf{u}_T^t$ the true state at time T, and M_T the nonlinear propagator of the numerical forecast model from 0 to time T; then, under the assumption of the perfect model, $\mathbf{u}_T^t = M_T(\mathbf{u}_0^t)$, where $\mathbf{u}_0^t$ is the true state at the initial time.

With a given prediction precision $\varepsilon > 0$, the maximum predictable time T_ε is defined as follows:

$$T_\varepsilon = \max\{\tau \mid ||M_T(\mathbf{u}_0) - \mathbf{u}_T^t||_2 \le \varepsilon, \ 0 \le T \le \tau\}. \quad (1)$$

Since the true value $\mathbf{u}_T^t$ cannot be obtained exactly, it is impossible to get T_ε by solving the nonlinear optimization problem (1). Inspired by the fact that the IC $\mathbf{u}_0$ is often provided by an analysis field, and the associated analysis error can generally be controlled in a specified range, Mu et al. (2002) reduced the maximum predictable time problem to the following LBMPT problem.

If we have an estimation of the uncertainty in the IC as follows,

$$\left\| \mathbf{u}_0 - \mathbf{u}_0^t \right\| \le \sigma, \quad (2)$$

then the LBMPT T_1 is defined as

$$T_1 = \min_{||\delta\mathbf{u}_0|| \le \sigma} \{T_{\mathbf{u}_0,\delta\mathbf{u}_0} \mid T_{\mathbf{u}_0,\delta\mathbf{u}_0} = \max\tau,$$

$$||M_t(\mathbf{u}_0 + \delta\mathbf{u}_0) - M_t(\mathbf{u}_0)|| \le \varepsilon, \ 0 \le t \le \tau\}, \quad (3)$$

where $\sigma > 0$ denotes the accuracy of the IC in terms of the norm $||\cdot||$, and $\delta\mathbf{u}_0$ is an initial perturbation superposed on the IC. According to Eq. (2), the true initial state is within the constraint region; we have

$$T_1 \le T_\varepsilon.$$

Problem 2: the upper bound of maximum prediction error

When a forecast is produced from an incorrect initial IC $\mathbf{u}_0$, the prediction error at the prediction time T is

$$E = ||M_T(\mathbf{u}_0) - \mathbf{u}_T^t||. \quad (4)$$

Similarly to problem 1, since the true value $\mathbf{u}_T^t$ cannot be obtained precisely, Mu et al. (2002) instead introduced the upper bound of the maximum prediction error within the given initial error limitation as follows:

$$E_{\mathbf{u}} = \max_{||\delta\mathbf{u}_0|| \le \sigma} ||M_T(\mathbf{u}_0 + \delta\mathbf{u}_0) - M_T(\mathbf{u}_0)||. \quad (5)$$

Note that $\mathbf{u}_0^t$ satisfies Eq. (2) and $\mathbf{u}_T^t = M_T(\mathbf{u}_0^t)$ under the assumption of perfect model; we have

$$E \le E_u.$$

Problem 3: the lower bound of maximum allowable initial error

Given the prediction time $T > 0$ and prediction precision $\varepsilon > 0$, the maximum allowable initial error is

$$\sigma_{\max} = \max\{\sigma \mid ||M_T(\mathbf{u}_0 + \delta\mathbf{u}_0) - \mathbf{u}_T^t|| \le \varepsilon, \ ||\delta\mathbf{u}_0|| \le \sigma\}. \quad (6)$$

Similar to problems 1 and 2, the above problem was reduced by Mu et al. (2002) to the following lower bound of maximum allowable initial error:

$$\overline{\sigma}_{\max} = \max\{\sigma \mid ||M_T(\mathbf{u}_0 + \delta\mathbf{u}_0) - M_T(\mathbf{u}_0)|| \le \varepsilon, \ ||\delta\mathbf{u}_0|| \le \sigma\}. \quad (7)$$

2.2 Conditional nonlinear optimal perturbation (CNOP)

In consideration of the nonlinearity impacts, Mu et al. (2003) introduced CNOPs into the study of predictability problems. Suppose the atmospheric or oceanic motions can be described by the following dynamic system:

$$\begin{cases} \dfrac{\partial \mathbf{U}}{\partial t} + \mathbf{F}(\mathbf{U}, t) = 0, \\ \mathbf{U}|_{t=0} = \mathbf{U}_0, \end{cases} \quad (8)$$

where $\mathbf{U}(\mathbf{x}, t) = (U_1(\mathbf{x}, t), U_2(\mathbf{x}, t), \cdots, U_n(\mathbf{x}, t))^T$ is the basic state, which is an n-dimensional vector; the superscript T represents the transpose, $\mathbf{U}_0$ is the initial basic state, and $\mathbf{x} = (x_1, x_2, \cdots, x_m)^T \in \Omega \subset \mathbf{R}^m$ and t are the spatial and temporal variables, respectively; $t = 0$ is the initial time; and $\mathbf{F}$ is a nonlinear partial differential operator.

Suppose M_τ is the nonlinear transmission propagator from the initial time $t = 0$ to the forecast time $t = \tau$; thus, the state of model (8) at time τ is

$$\mathbf{U}(\mathbf{x}, \tau) = M_\tau(\mathbf{U}_0). \quad (9)$$

If $\mathbf{u}_0$ stands for the initial perturbation of the basic state $\mathbf{U}(t)$ and $\mathbf{u}_I(\tau)$ is the development of $\mathbf{u}_0$ at time τ, that is,

$$\mathbf{u}_I(\tau) = M_\tau(\mathbf{U}_0 + \mathbf{u}_0) - M_\tau(\mathbf{U}_0), \quad (10)$$

then the initial perturbation $\mathbf{u}_0^*$ is called the conditional nonlinear optimal perturbation (CNOP) if and only if $\mathbf{u}_0^*$ is the solution of the following optimization problem:

$$J\left(\mathbf{u}_0^*\right) = \max_{\mathbf{u}_0 \in B_\sigma} ||M_\tau(\mathbf{U}_0 + \mathbf{u}_0) - M_\tau(\mathbf{U}_0)||, \quad (11)$$

where $B_\sigma = \{\mathbf{u}_0 \mid \mathbf{u}_0 \in \mathbf{R}^n, ||\mathbf{u}_0|| \le \sigma\}$ is the constraint domain on the initial perturbation. In terms of the L^2 norm, B_σ is a ball with the center at the origin and the radius σ. In addition, J is called the objective function in the context of optimal control theory.

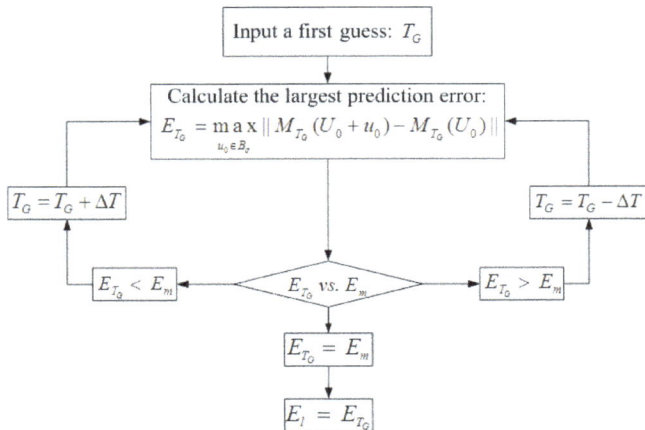

Figure 1. Flow chart of solving the LBMPT.

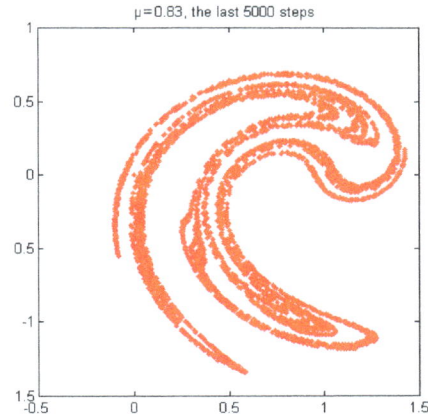

Figure 2. The distribution of solutions at the last 5000 steps when $\mu = 0.83$.

2.3 Estimation of the LBMPT

Duan and Luo (2010) designed a numerical method to calculate the LBMPT in their research of predictability (Fig. 1). It should be noted that CNOPs stand for the initial uncertainty in the given constrained domain which leads to the largest prediction error. Therefore, the maximum prediction error can be estimated by solving CNOPs.

In detail, for a given first guess value T_G of the prediction time, one can use a constraint nonlinear optimization algorithm to capture the CNOP so as to estimate the maximum prediction error E_{T_G} at time T_G caused by the initial error in a constraint domain B_σ.

If $E_{T_G} > E_m$ (E_m stands for the allowable prediction error), we try to reduce the integral time step $T_G = T_G - \Delta T$, where ΔT is a certain constant, and calculate the maximum prediction error at the reduced time. If $E_{T_G} < E_m$, we will increase the integral time step $T_G = T_G + \Delta T$ and calculate the maximum prediction error at the new T_G.

If T_G satisfies the conditions $E_{T_G + \Delta T} > E_m$ and $E_{T_G - \Delta T} \leq E_m$, then T_G is considered the lower bound of maximum predictable time which satisfies the prediction precision E_m under the given initial error. The operation flow chart is shown as Fig. 1.

2.4 The two-dimensional Ikeda model

The following two-dimensional Ikeda model is adopted as the prediction model:

$$\begin{cases} x_1(t+1) = 1 + \mu\,(x_1(t)\cos\theta_t - x_2(t)\sin\theta_t), \\ x_2(t+1) = \mu\,(x_1(t)\sin\theta_t + x_2(t)\cos\theta_t) \end{cases}, \quad (12)$$

$$\theta_t = a - \frac{b}{1 + x_1(t)^2 + x_2(t)^2}, \quad (13)$$

where $0 \leq \mu \leq 1$, $a = 0.4$ and $b = 6$.

From the expression of the model we find that the trigonometric functions appear in Eq. (12), and Eq. (13) is a frac-

tion whose denominator includes two quadratic components. Thus, the two-dimensional Ikeda model has fairly strong nonlinearity (Li and Zheng, 2016). The solutions of the model present different behaviors with the change in the model parameter μ. When the parameter varies from 0 to 1, the numerical solutions change from a point attractor to periodic solutions, then to chaos, and end up with a limit cycle (Li and Zheng, 2016). The predictability problems are always launching under chaos. According to the conclusions given by Li and Zheng (2016), the model solution appeared chaotic when $\mu \in [0.700, 0.902]$. Figure 2 shows the numerical solution of the last 5000 steps in 10 000 integral steps, while the initial value is set to $(x_0, y_0) = (0.25, -0.325)$ and the model parameter is $\mu = 0.83$.

3 The particle swarm optimization (PSO) algorithm

The PSO algorithm has recently got more and more attention (Eberhart and Shi, 2001; Banks et al., 2007, 2008; Poli et al., 2007). The PSO algorithm was originally proposed by social psychologists Kennedy and Eberhart (1995). It simulates the collective behavior of birds foraging. Each particle represents a potential solution in the PSO algorithm, flies with specific velocity, respectively, and adjusts its trajectories according to the flying experiences of its own and the companion's, finally finding the optimal location in the solution space. To avoid the particles rapidly flying out of the solving region and to improve the ability of PSO to search for a global optimal solution, Shi and Eberhart (1998) introduced the maximum velocity and the inertia weight into PSOs, so as to restrain the particle's behaviors in the searching process. Clerc and Kennedy (2002) proposed a limiting factor for the two acceleration coefficients after they analyzed theoretically the convergence of PSO.

The basic PSO algorithm consists of three processes, namely, generating particles' positions and velocities, assessing particles, and updating particles' positions and velocities.

The mathematical description of the classic PSO algorithm is as follows: in an n-dimensional search space, each particle of PSO represents a potential solution of the optimization problem. We denote M the swarm size, $\mathbf{X}_i(k) = (x_{i\,1}(k), x_{i\,2}(k), \cdots, x_{i\,n}(k))$ and $\mathbf{V}_i(k) = (v_{i1}(k), v_{i2}(k), \cdots, v_{i\,n}(k))$ the position and the velocity of the ith particle at the kth generation, respectively, $\mathbf{P}_i(k) = (p_{i\,1}(k), p_{i\,2}(k), \cdots, p_{i\,n}(k))$ the personal historical best position of the ith particle found so far, and $\mathbf{P}_g(k) = (p_{g\,1}(k), p_{g\,2}(k), \cdots, p_{g\,n}(k))$ the best position that the whole swarm attained so far; then the particle i's velocity and position in the next $k+1$ generation can be updated according to the following formula:

$$v_{id}(k+1) = w\,v_{id}(k) + c_1 r_1 \left(p_{id}(k) - x_{id}(k)\right)$$
$$+ c_2 r_2 \left(p_{gd}(k) - x_{id}(k)\right), \tag{14}$$
$$x_{id}(k+1) = x_{id}(k) + v_{id}(k+1), \tag{15}$$

where $i = 1, 2, \cdots, M$, $d = 1, 2, \cdots, n$, c_1 and c_2 are the acceleration coefficients, which make particles having the ability to self-summarize and learn from excellent particles among the group to approach their own and group historical optimal points. r_1 and r_2 are two random numbers that are subject to uniform distribution on the interval $[0, 1]$. ω is the inertia weight. It can be set as a fixed constant or a linear reduction function with the increase in the evolutional generations. The flow chart of the PSO algorithm is shown as Fig. 3.

When using a PSO to search CNOPs for the estimation of the LBMPT, the prediction error at the specified forecast time is the associated objective function J. The initial perturbation $\delta\mathbf{u}$, which is a two-dimensional vector in the search space in our situation, is the optimization variable.

4 Numerical experiments and their results analyses

4.1 The numerical experiments solving CNOPs by different optimization algorithms

In order to compare the performances of the ADJ-CNOP and PSO-CNOP in solving CNOPs, the CNOPs yielded by the filtering method are taken as the benchmark after fine-dividing the constraint domain of initial perturbations. The filtering method is implemented as follows. The corresponding circumscribed square of a constraint region of the CNOP is considered; four square meshes of a certain size are used to discretize the circumscribed square. For any mesh point outside the region, it is connected with the center of the region; the intersection point of this line with the boundary of the region is obtained. Integrating the Ikeda model from the initial basic state superimposed each of these intersection

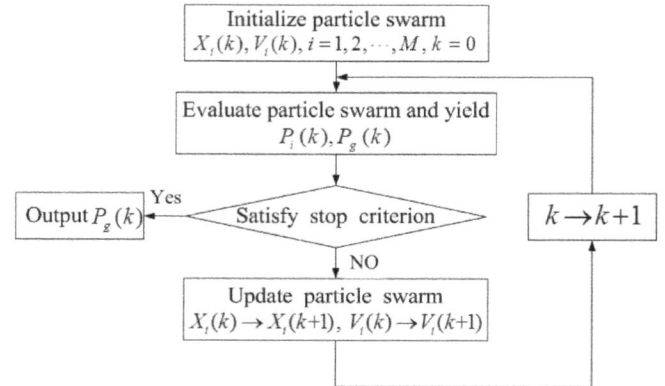

Figure 3. The flow chart of the PSO algorithm.

points and, for the mesh points inside the region, the prediction error caused by each initial error can be obtained. The CNOP refers to the mesh point which leads to the largest prediction error (Duan and Luo, 2010). Since the accuracy of the CNOP generated by the filtering method depends on the division size of the constraint region exclusively, the circumscribed squares of the constraint ball of CNOPs are separated into 1001×1001 small quadrate patches with very small side length 1.6402×10^{-5} in numerical experiments. For a detailed description of the operation of the filtering method, one can refer to Duan and Luo (2010) and Zheng et al. (2012).

In the numerical experiments, the initial basic state of the two-dimensional Ikeda model is $(x_0, y_0) = (0.25, -0.325)$, and the model parameter is $\mu = 0.83$. The population size of the PSO is $M = 60$, the maximum evolutional generation is set to 200, inertia weight is $\omega = 0.729$ and the accelerating factors are $C_1 = 2.05$ and $C_2 = 2.05$. The norms measuring IC errors and prediction errors are both the L^2 norm, and the radius σ of the constraint ball B_σ is 8.201×10^{-3}.

The particle swarm initialization scheme in PSO-CNOP is as follows:

$\mathbf{X}_i = (x_{i,1}(0), x_{i,2}(0))$ are random vectors obeying a uniform distribution on B_σ.

$\mathbf{V}_i(0) = (v_{i,1}(0), v_{i,2}(0)) = \mathbf{X}_i(0), \quad i = 1, 2, \cdots, M.$

The first guess of the perturbation $\delta\mathbf{u} = (x_0, y_0)$ for ADJ-CNOP is randomly picked from B_σ. The constrained optimization algorithm used in ADJ-CNOP is the SPG2.

With the prediction time increasing, there would appear simultaneously many CNOPs for the two-dimensional Ikeda model because of the impact of the strong nonlinearity. Hence, different forecast times are adopted to test the ability of ADJ-CNOP and PSO-CNOP to attack the nonlinearity obstacles. For each forecast time, the numerical experiment using ADJ-CNOP or PSO-CNOP to obtain CNOPs is conducted 40 times, respectively; 40 CNOPs are clustered by the fuzzy c-means clustering (FCM) method and the accuracy of the CNOPs is statistically analyzed.

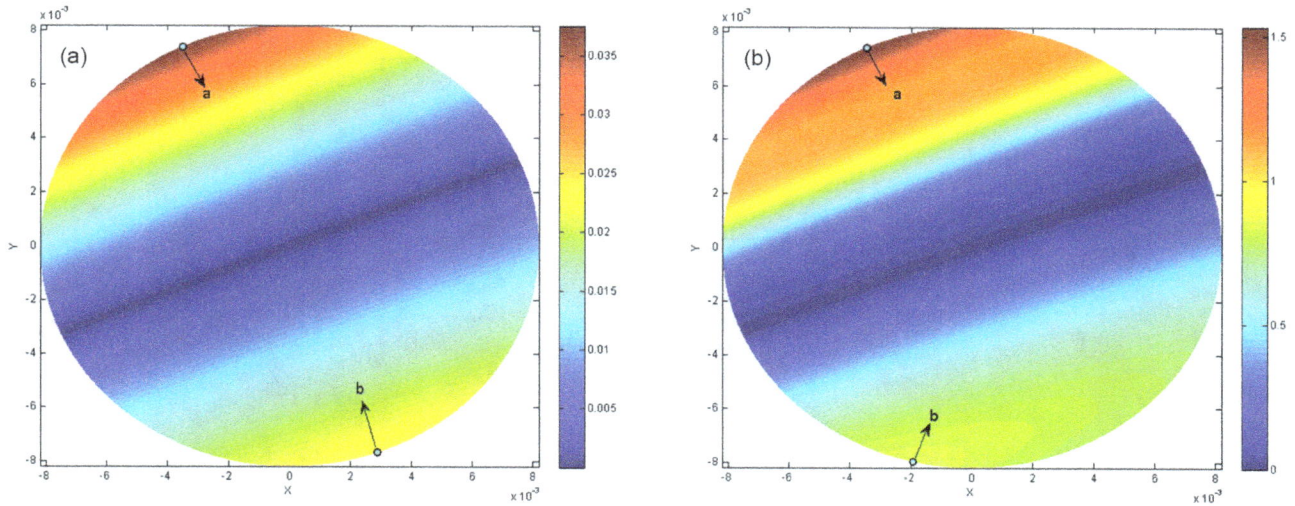

Figure 4. The distributions of OFVs at the prediction times $6\Delta t$ (left) and $13\Delta t$ (right), in which dots a and b are the global and local maximum points.

Table 1. Statistical analysis of CNOPs produced by different methods at $6\Delta t$.

Method	OFV	CNOP	Proportion
Filtering	3.7533×10^{-2}	$(-3.5066 \times 10^{-3}, 7.4131 \times 10^{-3})$	
ADJ-CNOP	3.7533×10^{-2}	$(-3.5068 \times 10^{-3}, 7.4130 \times 10^{-3})$	47.5 %
	2.3973×10^{-2}	$(2.9072 \times 10^{-3}, -7.6683 \times 10^{-3})$	52.5 %
PSO-CNOP	3.7533×10^{-2}	$(-3.5068 \times 10^{-3}, 7.4130 \times 10^{-3})$	100 %

The numerical experiment results show that when the integration time of the prediction model is short, the corresponding objective function (Eq. 11) presents good behavior with the change in the initial perturbation. It has only two extreme values in the constraint ball of the initial perturbation. One is the global maximum, and the extreme point corresponds to the global CNOP. The other is the local maximum, and the extreme point is a local CNOP. Figure 4 shows the distribution of objective function values (OFVs) when the prediction times are on 6th unit time steps, i.e., $6\Delta t$ (left) and $13\Delta t$ (right), respectively, in which the global maximum point is located at point a and point b is the position of the local maximum.

Tables 1 and 2 demonstrate the statistical analysis results of the CNOPs produced by ADJ-CNOP and PSO-CNOP when capturing CNOPs 40 times at the forecast times $6\Delta t$ and $13\Delta t$, and the related results generated by the filtering method. Through the FCM method, we find that CNOPs obtained by the ADJ-CNOP method are divided into two categories: one is related to the global CNOP that accounted for 47.5 % (70 %) for the forecast time $6\Delta t$ ($13\Delta t$) of the total; the other is the local CNOP that makes up 52.5 % (30 %) of the total. However, 40 CNOPs captured by PSO-CNOP are completely the same, and they are coincident with the CNOP yielded by the filtering method.

From Tables 1 and 2, we can see that although the prediction time is short, the ADJ-CNOP method still has a large probability of capturing local CNOPs, while PSO-CNOP can always catch the global CNOP. Actually, we can draw the same conclusion with the prediction time being increased to $13\Delta t$.

When the forecast time increases to $14\Delta t$, the 40 CNOPs yielded by ADJ-CNOP and PSO-CNOP are demonstrated in the following Fig. 5.

Figure 6 indicates all CNOPs generated by the filtering method with fine division of the constraint domain of initial perturbations (the circumscribed squares of the constraint ball of CNOPs are separated into 1001×1001 small quadrate patches with very small side length 1.6402×10^{-5}).

Figure 6 shows that when the prediction time reaches $14\Delta t$, there exist many global CNOPs, and all of them are located in a line. Based on this, we find that ADJ-CNOP not only gets global or local CNOPs, but also captures false CNOPs at the prediction time $14\Delta t$, since no matter how small an area we take around one of these "CNOPs", there always exists one point whose objective function value is larger than the objective function value of the "CNOP". According to the definition of CNOPs, these "CNOPs" are not true CNOPs. Hence, we call them false CNOPs.

Table 2. Same as Table 1, except at $13\Delta t$.

Method	OFV	CNOP	Proportion
Filtering	1.5301	$(-3.4679 \times 10^{-3}, 7.4313 \times 10^{-3})$	
ADJ-CNOP	1.5301	$(-3.4682 \times 10^{-3}, 7.4311 \times 10^{-3})$	70 %
	0.8516	$(-1.9285 \times 10^{-3}, -7.9706 \times 10^{-3})$	30 %
PSO-CNOP	1.5301	$(-3.4682 \times 10^{-3}, 7.4311 \times 10^{-3})$	100 %

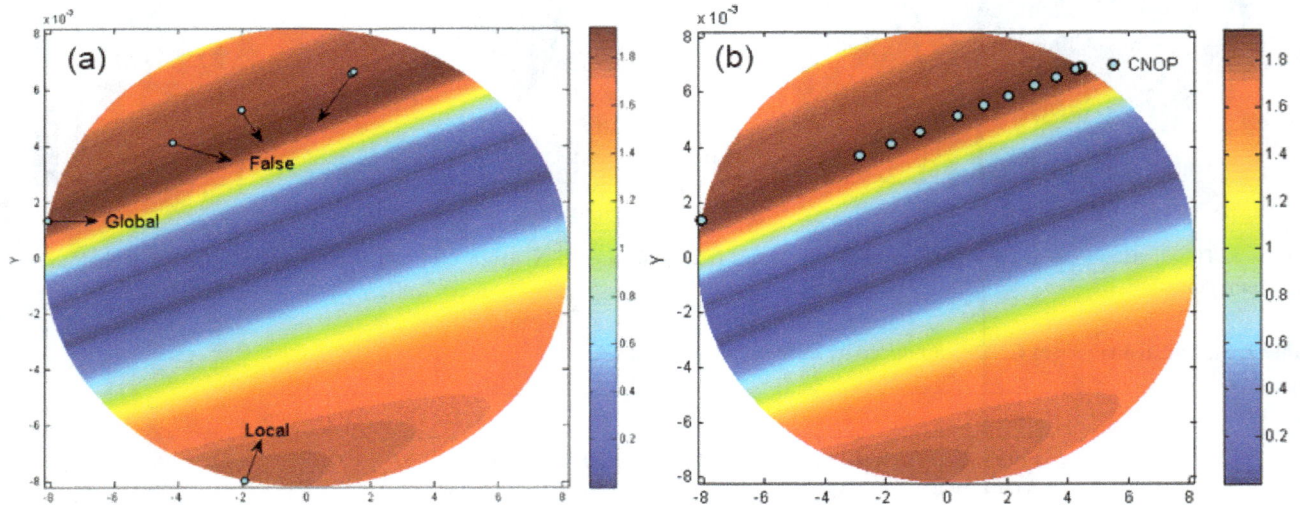

Figure 5. The distributions of OFVs at the prediction time $14\Delta t$, where the dots in the left (right) panel are the CNOPs produced by ADJ-CNOP (PSO-CNOP).

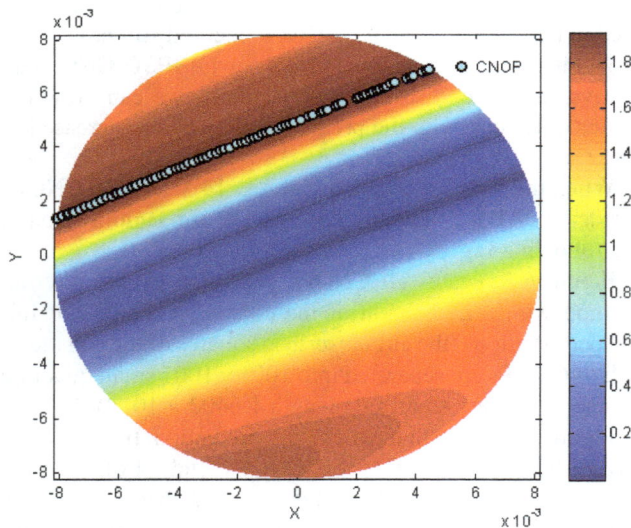

Figure 6. The distributions of OFVs at the prediction time $14\Delta t$, where dots are the CNOPs produced by the filtering method.

Additionally, comparing the right panel of Fig. 5 with Fig. 6, it is easy to know that although many CNOPs are produced by PSO-CNOP in 40 repeated numerical experiments, all of the CNOP points are located on the same line as the one presented by the filtering method. Therefore, the CNOPs yielded by PSO-CNOP are all global.

When we keep extending the prediction time, the behavior of the objective function will get much worse, and more extreme points will appear. In order to verify the performance of the PSO-CONP method in solving CNOPs in the strong nonlinear case, the mean value and variance of the OFVs of the 40 CNOPs at different forecast times are calculated and compared with the maximal OFV (MOFV) obtained by the filtering method.

According to Table 3, the OFVs of 40 CNOPs calculated by the PSO-CNOP method at each forecast time are almost consistent with the maximum of the objective function gotten by the filtering method at the same forecast time. Therefore, PSO-CNOP is still capable of solving global CNOPs of the two-dimensional Ikeda model for long forecast times. Figure 7 demonstrates the distributions of OFVs at the prediction times $15\Delta t$, $18\Delta t$ and $22\Delta t$, as well as the locations of all CNOPs generated by PSO-CNOP.

From the upper two panels of Fig. 7, we can see clearly that the CNOPs are all located in the maximal OFV region, which indicates that all of the CNOPs captured by the PSO-CNOP method are global. Because of the complexity of the

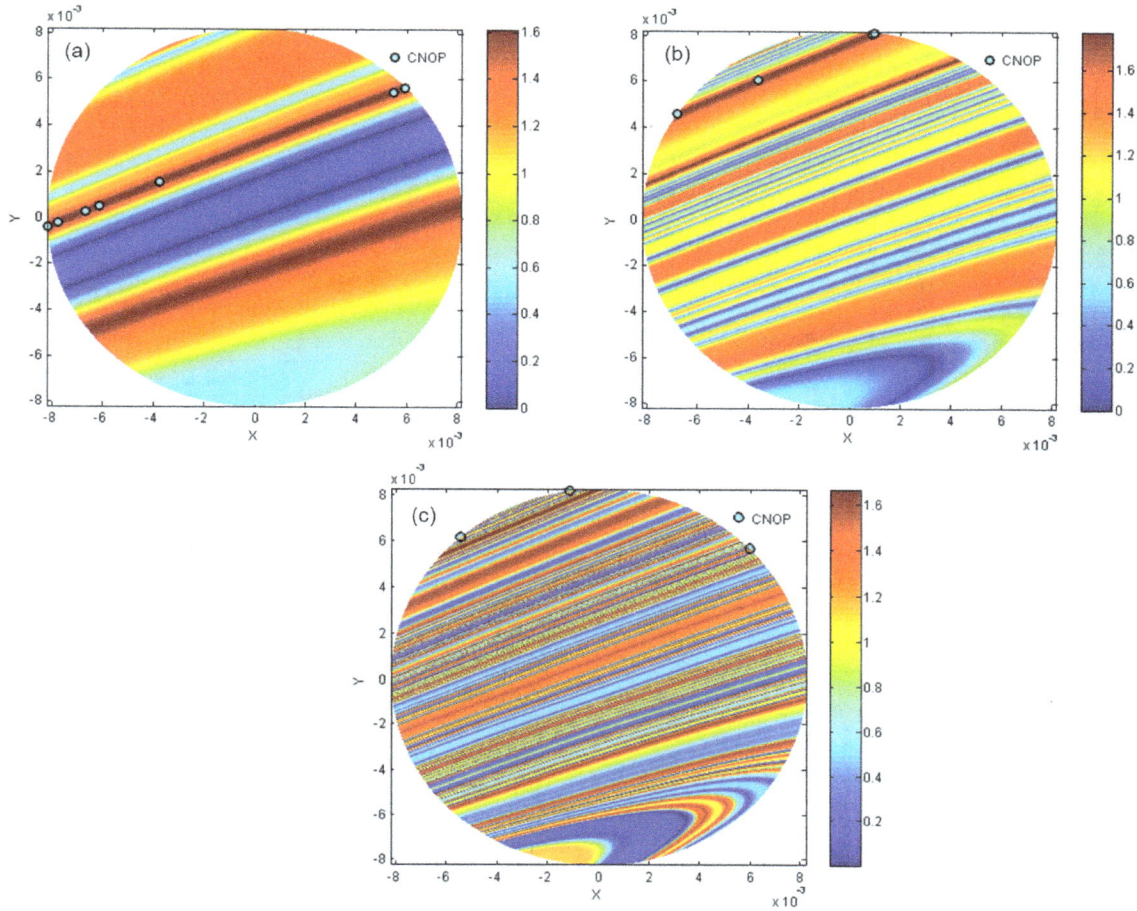

Figure 7. The distribution of OFVs at the prediction times $15\Delta t$ (the left of the upper panel), $18\Delta t$ (the right of the upper panel) and $22\Delta t$ (the lower panel), where dots denote CNOPs captured by PSO-CNOP.

Table 3. The precision analysis of the CNOPs produced by PSO-CNOP at different forecast times.

Prediction time	MOFV of the filtering method	CFV of PSO-CNOP	
		Mean value	Variance
$15\Delta t$	1.6106	1.6106	3.3142×10^{-16}
$16\Delta t$	1.3052	1.3052	8.6189×10^{-14}
$17\Delta t$	1.6521	1.6521	2.0092×10^{-16}
$18\Delta t$	1.7807	1.7807	7.6313×10^{-17}
$19\Delta t$	1.5401	1.5401	1.7300×10^{-16}
$20\Delta t$	1.3482	1.3482	8.8314×10^{-10}
$21\Delta t$	1.6980	1.6980	3.3546×10^{-10}
$22\Delta t$	1.6602	1.6593	4.6798×10^{-07}

distributions of OFVs at the prediction time $22\Delta t$, it cannot be confirmed directly from the lower panel of Fig. 7 whether or not the CNOPs are global. Therefore, we select one CNOP randomly from each cluster of the 40 CNOPs and zoom into the graph nearby the CNOP point to look at the OFV distri-

bution. Figure 8 gives one of the results, from which we can see that the CNOP is still located in the maximal OFV area.

With further increasing of the prediction time, the strong nonlinearity deteriorates the behavior of the objective function seriously. In this situation, a predictability study based on the CNOP method becomes no longer meaningful because the CNOPs are too dispersive.

4.2 Comparison between PSO-CNOP and GA-CNOP

In the following, the GA is adopted to capture CNOPs of the two-dimensional Ikeda model, and the results are compared with the ones obtained by the PSO-CNOP. The method using the GA to compute CNOP is called GA-CNOP. The relevant operational flow chart of the GA is shown in Fig. 9.

The configuration of the genetic operators and the relevant parameter are the same as in Zheng et al. (2014). A more detailed description of the GA and numerical experiment scheme of GA-CNOP can be found in Zheng et al. (2014).

The performances of PSO-CNOP and GA-CNOP in solving the CNOPs are tested for different population sizes; the

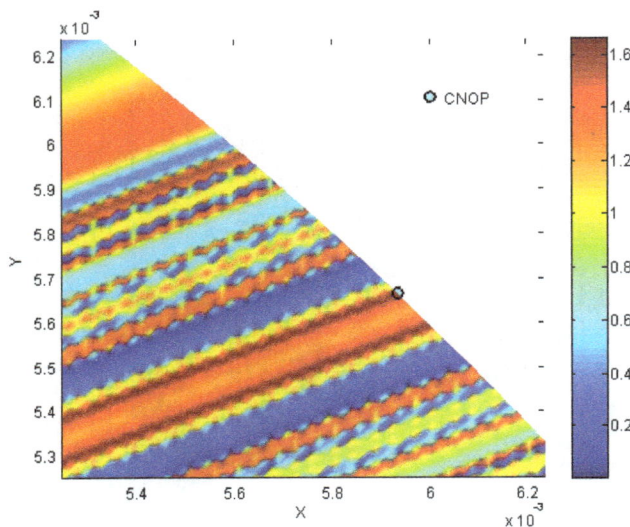

Figure 8. Local distribution of the OFV at the prediction time $22\Delta t$ nearby the CNOP point.

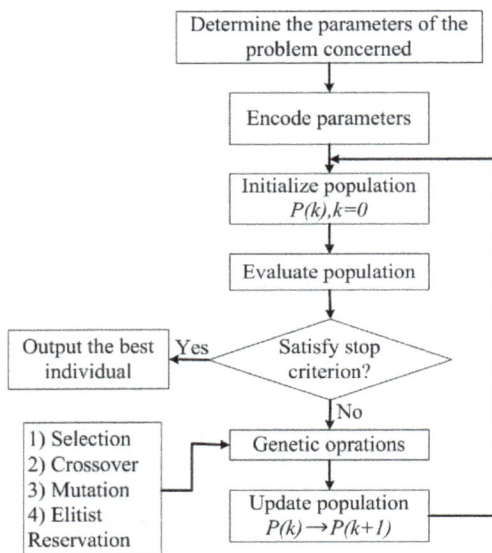

Figure 9. The flow chart of the GA.

Table 4. Mean CFV of 40 CNOPs produced by PSO-CNOP for every given population size.

Prediction time	Population size					
	5	10	15	30	45	60
$13\Delta t$	1.5301	1.5301	1.5301	1.5301	1.5301	1.5301
$19\Delta t$	1.5209	1.5396	1.5401	1.5401	1.5401	1.5401

Table 5. Same as Table 4, except that 40 CNOPs are produced by GA-CNOP.

Prediction time	Population size					
	16	26	36	40	50	60
$13\Delta t$	1.5299	1.5300	1.5300	1.5300	1.5300	1.5300
$19\Delta t$	1.5019	1.5295	1.5357	1.5400	1.5400	1.5400

From the numerical results we can see that the CFV of GA-CNOP is almost the same as PSO-CNOP; the GA is an effective optimal algorithm to obtain the optimal solution. However, the GA is more time consuming than PSO. At the same time, the computational time of PSO-CNOP and GA-CNOP is much greater than ADJ-CNOP, which is also the difference between stochastic searching algorithms and deterministic searching algorithms. Fortunately, in intelligent optimization algorithms, the parallel computation can be easily realized. The operators of different individuals in one generation are independent, and can be done in different CPUs, which thereby can take full advantage of fast developed parallel computation technology (Fang et al., 2009).

4.3 Estimation of the LBMPT

The CNOP method can be adopted to do a similar study on other two predictability sub-problems. Here we only focus on the estimation of the LBMPT to discuss the influence of nonlinearity. To demonstrate the effectiveness of PSO-CNOP in solving this problem, the filtering method, ADJ-CNOP and PSO-CNOP are used in the numerical experiments, respectively. In order to compare the estimation accuracy of the LBMPT generated by ADJ-CNOP and PSO-CNOP, the LBMPT computed by the filtering method with a fine division is taken as the benchmark. Different allowable prediction errors E_m, 0.5, 0.8, 1.1, 1.4, and 1.7, and a different constrained radius δ of initial perturbations, 0.01, 0.02, 0.03, 0.04, and 0.05, are employed to verify the performance of the ADJ-CNOP and PSO-CNOP methods. The lattice spacing of the filtering method is 0.001. Tables 7, 8 and 9 illustrate the LBMPTs computed by the filtering method, PSO-CNOP and ADJ-CNOP, respectively.

Comparing Table 7 with Table 8, we find that the LBMPTs estimated by the PSO-CNOP approach are completely the same as the ones computed by the filtering method. It is

results are statistically analyzed and presented in Tables 4 and 5.

It can clearly be seen from Tables 4 and 5 that the optimal population size of PSO-CNOP is about 15, but the optimal population size of GA-CNOP is about 40. Furthermore, the computational times and CFV calculated, respectively, by PSO-CNOP with a population size of 15 and GA-CNOP with a population size of 40, are compared with that of ADJ-CNOP. Table 6 illustrates the mean CFV and the average computation time obtained by different methods in their 40 numerical experiments.

Table 6. Analysis of CFV and average computational time by different methods.

Prediction time	$6\Delta t$		$13\Delta t$	
Method	Mean CFV	Time (s)	Mean CFV	Time (s)
ADJ-CNOP	3.0100×10^{-2}	5.0000×10^{-4}	1.3944	6.7500×10^{-4}
PSO-CNOP	3.7533×10^{-2}	0.0032	1.5301	0.0058
GA-CNOP	3.7533×10^{-2}	0.0299	1.5300	0.0358

Table 7. The LBMPT computed by the filtering method.

E_m	$\delta = 0.01$	$\delta = 0.02$	$\delta = 0.03$	$\delta = 0.04$	$\delta = 0.05$
0.5	9	7	7	6	6
0.8	10	9	7	7	6
1.1	10	10	9	7	7
1.4	10	10	9	7	7
1.7	12	12	12	11	10

Table 9. The LBMPT estimated by the ADJ-CNOP method.

E_m	$\delta = 0.01$	$\delta = 0.02$	$\delta = 0.03$	$\delta = 0.04$	$\delta = 0.05$
0.5	9	7	7	7	7
0.8	10	9	**8**	7	6
1.1	10	**11**	9	**9**	**8**
1.4	10	10	**10**	9	**9**
1.7	12	**13**	12	11	10

Table 8. The LBMPT estimated by the PSO-CNOP method.

E_m	$\delta = 0.01$	$\delta = 0.02$	$\delta = 0.03$	$\delta = 0.04$	$\delta = 0.05$
0.5	9	7	7	6	6
0.8	10	9	7	7	6
1.1	10	10	9	7	7
1.4	10	10	9	7	7
1.7	12	12	12	11	10

Table 10. Incorrect ratio in all 40 LBMPTs yielded by ADJ-CNOP.

E_m	$\delta = 0.01$	$\delta = 0.02$	$\delta = 0.03$	$\delta = 0.04$	$\delta = 0.05$
0.5	15 %	40 %	32.5 %	52.5 %	45 %
0.8	22.5 %	55 %	35 %	42.5 %	50 %
1.1	7.5 %	42.5 %	35 %	42.5 %	57.5 %
1.4	25 %	32.5 %	35 %	52.5 %	50 %
1.7	12.5 %	95 %	30 %	60 %	60 %

worth mentioning that with the prediction time extending, although the objective functions have multiple extreme values and PSO-CNOP would produce different CNOPs in different numerical experiments, each one of these CNOPs can be used to estimate the maximum prediction error, and the final LBMPTs obtained are the same since they are all global.

The bold numbers in Table 9 are the LBMPTs that are different from the ones computed by the filtering method. The LBMPTs yielded by the ADJ-CNOP method are generally larger. The reason is that the CNOP given by the ADJ-CNOP method is often local, even false. Therefore the estimation of the maximum prediction error based on the CNOP is usually questionable and untrusted.

To investigate the probability that the ADJ-CNOP method generates incorrect LBMPTs, we operate the numerical experiment shown in Table 9 40 times with various first guesses of the initial perturbations. The statistical analysis results are given in Table 10.

From Table 10, we can see that the ratio of incorrect LBMPTs based on ADJ-CNOP is high. The highest one even reaches 95 % when $\delta = 0.02$ and $E_m = 1.7$. This problem is serious for real weather forecasts since it can mislead forecasts with a large probability.

5　Conclusion and discussion

Since the two-dimensional Ikeda model has strong nonlinearity, when we utilize the ADJ-CNOP method to capture CNOPs, not only global or local CNOPs, but even false CNOPs, are obtained. The reason for this is that in the case of strong nonlinearity, the gradient provided by the adjoint model is incorrect. When the traditional optimization algorithm uses a wrong descent direction to search for extreme values of the objective function, false CNOPs are presented.

PSO is a heuristic search algorithm based on population. It can overcome the nonlinear influences and produce global CNOPs with high probability. In addition, the operation of the PSO algorithm is simple. This study applies the PSO algorithm to capture the CNOP of the two-dimensional Ikeda model. Numerical experiment results with different forecast times demonstrate that although the objective function has awful behavior and multiple extreme values, PSO-CNOP can still capture global CNOPs.

Furthermore, precision problems of using ADJ-CNOP to estimate the LBMPT are investigated. Results show that when the objective function has multiple extreme values, ADJ-CNOP has a large probability of producing the local CNOP, hence inducing false estimation of the LBMPT. As PSO-CNOP can always yield global CNOPs, therefore, the estimation of the LBMPT presented by PSO-CNOP is precise. It is consistent with the one yielded by the filtering method with fine division.

As we know, our numerical experiments focus on two-dimensional prediction models only. When considering high-dimensional and more complex models, whether or not the classic PSO algorithm used in this study can overcome the influence of high dimensions and the computation time meet the real requirement is still unknown. The problems of the curse of dimensionality and multimodal function are a big challenge for almost all intelligent optimization algorithms, also PSO. Whether it can be effective in higher-dimensional and more complicated models deserves further research. In short, the PSO-CNOP approach is an alternative method to study predictability problems in the case of strong nonlinearity.

Competing interests. The authors declare that they have no conflict of interest.

Acknowledgements. This work was supported by the National Natural Science Foundation of China (grant nos. 41430426 and 41331174).

Edited by: J. Duan

References

Banks, A., Vincent, J., and Anyakoha, C.: A review of particle swarm optimization. Part I: background and development, Nat. Comput., 6, 467–484, doi:10.1007/s11047-007-9049-5, 2007.

Banks, A., Vincent, J., and Anyakoha, C.: A review of particle swarm optimization. Part II: hybridisation, combinatorial, multicriteria and constrained optimization, and indicative applications, Nat. Comput., 7, 109–124, doi:10.1007/s11047-007-9050-z, 2008.

Birgin, E. G., Martinez, J. M., and Raydan, M.: Nonmonotone spectral projected gradient methods on convex sets, SIAM J. Optimiz., 10, 1196–1211, 2000.

Borges, M. D and Hartmann, D. L.: Barotropic Instability and Optimal Perturbations of Observed Nonzonal Flows, J. Atmos. Sci., 49, 335–353, 1992.

Buizza, R. and Palmer, T. N.: The Singular-Vector Structure of the Atmospheric Global Circulation, J. Atmos. Sci., 52, 1434–1456, 1995.

Clerc, M. and Kennedy, J.: The particle swarm: Explosion, stability, and convergence in a multi-dimensional complex space, IEEE T. Evolut. Comput., 6, 58–73, 2002.

Duan, W. S. and Luo, H. Y.: A New Strategy for Solving a class of Constrained Nonlinear Optimization Problems Related to Weather and Climate Predictability, Adv. Atmos. Sci., 27, 741–749, doi:10.1007/s00376-009-9141-0, 2010.

Duan, W. S., Mu, M., and Wang, B.: Conditional nonlinear optimal perturbation as the optimal precursors for ENSO events, J. Geophys. Res., 109, D23105, doi:10.1029/2004JD004756, 2004.

Duan, W. S., Xue, F., and Mu, M.: Investigating a nonlinear characteristic of El Niño events by conditional nonlinear optimal perturbation, Atmos. Res., 94, 10–18, 2008.

Eberhart, R. C. and Shi, Y.: Particle swarm optimization: developments, applications and resources, Congress on Evolutionary Computation, 37–30 May 2001, Korea, 1, 81–86, 2001.

Fang, C. L. and Zheng, Q.: The effectiveness of a genetic algorithm in capturing conditional nonlinear perturbation with parameterization "on-off" switches included by a model, J. Trop. Meteorol., 13, 13–19, 2009.

Farrell, B. F.: Small error dynamics and the predictability of atmospheric flows, J. Atmos. Sci., 47, 2409–2416, 1990.

Hassan R., Cohanim B., De Weck O, and Venter G.: A Comparison of Particle Swarm Optimization and the Genetic Algorithm [C], Proceedings of the 1st AIAA multidisciplinary design optimization specialist conference, 18–21 April 2005, Austin, Texas, 1897, 2005.

Ikeda, K.: Multiple-valued stationary state and its instability of the transmitted light by a ring cavity system, J. Opt. Commun., 30, 257–261, 1979.

Jiang, Z. N. and Wang, D. H.: A study on precursors to blocking anomalies in climatological flows by using conditional nonlinear optimal perturbations, Q. J. Roy. Meteor. Soc., 136, 1170–1180, 2010.

Kalnay, E.: Atmospheric modeling, data assimilation and predictability [M], Cambridge university press, UK, 2003.

Kennedy, J. and Eberhart, R. C.: Particle swarm optimization, in: Proceedings of the 1995 IEEE international conference on neural networks, Perth, Australia, Piscataway, NJ: IEEE Service Center, 27 November–1 December 1995, 1942–1948, 1995.

Lacarra, J. F. and Talagrand, O.: Short-range evolution of small perturbations in a baratropic model, Tellus, 40, 81–95, 1988.

Li, Q., Zheng, Q., and Zhou, S. Z.: Study on the dependence of the two-dimensional Ikeda model on the parameter, Atmos. Ocean. Sci. Lett., 9, 1–6, doi:10.1080/16742834.2015.1128694, 2016.

Liu, D. C. and Nocedal, J.: On the limited memory BFGS method for large scale optimization, Math. Program., 45, 503–528, 1989.

Lorenz, E. N.: Climate predictability. The Physical Basis of Climate Modeling, WMO GARP Publ. Ser. No. 16, WMO, Sweden, 132–136, 1975.

Mu, M. and Duan, W. S.: A new approach to studying ENSO predictability: Conditional nonlinear optimal perturbation, Chinese Sci. Bull., 48, 1045–1047, 2003.

Mu, M. and Zhang, Z. Y.: Conditional nonlinear optimal perturbation of a two-dimensional quasigeostrophic model, J. Atmos. Sci., 63, 1587–1604, 2006.

Mu, M., Duan, W. S., and Wang, J. C.: The predictability problems in numerical weather and climate prediction, Adv. Atmos. Sci., 19, 191–204, 2002.

Mu, M., Wang, H. L., and Zhou, F. F.: A preliminary application of conditional nonlinear optimal perturbation to adaptive observation, J. Atmos. Sci., 31, 1102–1112, 2007.

Mu, M., Zhou, F. F., and Wang, H. H.: A method for identifying the sensitive areas in targeted observations for tropical cyclone prediction: Conditional nonlinear optimal perturbation, Mon. Weather Rev., 137, 1623–1639, 2009b.

Conditional nonlinear optimal perturbations based on the particle swarm optimization and their applications...

193

Poli, R., Kennedy, J., and Blackwell, T.: Particle swarm optimization, Swarm Intell., 1, 33–57, 2007.

Powell, M. J. D.: VMCWD: A FORTRAN subroutine for constrained optimization, DAMTP Report 1982/NA4, University of Cambridge, NY, USA, 1982.

Riviere, O., Lapeyre, G., and Talagrand, O.: A novel technique for nonlinear sensitivity analysis: application to moist predictability, Q. J. Roy. Meteorol. Soc., 135, 1520–1537, 2010.

Shi, Y. and Eberhart, R. C.: A Modified Particle Swarm Optimizer, in: The 1998 Conference of Evolutionary Computation, Piscataway, IEEE Press, 4–9 May, Anchorage, AK, 69–73, 1998.

Terwisscha van Scheltinga, A. D. and Dijkstra, H. A.: Conditional nonlinear optimal perturbations of the double-gyre ocean circulation, Nonlin. Processes Geophys., 15, 727–734, doi:10.5194/npg-15-727-2008, 2008.

Wang, Q., Mu, M., and Dijikstra, H. A.: Application of the conditional nonlinear optimal perturbation method to the predictability study of the Kuroshio large meander, Adv. Atmos. Sci, 29, 118–134, 2012.

Wang, Q., Mu, M., and Dijkstra, H. A.: Effects of nonlinear physical processes on optimal error growth in predictability experiments of the Kuroshio Large Meander, J. Geophys. Res.-Oceans, 118, 6425–6436, doi:10.1002/2013JC009276, 2013.

Yu, Y. S., Mu, M., and Duan, W. S.: Does model error cause a significant spring predictability barrier for El Nino events in the Zebiak-Cane model?, J. Climate, 25, 1263–1277, 2012.

Zheng, Q., Dai, Y., Zhang, L., Sha, J. X., and Lu, X. Q.: On the application of a genetic algorithm to the predictability problems involving "on/off" switches, Adv. Atmos. Sci., 29, 422–434, doi:10.1007/s00376-011-1054-z, 2012.

Zheng, Q., Sha, J. X., Shu, H., and Lu, X. Q.: A variant constrained genetic algorithm for solving conditional nonlinear optimal perturbations, Adv. Atmos. Sci., 31, 219–229, doi:10.1007/s00376-013-2253-6, 2014.

Regularization destriping of remote sensing imagery

Ranil Basnayake[1,2], Erik Bollt[1,2], Nicholas Tufillaro[3], Jie Sun[1,2], and Michelle Gierach[4]

[1]Department of Mathematics, Clarkson University, Potsdam, NY 13699, USA
[2]Clarkson Center for Complex Systems Science (C^3S^2), Potsdam, NY 13699, USA
[3]College of Earth, Ocean, and Atmospheric Sciences, Oregon State University, Corvallis, OR 97331, USA
[4]Jet Propulsion Laboratory, California Institute of Technology, Pasadena, CA, 91109, USA

Correspondence to: Ranil Basnayake (rbasnaya@clarkson.edu)

Abstract. We illustrate the utility of variational destriping for ocean color images from both multispectral and hyperspectral sensors. In particular, we examine data from a filter spectrometer, the Visible Infrared Imaging Radiometer Suite (VIIRS) on the Suomi National Polar Partnership (NPP) orbiter, and an airborne grating spectrometer, the Jet Population Laboratory's (JPL) hyperspectral Portable Remote Imaging Spectrometer (PRISM) sensor. We solve the destriping problem using a variational regularization method by giving weights spatially to preserve the other features of the image during the destriping process. The target functional penalizes "the neighborhood of stripes" (strictly, directionally uniform features) while promoting data fidelity, and the functional is minimized by solving the Euler–Lagrange equations with an explicit finite-difference scheme. We show the accuracy of our method from a benchmark data set which represents the sea surface temperature off the coast of Oregon, USA. Technical details, such as how to impose continuity across data gaps using inpainting, are also described.

1 Introduction

Striping is a persistent artifact in remote sensing images and is particularly pronounced in visible–near infrared (VNIR) water-leaving radiance products such as those produced by operational sensors including NPP VIIRS, Landsat 8 Operational Land Imager (OLI), and Geostationary Ocean Color Imager (GOCI), as well airborne instruments such as NASA's JPL PRISM sensor. These sensors cover a temporal sampling range from daily (VIIRS) to hourly (GOCI) and spectral sampling from multi-spectral (VIIRS, GOCI) to hyperspectral

(PRISM). Striping is pronounced in products from all these sensors because atmospheric correction for ocean color products typically removes at least 90 % of the signal recorded at the top of the atmosphere (TOA). Put another way, any artifacts in the TOA signal are amplified by at least a factor of 10 in any derived water products such as normalized water-leaving radiance of a specific spectral band ($nLw(\lambda)$), or in product fields such as total suspended sediment (TSS) concentration maps.

Striping is ubiquitous and difficult to remove because it has many possible origins. The detectors themselves are subjected to small amplitude variations in both sensitivity and calibration. The view angles (azimuthal and zenith) also vary from detector to detector and from pixel to pixel. Other differences in the instrument's optical path, components (e.g., mirrors), asynchronous readout, and so on also cause striping. Not unexpectedly, the magnitude of the striping varies from image to image. Striping is particularly problematic when comparing a sequence of images, since any difference in computations between images produces spurious results in the neighborhood of stripes.

Ocean products from NPP VIIRS have shown problematic striping since its launch, which has led to focus efforts at both NASA and NOAA to find correction methods. NASA created a vicarious destriping method for VIIRS images based on a collection of long-term on-orbit image data, including solar and lunar calibrations. NASA's Ocean Biology Processing Group (OBPG) began serving operational products with their vicarious calibrations and destriping for VIIRS in 2014 Eplee et al. (2012). In contrast, a method for hierarchically destriping VNIR images based on a single scene was proposed in Bouali (2010). This particular variational method was more recently augmented with filtering using a hierarchical image decomposition (Bouali and Ignatov, 2013), and that algorithm has also been implemented by scientists at NOAA for operations with images from VIIRS (Mikelsons et al., 2015, 2014). Scene-based processing methods are advantageous for sensors which do not have dedicated calibration subsystems such as a solar diffuser or where the data sets are limited in scope (such as in the case of airborne sensors) and do not include uniform scenes for vicarious calibration.

2 Regularization destriping: the functional and its minimization

The method described here is closely related to the destriping functional described in Bouali (2010). Our work differs in its exact functional form and its principle of solution. In particular, we formulate a solution for destriping in an inverse-problem framework, and keep only the part of the functional in Bouali (2010) that smooths the stripes. This new formulation allows us to provide an explicit numerical solution instead of an iterative one, the former being more efficient and thus better suited for operational codes. We explicitly introduce a regularization parameter that controls the relative balance between the data term ("fitting the original image") and the regularity term ("smoothing out the stripes"). Such regularization is a common practice in inverse problems (Vogel, 2002), and fall under the rubric of Tikhonov regularization theory. As a further refinement of the destriping functional, we specifically define weights for the regularization term so that the algorithm applies only to the stripes while preserving the other features.

Assuming that the stripes are parallel to one another in the image plane, we take the direction of the stripes as the x (horizontal) direction. Thus the data term representing the horizontal gradient difference between the original and the destriped images is given as

$$E_D(u) = \int_\Omega \left(\frac{\partial}{\partial x}(u - f) \right)^2 d\Omega, \tag{1}$$

where Ω is the image domain on xy plane, $f(x, y)$ is the original image with stripes, and $u(x, y)$ is the destriped image.

The regularization term emphasizes the smoothness in the vertical direction, which is assumed to be free of stripes. This regularization term is given by

$$E_R(u) = \int_\Omega \left(\frac{\partial u}{\partial y} \right)^2 d\Omega. \tag{2}$$

The regularization parameter $\alpha > 0$ balances the data term and the regularization term. The resulting destriping functional is

$$E_C(u) = \int_\Omega \left(\frac{\partial}{\partial x}(u - f) \right)^2 d\Omega + \alpha \int_\Omega \left(\frac{\partial u}{\partial y} \right)^2 d\Omega. \tag{3}$$

This is the x-directional destriping functional proposed by Bouali (2010), and it is equivalent to the basic form of our destriping functional when $\alpha = 1$. The choice of α, as we show later, is key to achieving the balance between matching the original image and removing stripes. However, our approach differs from Bouali (2010) and our goal is to develop a destriping method that is easy to implement while preserving the other features of the image.

A drawback of scene-based destriping is unintended changes in the values of all the pixels and not just the stripes. If we apply the regularization term for the whole image as it is in Eq. (3), the entire image is effected. This could modify the original features of the image, in addition to recovering stripes. Therefore, we further develop our functional in Eq. (3) to regularize only the stripes. We introduce a mask (L) to the regularization term, to limit the smoothing effects on the stripes. To obtain L from the image, we first compute the slope of the image transverse to the stripes using first-order finite differences. Then we sum the absolute differences parallel to the stripes. This yields the total value (S) corresponding to each row. From the peaks of graph S versus r, where r is the row index, we can identify the stripes and select a "threshold" to separate the stripes from the other features. To compute the S curve, we do not need to consider the complete image and instead we can consider a vertical image segment that contains some part of all the biased data. In this case, we can preserve the actual features such as roads or any other real signals that are exactly aligned with stripes.

The mathematical expression for the computation of S for an image f, of size $m \times n$, with a suitable boundary condition, can be written as

$$S(r) = \sum_{c=1}^{n} |f(r, c) - f(r + 1, c)|, \tag{4}$$

where $r = 1, 2, \ldots, m$ and $f(m + 1, c)$ is the introduced boundary row. Now defining the "threshold" value from the S curve, we obtain the sparse matrix $\mathbf{L}$ by assigning "ones" for the locations of the stripes. We will explain the procedure of selecting an appropriate "threshold" value using examples in Sect. 3. Any row r, where $r = 1, 2, \ldots, m$ of matrix $\mathbf{L}$ with size $m \times n$ can be defined as

$$\mathbf{L}(r, c) = \begin{cases} 1, & \text{if } S(r) \geq \text{threshold} \\ 0, & \text{otherwise}, \end{cases} \tag{5}$$

where $c = 1, 2, \ldots, n$.

Then the new destriping functional, with the spatially weighted regularization term, is written as

$$E(u) = \int_\Omega \left(\frac{\partial}{\partial x}(u - f) \right)^2 d\Omega + \alpha \int_\Omega L \left(\frac{\partial u}{\partial y} \right)^2 d\Omega. \tag{6}$$

The destriped image is obtained by minimizing the functional after choosing an appropriate regularization parameter. Note that the functional $E(u)$ is invariant under constant shift. That is, $E(u+a) = E(u)$ for any constant a, implying that minimization of $E(u)$ leads to an infinite number of solutions. In this work, we impose appropriate boundary conditions (as discussed below) that ensure uniqueness of the solution.

We create a destriped image by minimizing the energy functional in Eq. (6) using the Euler–Lagrange equation. For a functional of the form

$$J(u) = \int_\Omega F\left(x, y, u, u_x, u_y\right) d\Omega$$

on the bounded domain Ω, the Euler–Lagrange equation is given as

$$\frac{\partial F}{\partial u} - \frac{\partial}{\partial x}\left(\frac{\partial F}{\partial u_x}\right) - \frac{\partial}{\partial y}\left(\frac{\partial F}{\partial u_y}\right) = 0. \tag{7}$$

As explained in Vogel (2002) and Basnayake and Bollt (2014), the Euler–Lagrange equation,

$$u_{xx} + \alpha L u_{yy} = f_{xx} \tag{8}$$

is obtained by applying Eq. (7) to Eq. (6), where subscripts represent the argument variable(s) of the partial derivatives. We can rewrite Eq. (8) as

$$\left(D_{xx} + \alpha L D_{yy}\right) u = D_{xx} f, \tag{9}$$

where the operators $D_{\bullet\bullet}$ are two-dimensional arrays of size $k \times k$ used to compute the partial derivatives of a given vector of size $k \times 1$ with respect to the indices $\bullet\bullet$.

We use finite-difference approximations with suitable boundary conditions for each derivative to directly represent these differential operators. In this work, we apply "reflexive" boundary conditions parallel to the stripes and "zero" boundary conditions transverse to the stripes when we the generate derivative operators. These boundary conditions lead $\left(D_{xx} + \alpha L D_{yy}\right)$ to a full rank operator and hence we reach a unique solution. In Eq. (9) we stack the given image of size $p \times q$ into $k \times 1$ vector, where $k = pq$. We can now use the differential operators to write the linear Euler–Lagrange equation in the form of $Au = b$, and solve for u using an appropriate numerical method rather than solving the Euler–Lagrange equation for u from an iterative scheme such as the gradient descent method.

2.1 Construction of the differential operator

We construct the operator D_{xx} using finite-difference approximations. The operator D_{yy} is built by taking the transpose of the finite-difference stencil. Suppose we have a function $M(x, y) \in \mathbb{R}^{p \times q}$. We need to compute the second-order

partial derivative of M with respect to x. We use a fourth-order finite-difference approximation and compute the pointwise second partial derivative of the array M with respect to x.

As an example, take an array $M(x, y)$ of size 3×5 where we want to compute $M_{xx}(x, y)$. We index the elements in the array in the form of a column vector as shown in Table 1, with two added boundary columns for each side.

The boundary points are highlighted in bold. If we compute the partial derivative of m_1 with respect to x, the resulting approximation is

$$\frac{\partial^2 m_1}{\partial x^2} = \frac{1}{12h^2}\left[-\mathbf{m_4} + 16\mathbf{m_1} - 30m_1 + 16m_4 - m_7\right]$$

$$= \frac{1}{12h^2}\left[-14m_1 + 15m_4 - m_7\right].$$

Computing the finite-difference approximations for each element in the array, we can obtain the differential operator D_{xx} corresponding to the 3×5 array in Table 1. The resulting operator is a sparse matrix with only five non-zero diagonals, and it is shown in Table 2 with a multiplication factor of $12h^2$.

2.2 Solution to the Euler–Lagrange equation

Now we can determine the solution to Eq. (9). We rewrite the Eq. (9) as

$$Au = b, \tag{10}$$

where $A = D_{xx} + \alpha L D_{yy}$ and $b = D_{xx} f$. If the size of the given striped image f is $p \times q$, then A is a $k \times k$ sparse array and b is a $k \times 1$ array, where $k = pq$.

Using a suitable value for the regularization parameter, Eq. (10) can be solved as a linear system. The system is sparse, and hence the computation time for an image with n pixels is of $O(n)$ for each iteration. Clearly, at this stage, for a given α, Eq. (10) is straightforward to solve; however, in terms of the image processing, the specific choice of α plays an important role. To achieve "the most appropriate solution," we need to determine the best regularization parameter α.

2.3 Selection of the regularization parameter

The condition number of the resulting matrix quantifies the amplification of computational errors seen while solving the problem by direct computation. The condition number may be computed as the ratio between the largest singular value and the smallest singular value of the coefficient matrix. If the condition number is large, then the coefficient matrix is said to be ill-conditioned and hence the corresponding system is ill-posed. In an ill-posed system, the solution is highly sensitive to perturbations of the input data. Regularizing an ill-posed system, which emphasizes a desired property of the

Table 1. An array of 3×5 with boundary points in bold.

$\mathbf{m_4}$	$\mathbf{m_1}$	m_1	m_4	m_7	m_{10}	m_{13}	$\mathbf{m_{13}}$	$\mathbf{m_{10}}$
$\mathbf{m_5}$	$\mathbf{m_2}$	m_2	m_5	m_8	m_{11}	m_{14}	$\mathbf{m_{14}}$	$\mathbf{m_{11}}$
$\mathbf{m_6}$	$\mathbf{m_3}$	m_3	m_6	m_9	m_{12}	m_{15}	$\mathbf{m_{15}}$	$\mathbf{m_{12}}$

Table 2. A discretized derivative operator $D_{xx} \times 12h^2$ for a 3×5 matrix.

	m_1	m_2	m_3	m_4	m_5	m_6	m_7	m_8	m_9	m_{10}	m_{11}	m_{12}	m_{13}	m_{14}	m_{15}
m_1	-14	0	0	15	0	0	-1	0	0	0	0	0	0	0	0
m_2	0	-14	0	0	15	0	0	-1	0	0	0	0	0	0	0
m_3	0	0	-14	0	0	15	0	0	-1	0	0	0	0	0	0
m_4	15	0	0	-30	0	0	16	0	0	-1	0	0	0	0	0
m_5	0	15	0	0	-30	0	0	16	0	0	-1	0	0	0	0
m_6	0	0	15	0	0	-30	0	0	16	0	0	-1	0	0	0
m_7	-1	0	0	16	0	0	-30	0	0	16	0	0	-1	0	0
m_8	0	-1	0	0	16	0	0	-30	0	0	16	0	0	-1	0
m_9	0	0	-1	0	0	16	0	0	-30	0	0	16	0	0	-1
m_{10}	0	0	0	-1	0	0	16	0	0	-30	0	0	15	0	0
m_{11}	0	0	0	0	-1	0	0	16	0	0	-30	0	0	15	0
m_{12}	0	0	0	0	0	-1	0	0	16	0	0	-30	0	0	15
m_{13}	0	0	0	0	0	0	-1	0	0	15	0	0	-14	0	0
m_{14}	0	0	0	0	0	0	0	-1	0	0	15	0	0	-14	0
m_{15}	0	0	0	0	0	0	0	0	-1	0	0	15	0	0	-14

problem, introduces a stable way to define a desirable solution (Vogel, 2002; Hansen et al., 2006; Hansen, 2010). This is the standard trade-off between regularity and stability in Tikhonov regularization terms.

We regularize our computed solution by emphasizing the expected physics. To damp the accumulated errors from the residuals, we must make sure that we add sufficient regularity. The balance between the data term and the regularization term is very important: if we add too much regularity, it will divert the solution from the desired solution. Stated in terms of Tikhonov regularization, α serves the role to select a unique optimizer u, from what would be an otherwise ill-posed system had only the data fidelity had been chosen. In terms of the images, the data fidelity states that the optimizer image u should "appear as" the original data image, measured here in terms of along the stripes, but "smoothed" according to the regularizer term, in this case transverse to the stripes. The question then is how to balance these two terms.

Some of the common methods to determine the regularization parameter in inverse problems are the L-curve method of Hansen (1992, 2000) and U-curve method of Krawczyk-StańDo and Rudnicki (2007) and Krawczyk-Stado and Rudnicki (2008). After trials with both methods we selected the U-curve method. Starting with the functional

$$J(u) = \underset{u}{\arg\min} \|Bu - v\|_2^2 + \alpha \|u\|_2^2, \tag{11}$$

where B is a $k \times k$ array and u, v are $k \times 1$ arrays for an integer k, a regularized solution is computed for a fixed α. Then the norm of the residuals and solution are written as

$$x(\alpha) = \|Bu_\alpha - v\|_2^2 \text{ and}$$
$$y(\alpha) = \|u_\alpha\|_2^2. \tag{12}$$

In general, $\alpha \in (0, 1)$ and in this work we picked 100 equally spaced α values in the interval of $(10^{-12}, 1)$. Considering all the α values, and computing the sum of the reciprocals, results in a U-curve,

$$U(\alpha) = \frac{1}{x(\alpha)} + \frac{1}{y(\alpha)}. \tag{13}$$

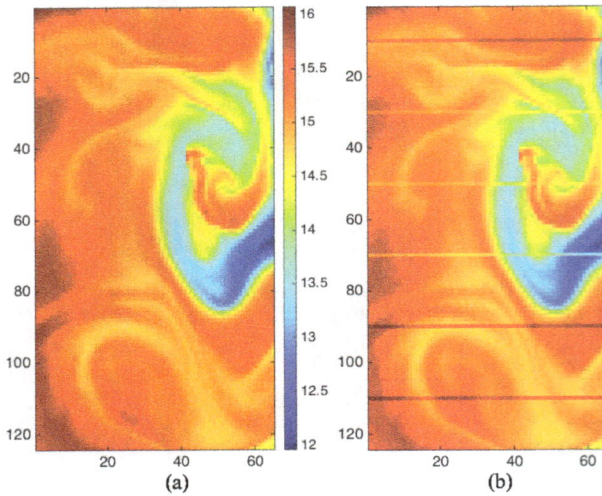

Figure 1. Panel **(a)** shows (simulated) variations of sea surface temperature off the coast of Oregon, USA, on 1 August 2002. The image was generated from a Regional Ocean Model System (ROMS, Courtesy of John Osborne, Oregon State University), using the data assimilation from the Geostationary Operational Environmental Satellite (GOES). Panel **(b)** is created by adding artificial stripes to every 20th row.

The "best" α value corresponds to the minimum of $U(\alpha)$. The "best" α value must be in the interval $\alpha \in \left(\sigma_r^{2/3}, \sigma_1^{2/3}\right)$, where σ_1 is the largest singular value of the operator B, and σ_r is the smallest non-zero singular value of the operator B (Krawczyk-StańDo and Rudnicki, 2007).

2.4   Inpainting data gaps

Unlike terrestrial images, which can show sharp edges, ocean color images typically appear continuous. This is because the water tends to diffuse any color agents in the water column, and the spatial resolution of the sensor is usually finer than those diffusive features. The same holds spectrally if the sampling wavelength is less than the autocorrelation function of the spectra, as is the case for hyperspectral images. However, this continuity is broken if gaps appear in the data.

Clouds are very bright and often saturate the sensor. In normal processing, clouds are typically masked from the data since their large radiance values obscure the (relatively dark) ocean. These types of processing masks also cause large, irregular data gaps. There are other sources of data gaps as well, as we discuss here, that can be inherent in the sensor design.

As an example of data gaps introduced by system design, consider the VIIRS sensor, which uses 16 detector elements to generate each spatial image. The spatial footprint of each adjacent sensor element overlaps at the detector edges of Cao et al. (2013). To reduce data transmission from the satellite to ground stations, the overlapping regions are not transmit-

ted, causing the so-called "bow-tie" effect. This appears as visible horizontal stripes in the Level 1 or Level 2 unmapped pixel values. These gaps are removed during projection to ground-based coordinates, so-called Level-3 data (L3). However, stripes which are linear in the "unmapped" data become nonlinear after mapping, and are more difficult to remove. Therefore we work with the Level 1 and Level 2 data where the stripes are linear, and often aligned with the focal plane array or detector elements.

We need to preprocess the image data in such a way as to ensure continuity across any data gaps. An obvious way to fix the gaps is to use "inpainting", a technique from image processing – rather than infer what the actual missing data might be, inpainting simply imposes continuity across the whole image when gaps are present. In a museum setting, inpainting refers to the process whereby a painter–restorer interprets a damaged painting by artistically filling in damaged or missing parts of a painting, smoothly bleeding in the colors of the painting that surrounds the damage (Bertalmio et al., 2011). In modern-day computational parlance, inpainting refers to an image-processing algorithm that mimics this idea. A wide variety of methods have been implemented (Bertalmio et al., 2000).

Missing data are filled in by solving the Laplace equation with Dirichlet boundary conditions, $\nabla u = 0$ in Ω and $u = f$ on $\partial\Omega$; specifically, we use the MATLAB routine roifill. Because the algorithm approximates the partial derivatives using finite-difference approximations, inpainting should be done before applying the destriping algorithm on images.

3   Results and discussion

In this section, we first apply the destriping method to a simulated image of sea surface temperature (SST), and then apply the destriping algorithm to two real-world images, one from the multi-spectral NOAA imager VIIRS, and the second from the JPL PRISM hyperspectral sensor.

3.1   Benchmark data: sea surface temperature

A benchmark data set – sea surface temperature data off the Oregon coast – was used to test the codes. The original image data is shown in Fig. 1a, where red represents high temperature and blue represents the low temperature. This simulated image is from a three-dimensional Regional Ocean Model System (ROMS) showing the Oregon Coast on 1 August 2002 (Osborne et al., 2011). The area covered is 41–$46°$ N, -124–$-125°$ W.

The image (b) in Fig. 1 shows the artificially added stripes on the image shown in Fig. 1a. The intensities of stripes are assigned as the absolute values of difference between the original intensity and the average of each striped row. However, stripes at the 10th, 90th and 110th rows have some

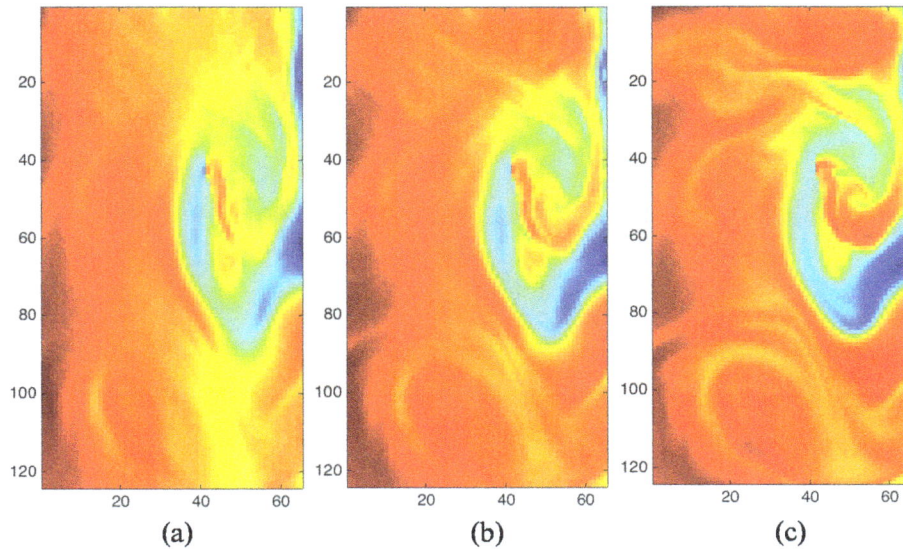

Figure 2. These three images represent the destriped versions of the image, shown in Fig. 1b, in three different ways. Panels **(a)** and **(b)** are obtained with $\alpha = 1$ and $\alpha = 3 \times 10^{-1}$ from the functional in Eq. (3). Panel **(c)** is obtained with $\alpha = 7 \times 10^{-1}$ from the functional with spatially weighted regularization term, as shown in Eq. (6).

added noise intensity values so that they can be visualized properly.

Our next step is to apply the destriping method to the SST data and check the accuracy of the algorithm. There, we compare the solutions with regularization parameters $\alpha = 1$ and $\alpha = 3 \times 10^{-1}$ using the functional with an unweighted regularization term, as shown in Eq. (3) and $\alpha = 7 \times 10^{-3}$ with a spatially weighted regularization term, as shown in Eq. (6). The idea of this comparison is to show the importance of the selection of our regularization method and how the weighted regularization term improves the destriping results. The destriped image from the unweighted regularization term is shown in Fig. 2a, and clearly it is overly smooth as most of the original features are destroyed from the destriping. Therefore, we chose an appropriate α value to balance the data term and the unweighted regularization term from the U-curve method. The U-curve solution is $\alpha = 3 \times 10^{-1}$ and the resulting destriped image is shown in Fig. 2b. However, it can be observed that some features of the destriped image have undergone smoothing at some places in addition to the stripes during the destriping process. Hence, it is an important task to preserve the other features where there are no stripes while applying destriping. We achieve this by improving the regularization term by incorporating different weights to the places where stripes are available and to where they are not available. The weighted matrix L can be obtained from the expression in Eq. (5) by giving zeros to the places where no stripes. In this manner, we can preserve the features of the original image from smoothing, and the resulting image is shown in Fig. 2c.

The next task is to check the accuracy of the estimated values for the stripes. The stripe at the 110th row was randomly selected for this purpose. There we plot the intensity values of the stripe (black), reconstruction (red) and the actual values as they were against the column index. The results from the functional in Eq. (3) with $\alpha = 1$ is shown in the graph (c) of Fig. 3. The reconstructions of the first 20 pixels and the last 10 pixels are closer to the original values, but the rest have significant differences compared to the actual value. The graph (b) in Fig. 3 shows destriped results of the 110th row from the functional in Eq. (3) using the U-curve solution, $\alpha = 3 \times 10^{-1}$. In this case, the reconstruction of some middle pixels (18–28 and 31–37) and the last 10 pixels are off by some extent, but the rest is much closer to the actual values than the reconstruction with $\alpha = 1$. The graph (c) in Fig. 3 shows the best solution among these approaches, obtained from the weighted regularization term with $\alpha = 7 \times 10^{-1}$. This reconstruction is very close to the actual values and hence it is the most accurate reconstruction. These are the sort of results that we expect from a destriping algorithm. More importantly, if we compare "a stripe-like feature" at the 18th row, we can confirm the importance of giving weights to the regularization term, as the weighted regularization term focus only the stripes.

In addition to the reconstruction of stripes, we need to pay attention to the rest of the features of the image. The idea of destriping is to remove artificial stripes while preserving the other original features of the image. Therefore, we randomly select the 67th row of the image to compare before and after effects of destriping at a place where there is no actual stripe. The graphs (a) and (b) in Fig. 4 are from the func-

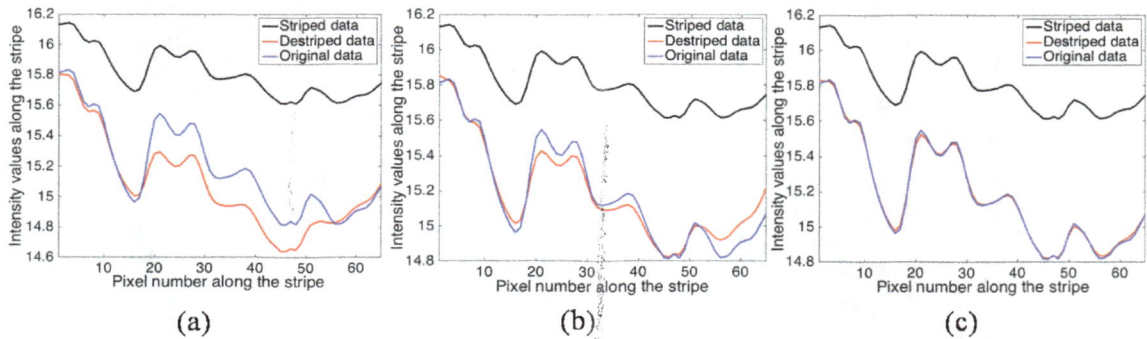

Figure 3. This figure presents three different reconstructions of the stripe at the 110th row of the image shown in Fig. 1b. The graphs show the actual data in blue, striped data in black and the destriped data in red. Graphs **(a)** and **(b)** represent the reconstructions from the functional in Eq. (3) with $\alpha = 1$ and $\alpha = 3 \times 10^{-1}$, respectively. The graph **(c)** shows the reconstruction from the functional in Eq. (6) with $\alpha = 7 \times 10^{-1}$.

tional in Eq. (3) with $\alpha = 1$ and $\alpha = 3 \times 10^{-1}$, respectively, and the graph (c) in Fig. 4 from the functional in Eq. (6) with $\alpha = 7 \times 10^{-1}$. The graphs (a) and (b) in Fig. 4 conclude that the destriping has affected the other features of the image when the functional in Eq. (3) is applied regardless of the value of the regularization parameter. However, the graph (c) in Fig. 4 shows that the functional with the spatially weighted regularization term in Eq. (6) does not affect the other features of the image.

The S curve corresponding to the SST image is shown in graph (a) in Fig. 5. The "threshold" value for this problem is also shown on the same graph and it is 18 in this case. If we consider the units of the sea surface temperature data, the threshold value can written as 842 °C longitude per latitude. To determine the best "threshold" value that does not affect the image regions with no stripes, we begin the procedure by assigning the "threshold" value to the maximum peak of the S curve. Then we apply the destriping algorithm and check the solution. If at least one stripe is still visible, we assign the next highest peak of the S curve as the "threshold" value and check for the stripes. Continuing in this manner until all the stripes are removed, the best "threshold" value can be determined. The peak points that have crossed the threshold value are the stripes. As a comparison of the full destriping image, the absolute percentage error between the striped and destriped image is shown in the image (b) in Fig. 5. This is computed by

$$\text{Absolute Percentage Error} = \frac{|f - u|}{f} \times 100,$$

where f and u are striped and destriped images respectively. The graph of the absolute percentage error shows that not only the 67th row but also all the other non-stripe rows are not affected from this spatially weighted regularization method.

3.2 Example 2 – VIIRS images

A good example of VIIRS stripping is shown (Fig. 6) in a chlorophyll concentration map near the Santa Monica region in southern California on 7 November 2014 NAS (2014). The chlorophyll concentration is given in milligrams per cubic meter. Green is the land and the dark blue is the dropped data and the missing data. The full VIIRS data granule covers $-122.09°$ W to $-116.90°$ E and $34.2°$ N to $31.6°$ S (Fig. 6). We consider a subsetted (cropped) region of interest to highlight the stripes in Fig. 7a.

The first step of applying this destriping method is to determine the threshold value to separate the neighborhood of stripes and the rest of the features. However, when we deal with real data, we may not always get nice and smooth images. For instance, if we compute the S curve values using the whole image, we are unable to get any evidence to determine the threshold value, as the sum of the magnitude of forward differences in some other rows are higher than that of the stripes. Therefore, we need to carefully pick a subregion of the image that includes all the rows with some selected columns. The sum of the magnitude of forward differences in non-stripe rows should be less than the that of stripes in this region. Then the stripes can be easily highlighted. For this example, we select the region using the columns from 137 to 148, and it is shown in Fig. 7b with the S curve. The threshold value can be determined from the S curve, and it is 0.558 for this image. If we include the proper units of the image data, the threshold value is 86.04 mg m^{-3}.

The effects of spatially weighted regularizing destriping are shown in Fig. 7e. We apply our algorithm to the image shown in Fig. 7a. In this case, the regularization parameter was $\alpha = 10^{-2}$ with the weighted regularization term. The intensity of the destriped image now varies smoothly. Therefore, the image is smooth enough to further post-process for other applications such as computing optical flow.

The approach is proposed in Bouali (2010) without the transverse direction functional, effectively setting $\alpha = 1$ in

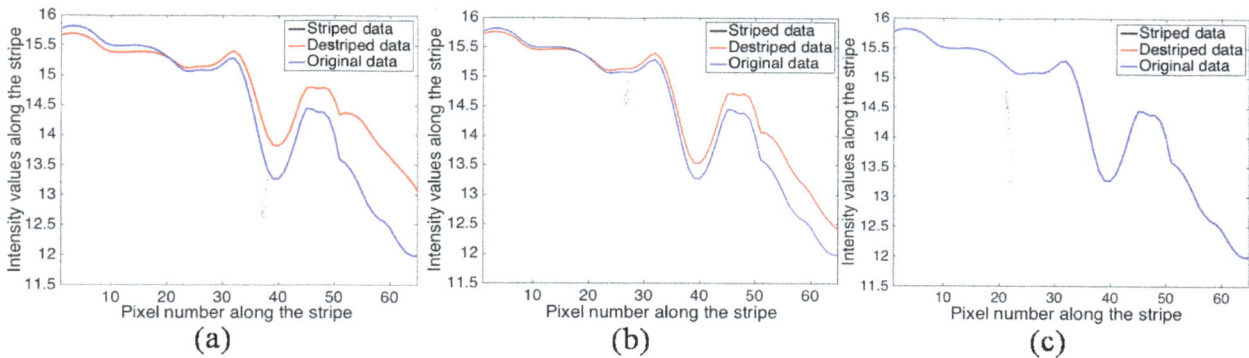

Figure 4. This figure shows the effects of destriping on the places where there are "no stripes". We randomly picked the 67th row for this comparison. When $\alpha = 1$ in the unweighted regularization functional, the image is more smoothed and affects destriping to the whole image. Much better results can be obtained from the unweighted regularization functional with $\alpha = 3 \times 10^{-1}$, which is the U-curve solution and is shown in graph (**b**). A spatially weighted regularization term with $\alpha = 7 \times 10^{-1}$ has less of an effect on the other features of the image and we can observe that from the graph (**c**).

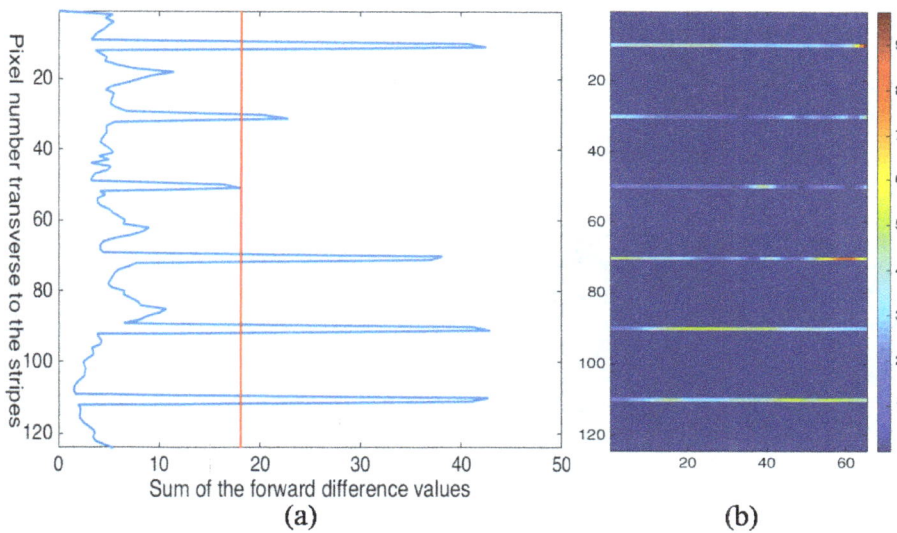

Figure 5. Panel (**a**) shows the S function values against the column numbers. The peak points represent the stripes. When "threshold" is set at 18, only stripes can be included for the regularization but excludes all the other features of the image. With the units of the image data, the threshold value can written as 842 °C longitude per latitude. Then the percentage error between the striped and destriped image are shown in the panel (**b**). The error where there are no stripes is always closer to the zero.

our functional Eq. (3). However, our destriping functional as shown in Eq. (6) has weight matrix **L** inside the regularization term and hence the two approaches are not the same. Panel (c) in Fig. 7 represents the destriped image when $\alpha = 1$ without the weighted regularization term, where image (a) is the original image. The destriped image is blurred and we lose some information from the original image due to "over regularization". In this case, by over regularization we mean that continuity in the transverse direction to the "stripes" has been emphasized so strongly in functional Eq. (3), when α is close to 1, that it is past balance against the need to also emphasize the image data in the first term, called "data fidelity".

On the other hand, image (d) is the destriped image of image (a) with $\alpha = 10^{-5}$ in weighted regularization terms. In this image, we still can observe some stripes due to less regularity to smooth the stripes. Hence, it is clearly observed that a given weighting factor to the regularization term is necessary, and choosing an appropriate regularization parameter is also very important. We can observe this scenario when we compare the images (c), (d) and (e) in Fig. 7 as α varies from 10^{-5} to 1. Therefore, once we have an appropriate α, we do not need an extra functional to reduce the blurring effects as explained in Bouali (2010), and hence our approach provides a simple algorithm to remove the stripes. The appropriate α is

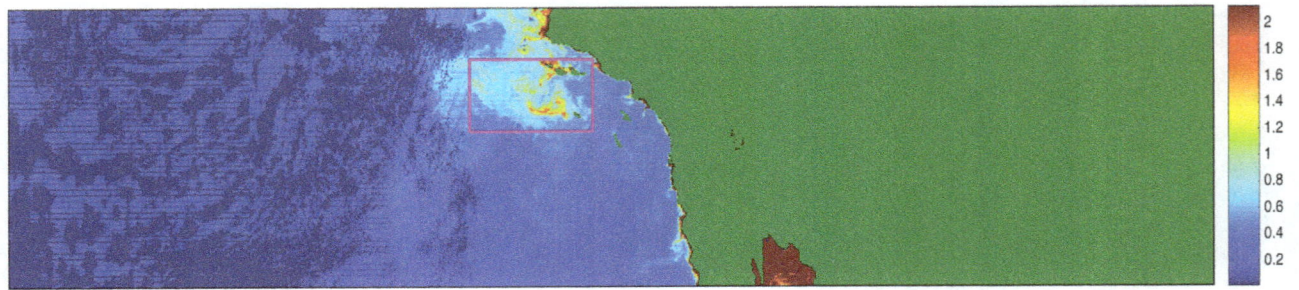

Figure 6. The image shows the chlorophyll concentration in milligrams per cubic meter near the Santa Monica region in southern California as viewed by VIIRS on 7 November 2014. Green represents the land and dark blue represents the dropped data due to bow-tie effects and missing data due to clouds. For a detailed discussion, we next consider the subset of the image that is covered by the pink square in the image.

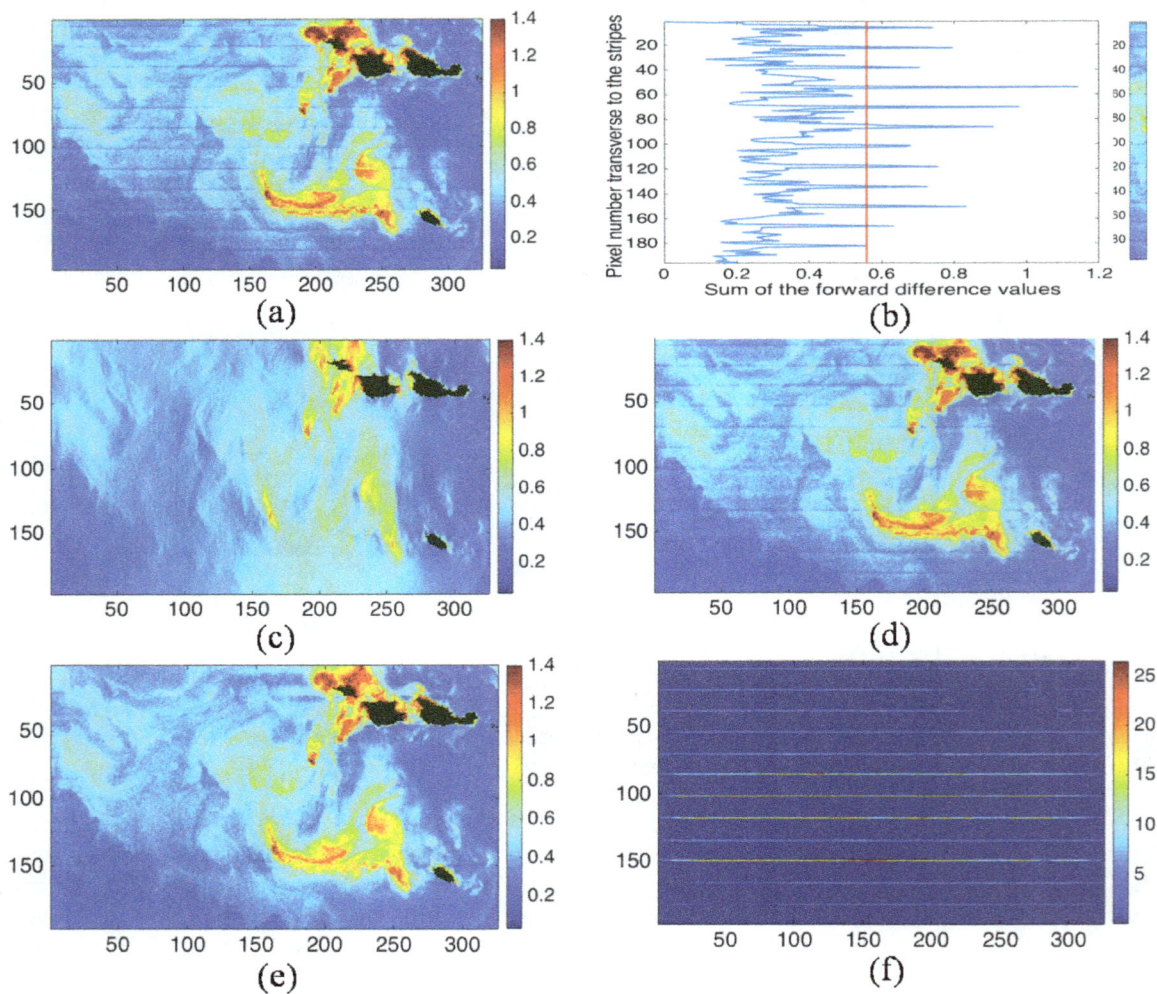

Figure 7. Panel (**a**) shows the cropped region shown in Fig. 6 and the graph (**b**) shows the S curve with the threshold and the image piece (columns 137 to 148) that we used compute the values of S curve. The panels (**c**), (**d**) and (**e**) represent the destriped images of (**a**) with $\alpha = 1$ with unweighted regularization term and $\alpha = 10^{-5}$ and $\alpha = 10^{-2}$ with weighted regularization term, respectively. Panel (**e**) provides the best solution for the destriped image. Panel (**c**) is over regularized whereas panel (**d**) is not sufficiently regularized. Panel (**f**) represents the percentage error between panels (**a**) and (**e**).

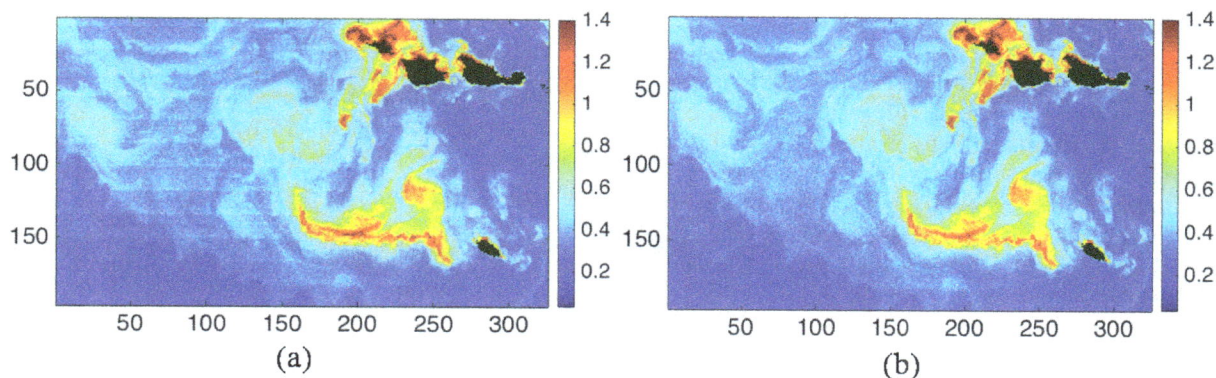

Figure 8. Panel **(a)** shows the destriped image scene of the panel **(a)** in Fig. 7 from NASA's vicarious calibration of The L2 (*.nc) product method. While NASA's vicarious calibration of The L2 (*.nc) product method does improve over the raw image, there are still stripe artifacts present. The following image **(b)** represents the destriped image of the image **(a)** from our method.

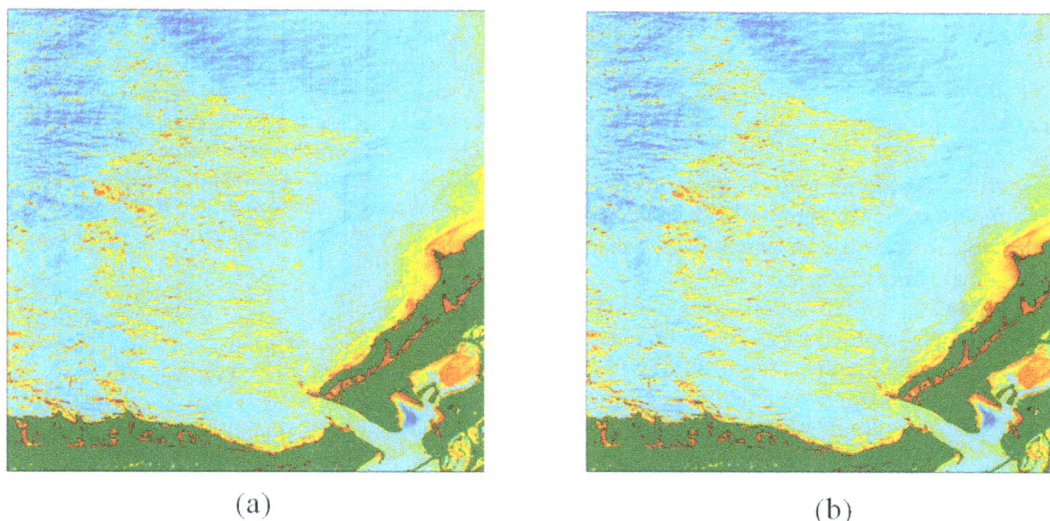

Figure 9. Panel **(a)** is band 22 ($\sim 410\,\text{nm}$) of a hyperspectral image which was taken from JPL PRISM JPL (2015). The stripe pattern is vertical and the destriped images are shown in image **(b)**. Green represents the land of the observed region.

selected from the U-curve method, as explained in Sect. 2.3. Panel (f) in Fig. 7 shows the percentage error between the striped and destriped images. This is proof that the effects from the destriping on the image where there are no stripes is negligible.

Image-based methods such as the variational destriping can be used together with other destriping methods. For instance, NASA's vicarious calibration of the L2 (*.nc) product method uses a monthly moon calibration to monitor the striping and calibrations and create up-to-date corrections using the entire image collection. Figure 8a shows the VIIRS image in Fig. 7a after the NASA correction, and the data are publicly available at NAS (2014). However, the striping artifacts are still visible in that image. Fig. 8b shows the application of the weighted regularizing destriping to the image

from NASA's vicarious calibration of The L2 (*.nc) product method. The product from the applications of both corrections is superior to either individual correction. The regularization parameter for this image was 5×10^{-3}, and the destriped image is shown in Fig. 8b. The threshold value for the columns 137 to 148 was 0.55. If we include the proper units of the chlorophyll data, the threshold value is $84.81\,(\text{mg})\,\text{longitude}/(\text{m}^3)\,\text{latitude}$.

When we compare the images (a) and (b) in Fig. 8, it can be observed that our method can be used to further improve upon the destriped images, after NASA's vicarious calibration of the L2 (*.nc) product method has already been applied.

3.3 Example 3 – JPL PRISM images

In the last example, we apply the variational destriping algorithm to another data set from the JPL PRISM hyperspectral imager, where the data is publicly available at JPL (2015). Stripes arise in this sensor because of the focal-plane array read-out mechanism has cross-talk artifacts. An image from the band centered at 410 nm (band 22), which was taken from an airborne campaign around Monterey, CA, near the Elkhorn Slough (Pantazis et al., 2015), is shown in Fig. 9.

The regularization parameter to destripe the image (a) in Fig. 9 was $\alpha = 5 \times 10^{-3}$. The destriped image is shown in image (b) in Fig. 9. We used the columns from 50 to 70 to determine the threshold value to assign the weights in the regularization term, and this value was 1.7 for this image. It can be clearly observed that the stripes are smoothed from the destriping method while preserving the other features of the image.

4 Conclusions

We present a variational destriping method by explicitly including a tunable regularization parameter with a weighted regularization term to a part of the destriping functional in Bouali (2010). In other words, we modeled one piece of the destriping functional in Bouali (2010) into a standard variational-based approach employing the Tikhonov regularization theory of Vogel (2002). According to the Tikhonov regularization theory, the tuning parameter allows us to properly balance the effects of optimizing the data fidelity and the smoothing effects of regularization term with the help of given weights. The introduction of the weighted regularization term avoids the effects on the original features of the image during the process of destriping. As a preprocessing step, we apply an image-processing technique to avoid the error accumulation from the "bow-tie" effects and missing data due to clouds. We also demonstrate alternative numerical aspects to implementing these methods, which allows us to write the solver in terms of common matrix manipulations. Lastly, we show that applying a scene-based method to the destriped VIIRS L2 product from NASA calibration method results in additional improvements in scene uniformity.

Author contributions. R. Basnayake, E. Bollt and J. Sun developed the variational-based destriping algorithm. N. Tufillaro collected raw data from reliable sources and processed the data so that it could be readable in MATLAB. M. Gierach provided the JPL-PRISM data.

Competing interests. The authors declare that they have no conflict of interest.

Acknowledgements. The authors Ranil Basnayake, Erik Bollt, Nicholas Tufillaro and Jie Sun were supported by the National Geospatial Intelligence Agency under grant number HM02101310010. Erik Bollt was also supported by the Office of Naval Research, PI, N00014-15-1-2093, thanks to the Office of Naval Research grant no. N00014-16-1-2492.

Edited by: Stephen Wiggins

References

Basnayake, R. and Bollt, E. M.: A Multi-time Step Method to Compute Optical Flow with Scientific Priors for Observations of a Fluidic System, in: Ergodic Theory, Open Dynamics, and Coherent Structures, Springer, 59–79, 2014.

Bertalmio, M., Sapiro, G., Caselles, V., and Ballester, C.: Image inpainting, in: Proceedings of the 27th annual conference on Computer graphics and interactive techniques, ACM Press/Addison-Wesley Publishing Co., 417–424, 2000.

Bertalmio, M., Caselles, V., Masnou, S., and Sapiro, G.: Inpainting, Encyclopedia of Computer Vision, 2011.

Bouali, M.: A simple and robust destriping algorithm for imaging spectrometers: Application to MODIS data, in: Proceedings of ASPRS 2010 Annual Conference, San Diego, CA, USA, vol. 2630, 84–93, 2010.

Bouali, M. and Ignatov, A.: Adaptive Reduction of Striping for Improved Sea Surface Temperature Imagery from Suomi National Polar-Orbiting Partnership (S-NPP) Visible Infrared Imaging Radiometer Suite (VIIRS), J. Atmos. Ocean. Tech., 31, 150–163, https://doi.org/10.1175/JTECH-D-13-00035.1, 2013.

Cao, C., Xiong, X., Wolfe, R., De Luccia, F., Liu, Q., Blonski, S., Lin, G., Nishihama, M., Pogorzala, D., Oudrari, H., and others: Visible Infrared Imaging Radiometer Suite (VIIRS) Sensor Data Record (SDR) User's Guide. Version 1.2, 10 September 2013, NOAA Technical Report NESDIS, 142, 2013.

Eplee, R. E., Turpie, K. R., Fireman, G. F., Meister, G., Stone, T. C., Patt, F. S., Franz, B. A., Bailey, S. W., Robinson, W. D., and McClain, C. R.: VIIRS on-orbit calibration for ocean color data processing, vol. 8510 of Proc. SPIE, 85101G–85101G–18, 2012.

Hansen, P. C.: Analysis of discrete ill-posed problems by means of the L-curve, SIAM Rev., 34, 561–580, 1992.

Hansen, P. C.: The L-Curve and its Use in the Numerical Treatment of Inverse Problems, in: Computational Inverse Problems in Electrocardiology, Advances in Computational Bioengineering, edited by: Johnston, P., WIT Press, 119–142, 2000.

Hansen, P. C.: Discrete inverse problems: insight and algorithms, vol. 7, SIAM, Society for Industrial and Applied Mathematics, 2010.

Hansen, P. C., Nagy, J. G., and O'leary, D. P.: Deblurring images: matrices, spectra, and filtering, vol. 3, Siam, Society for Industrial and Applied Mathematics, 2006.

JPL: Portable Remote Imaging Spectrometer, available at: http://prism.jpl.nasa.gov/prism_data.html, last access: 20 May 2015.

Krawczyk-Stado, D. and Rudnicki, M.: The use of L-curve and U-curve in inverse electromagnetic modelling, Stud. Comp. Intell., 119, 73–82, 2008.

Krawczyk-StańDo, D. and Rudnicki, M.: Regularization parameter selection in discrete ill-posed problems—the use of the U-curve, Int. J. Appl. Math. Comp., 17, 157–164, 2007.

Mikelsons, K., Wang, M., Jiang, L., and Bouali, M.: Destriping algorithm for improved satellite-derived ocean color product imagery, Opt. Express, 22, 28058–28070, 2014.

Mikelsons, K., Ignatov, A., Bouali, M., and Kihai, Y.: A fast and robust implementation of the adaptive destriping algorithm for SNPP VIIRS and Terra/Aqua MODIS SST, vol. 9459 of Proc. SPIE, 94590R–94590R–13, 2015.

NASA Ocean Color, available at: http://oceancolor.gsfc.nasa.gov/cgi/browse.pl?sen=am, last access: 1 June 2014.

Osborne, J., Kurapov, A., Egbert, G., and Kosro, P.: Spatial and temporal variability of the M 2 internal tide generation and propagation on the Oregon shelf, J. Phys. Oceanogr., 41, 2037–2062, 2011.

Pantazis, M., Gorp, B. V., Dierseen, H., Blerach, M., and Fichot, C.: The Portable Remote Imaging Spectrometer (PRISM): Recent Campaigns and Developments, Fourier Transform Spectroscopy and Hyperspectral Imaging and Sounding of the Environment, HM4B.5, OSA, 2015.

Vogel, C. R.: Computational Methods for Inverse Problems, Frontiers in Mathematics, SIAM, Society for Industrial and Applied Mathematics, 2002.

PERMISSIONS

All chapters in this book were first published in NPG, by Copernicus Publications; hereby published with permission under the Creative Commons Attribution License or equivalent. Every chapter published in this book has been scrutinized by our experts. Their significance has been extensively debated. The topics covered herein carry significant findings which will fuel the growth of the discipline. They may even be implemented as practical applications or may be referred to as a beginning point for another development.

The contributors of this book come from diverse backgrounds, making this book a truly international effort. This book will bring forth new frontiers with its revolutionizing research information and detailed analysis of the nascent developments around the world.

We would like to thank all the contributing authors for lending their expertise to make the book truly unique. They have played a crucial role in the development of this book. Without their invaluable contributions this book wouldn't have been possible. They have made vital efforts to compile up to date information on the varied aspects of this subject to make this book a valuable addition to the collection of many professionals and students.

This book was conceptualized with the vision of imparting up-to-date information and advanced data in this field. To ensure the same, a matchless editorial board was set up. Every individual on the board went through rigorous rounds of assessment to prove their worth. After which they invested a large part of their time researching and compiling the most relevant data for our readers.

The editorial board has been involved in producing this book since its inception. They have spent rigorous hours researching and exploring the diverse topics which have resulted in the successful publishing of this book. They have passed on their knowledge of decades through this book. To expedite this challenging task, the publisher supported the team at every step. A small team of assistant editors was also appointed to further simplify the editing procedure and attain best results for the readers.

Apart from the editorial board, the designing team has also invested a significant amount of their time in understanding the subject and creating the most relevant covers. They scrutinized every image to scout for the most suitable representation of the subject and create an appropriate cover for the book.

The publishing team has been an ardent support to the editorial, designing and production team. Their endless efforts to recruit the best for this project, has resulted in the accomplishment of this book. They are a veteran in the field of academics and their pool of knowledge is as vast as their experience in printing. Their expertise and guidance has proved useful at every step. Their uncompromising quality standards have made this book an exceptional effort. Their encouragement from time to time has been an inspiration for everyone.

The publisher and the editorial board hope that this book will prove to be a valuable piece of knowledge for researchers, students, practitioners and scholars across the globe.

LIST OF CONTRIBUTORS

Claudia Cherubini
Department of Mechanical, Aerospace & Civil Engineering, Brunel University London, Uxbridge, UB8 3PH, UK
School of Civil Engineering, The University of Queensland, Queensland, Australia

Nicola Pastore, Concetta I. Giasi, and Nicoletta Maria Allegretti
DICATECh, Department of Civil, Environmental, Building Engineering, and Chemistry, Politecnico di Bari, Bari, Italy

Odim Mendes, Ezequiel Echer and Varlei Everton Menconi
Space Geophysics Division (DGE/CEA), Brazilian Institute for Space Research (INPE), São José dos Campos, São Paulo, Brazil

Margarete Oliveira Domingues
Associated Laboratory of Computation and Applied Mathematics (LAC/CTE), Brazilian Institute for Space Research (INPE), São José dos Campos, São Paulo, Brazil

Rajkumar Hajra
Laboratoire de Physique et Chimie de l'Environnement et de l'Espace (LPC2E), CNRS, Orléans 45100, France

Jobst Heitzig and Kirsten Thonicke
Potsdam Institute for Climate Impact Research, Telegrafenberg A31, 14473 Potsdam, Germany

Finn Müller-Hansen and Jürgen Kurths
Potsdam Institute for Climate Impact Research, Telegrafenberg A31, 14473 Potsdam, Germany
Department of Physics, Humboldt University Berlin, Newtonstraße 15, 12489 Berlin, Germany

Manoel F. Cardoso and Eloi L. Dalla-Nora
Center for Earth System Science, National Institute for Space Research, Rodovia Presidente Dutra 40, 12630-000 Cachoeira Paulista, São Paulo, Brazil

Jonathan F. Donges
Potsdam Institute for Climate Impact Research, Telegrafenberg A31, 14473 Potsdam, Germany
Stockholm Resilience Center, Stockholm University, Kräftriket 2B, 114 19 Stockholm, Sweden

Carmelo Alonso
Earth Observation Systems, Indra Sistemas S.A., Madrid, Spain
Grupo de Sistemas Complejos, U.P.M, Madrid, Spain

Rosa M. Benito
Grupo de Sistemas Complejos, U.P.M, Madrid, Spain

Ana M. Tarquis
Grupo de Sistemas Complejos, U.P.M, Madrid, Spain
CEIGRAM, E.T.S.I.A.A.B., U.P.M, Madrid, Spain

Ignacio Zúñiga
Dpt. Física Fundamental, Facultad de Ciencias, Universidad Nacional de Educación a Distancia (UNED), Madrid, Spain

Pedro Monroy, Emilio Hernández-García, and Cristóbal López
IFISC, Instituto de Física Interdisciplinar y Sistemas Complejos (CSIC-UIB), 07122 Palma de Mallorca, Spain

Vincent Rossi
IFISC, Instituto de Física Interdisciplinar y Sistemas Complejos (CSIC-UIB), 07122 Palma de Mallorca, Spain
Mediterranean Institute of Oceanography (UM 110, UMR 7294), CNRS, Aix Marseille Univ., Univ. Toulon, IRD, 13288, Marseille, France

Sergey V. Prants, Maxim V. Budyansky and Michael Yu. Uleysky
Laboratory of Nonlinear Dynamical Systems, Pacific Oceanological Institute of the Russian Academy of Sciences, 43 Baltiyskaya st., 690041 Vladivostok, Russia

Hazuki Arakida, and Shunji Kotsuki
RIKEN Advanced Institute for Computational Science, Kobe, 650-0047, Japan

Takemasa Miyoshi
RIKEN Advanced Institute for Computational Science, Kobe, 650-0047, Japan
Department of Atmospheric and Oceanic Science, University of Maryland, College Park, MD 20742, USA
Application Laboratory, Japan Agency for Marine-Earth Science and Technology, Yokohama, 236-0001, Japan

Takeshi Ise
Field Science Education and Research Center, Kyoto University, Kyoto, 606-8502, Japan

Shin-ichiro Shima
RIKEN Advanced Institute for Computational Science, Kobe, 650-0047, Japan

Graduate School of Simulation Studies, University of Hyogo, Kobe, 650-0047, Japan

Sergei I. Badulin
P. P. Shirshov Institute of Oceanology of the Russian Academy of Sciences, Moscow, Russia
Laboratory of Nonlinear Wave Processes, Novosibirsk State University, Novosibirsk, Russia

Vladimir E. Zakharov
P. P. Shirshov Institute of Oceanology of the Russian Academy of Sciences, Moscow, Russia
Laboratory of Nonlinear Wave Processes, Novosibirsk State University, Novosibirsk, Russia
Department of Mathematics, University of Arizona, Tucson, USA
P. N. Lebedev Physical Institute of the Russian Academy of Sciences, Moscow, Russia
Waves and Solitons LLC, Phoenix, Arizona, USA

Benjamin K. B. Abban, Christopher G. Wilson and Christos P. Giannopoulos
Hydraulics and Sedimentation Lab, Department of Civil&Environmental Engineering, University of Tennessee – Knoxville, Knoxville, TN 37996, USA

A. N. (Thanos) Papanicolaou
Hydraulics and Sedimentation Lab, Department of Civil&Environmental Engineering, University of Tennessee – Knoxville, Knoxville, TN 37996, USA
Tennessee Water Resources Center, Knoxville, TN 37996, USA

Dimitrios C. Dermisis
College of Engineering, Department of Chemical, Civil&Mechanical Engineering, McNeese State University, Lake Charles, LA 70605, USA

Kenneth M. Wacha
USDA-ARS – National Laboratory for Agriculture and the Environment, Ames, IA 50011, USA

Mohamed Elhakeem
Abu Dhabi University, Abu Dhabi, P.O. Box 59911, Abu Dhabi, United Arab Emirates

Sergey B. Turuntaev
Institute of Geosphere Dynamics, Russian Academy of Sciences, Moscow, 119334, Russian Federation
Moscow Institute of Physics and Technology, Moscow, 141701, Russian Federation
All-Russian Research Institute of Automatics, Moscow, 127055, Russian Federation

Vasily Y. Riga
Moscow Institute of Physics and Technology, Moscow, 141701, Russian Federation

Aaron Coutino and Marek Stastna
Department of Applied Mathematics, University of Waterloo, Waterloo, Canada

Vladimir Zakharov and Andrei Pushkarev
Department of Mathematics, University of Arizona, Tucson, AZ 85721, USA
Lebedev Physical Institute RAS, Leninsky 53, Moscow 119991, Russia
Novosibirsk State University, Novosibirsk, 630090, Russia
Waves and Solitons LLC, 1719 W. Marlette Ave., Phoenix, AZ 85015, USA

Donald Resio
Taylor Engineering Research Institute, University of North Florida, Jacksonville, FL, USA

Víctor José García-Garrido and Ana María Mancho
Instituto de Ciencias Matemáticas, CSIC-UAM-UC3M-UCM, C/ Nicolás Cabrera 15, Campus de Cantoblanco UAM, 28049 Madrid, Spain

Jezabel Curbelo
Instituto de Ciencias Matemáticas, CSIC-UAM-UC3M-UCM, C/ Nicolás Cabrera 15, Campus de Cantoblanco UAM, 28049 Madrid, Spain
Departamento de Matemáticas, Facultad de Ciencias, Universidad Autonóma de Madrid, Madrid, Spain

Carlos Roberto Mechoso
Department of Atmospheric and Oceanic Sciences, University of California at Los Angeles, Los Angeles, CA, USA

Stephen Wiggins
School of Mathematics, University of Bristol, Bristol, UK

Qin Zheng, Zubin Yang, and Jun Yan
Institute of Science, PLA University of Science and Technology, Nanjing, 211101, China

Jianxin Sha
Troop 94906, People's Liberation Army, Suzhou, 215157, China

Ranil Basnayake, Erik Bollt and Jie Sun
Department of Mathematics, Clarkson University, Potsdam, NY 13699, USA
Clarkson Center for Complex Systems Science (C3S2), Potsdam, NY 13699, USA

Nicholas Tufillaro
College of Earth, Ocean, and Atmospheric Sciences, Oregon State University, Corvallis, OR 97331, USA

Michelle Gierach
Jet Propulsion Laboratory, California Institute of Technology, Pasadena, CA, 91109, USA

Index

www.ingramcontent.com/pod-product-compliance
Lightning Source LLC
Chambersburg PA
CBHW080646200326
41458CB00013B/4753